Remote Sensing Technology in Modern Agriculture

Remote Sensing Technology in Modern Agriculture

Edited by Georgie Johnson

New York

Published by Syrawood Publishing House,
750 Third Avenue, 9th Floor,
New York, NY 10017, USA
www.syrawoodpublishinghouse.com

Remote Sensing Technology in Modern Agriculture
Edited by Georgie Johnson

International Standard Book Number: 978-1-64740-416-1 (Hardback)

Cataloging-in-publication Data

Remote sensing technology in modern agriculture / edited by Georgie Johnson.
p. cm.
Includes bibliographical references and index.
ISBN 978-1-64740-416-1
1. Agriculture--Remote sensing. 2. Plants, Cultivated--Remote sensing.
3. Agricultural innovations. I. Johnson, Georgie.
S494.5.R4 R463 2023
577.82--dc23

TABLE OF CONTENTS

PREFACE

Remote sensing technology encompasses the systems that use digital cameras and remote control systems for obtaining photographs and images from hard-to-access places. It has multiple applications in agriculture, such as obtaining information about crops, soil, and surrounding environment. This technology enables the collection of high-resolution information from agricultural land and assists the farmers and crop scientists in making various decisions. Different types of image capturing devices used in remote imaging include drones, satellites, in-field devices and digital cameras integrated in wireless sensor networks on the ground. Remote sensing is able to generate different types of spectral information such as multispectral images, standard RGB images, and hyperspectral images. Some major applications of remote sensing technology include plant monitoring, automatic harvesting, mapping and detection of plants, water management, and yield estimation. This book outlines the use of remote sensing technology in modern agriculture in detail. It presents researches and studies performed by experts across the globe. This book aims to equip students and experts with the latest concepts in this area.

This book unites the global concepts and researches in an organized manner for a comprehensive understanding of the subject. It is a ripe text for all researchers, students, scientists or anyone else who is interested in acquiring a better knowledge of this dynamic field.

I extend my sincere thanks to the contributors for such eloquent research chapters. Finally, I thank my family for being a source of support and help.

Editor

Applications of Remote Sensing in Precision Agriculture

Rajendra P. Sishodia [1,*], **Ram L. Ray** [1] and **Sudhir K. Singh** [2]

[1] College of Agriculture and Human Sciences, Prairie View A&M University, Prairie View, TX 77446, USA; raray@pvamu.edu
[2] K. Banerjee Centre of Atmospheric & Ocean Studies, IIDS, Nehru Science Centre, University of Allahabad, Prayagraj 211002, India; sudhirinjnu@gmail.com
* Correspondence: rpsishodia@pvamu.edu.

Abstract: Agriculture provides for the most basic needs of humankind: food and fiber. The introduction of new farming techniques in the past century (e.g., during the Green Revolution) has helped agriculture keep pace with growing demands for food and other agricultural products. However, further increases in food demand, a growing population, and rising income levels are likely to put additional strain on natural resources. With growing recognition of the negative impacts of agriculture on the environment, new techniques and approaches should be able to meet future food demands while maintaining or reducing the environmental footprint of agriculture. Emerging technologies, such as geospatial technologies, Internet of Things (IoT), Big Data analysis, and artificial intelligence (AI), could be utilized to make informed management decisions aimed to increase crop production. Precision agriculture (PA) entails the application of a suite of such technologies to optimize agricultural inputs to increase agricultural production and reduce input losses. Use of remote sensing technologies for PA has increased rapidly during the past few decades. The unprecedented availability of high resolution (spatial, spectral and temporal) satellite images has promoted the use of remote sensing in many PA applications, including crop monitoring, irrigation management, nutrient application, disease and pest management, and yield prediction. In this paper, we provide an overview of remote sensing systems, techniques, and vegetation indices along with their recent (2015–2020) applications in PA. Remote-sensing-based PA technologies such as variable fertilizer rate application technology in Green Seeker and Crop Circle have already been incorporated in commercial agriculture. Use of unmanned aerial vehicles (UAVs) has increased tremendously during the last decade due to their cost-effectiveness and flexibility in obtaining the high-resolution (cm-scale) images needed for PA applications. At the same time, the availability of a large amount of satellite data has prompted researchers to explore advanced data storage and processing techniques such as cloud computing and machine learning. Given the complexity of image processing and the amount of technical knowledge and expertise needed, it is critical to explore and develop a simple yet reliable workflow for the real-time application of remote sensing in PA. Development of accurate yet easy to use, user-friendly systems is likely to result in broader adoption of remote sensing technologies in commercial and non-commercial PA applications.

Keywords: big data analysis; disease and pest management; nutrient management; satellite remote sensing; UAV; vegetation indices; water management

---

1. Introduction

Agriculture, an engine of economic growth for many nations, provides the most basic needs of humankind: food and fiber [1,2]. Technological changes during the past century, such as the Green Revolution, have transformed the face of agriculture [3]. The improved crop varieties, synthetic fertilizers, pesticides, and irrigation during the 1960s–1980s, known as the Green Revolution or third agricultural revolution, enhanced crop productivity and food security, especially in developing nations [4]. Consequently, despite the doubling population and tripling food demand since the 1960s, global agriculture has been able to meet the demands with only a 30% expansion in the cultivated area [4,5]. The demand for food and agricultural products is projected to further increase by more than 70% by 2050 [6]. Given the limited availability of arable land, a significant part of this increased demand will be met through agricultural intensification, i.e., increased use of fertilizers, pesticides, water, and other inputs.

However, intensified use of agricultural inputs also causes environmental degradation, including groundwater depletion, reduced surface flows, and eutrophication [7–11]. Excessive and/or inefficient use of natural resources (e.g., soil and water), fertilizers, and pesticides for agricultural production cause economic losses as well as increased water and nutrient losses from agriculture that lead to environmental degradation [12]. For an economically and environmentally sustainable production system, there is a need to develop techniques that can increase crop production through increased efficiency of inputs use and reduced environmental losses [13].

Precision agriculture (PA) is a key component of sustainable agricultural systems in the 21st century [13,14]. PA has been defined in multiple ways, yet the underlying concept remains the same [15]. PA entails a management strategy that uses a suite of advanced information, communication, and data analysis techniques in the decision making process (e.g., application of water, fertilizer, pesticide, seed, fuel, labor, etc.), which helps enhancing crop production and reducing water and nutrient losses and negative environmental impacts [16–20]. Information-based management, site-specific crop management, target farming, variable rate technology, and grid farming are some other names used synonymously for PA [15,21]. In addition to crop production, PA has been used in viticulture, horticulture, pasture, and livestock production and management [18,22].

Presently, agriculture can be considered to be going through a fourth revolution facilitated mainly by advances in information and communication technologies [23]. Emerging technologies, such as remote sensing, global positioning systems (GPS), geographic information systems (GIS), Internet of Things (IoT), Big Data analysis, and artificial intelligence (AI) are promising tools being utilized to optimize agricultural operations and inputs aimed to enhance production and reduce inputs and yield losses [13,24,25]. Several IoT technology systems utilizing cloud computing, wireless sensors networks, and big data analysis have been developed for smart farming operations such as automated wireless-controlled irrigation systems and intelligent disease and pest monitoring and forecasting systems [24,25]. AI techniques, including machine learning (e.g., artificial neural networks) have been used to estimate ET, soil moisture, and crop predictions for automated and precise application of water, fertilizer, herbicide, and insecticides [23]. These technologies and tools enable farmers to characterize spatial variability (e.g., soils) among farms and large crop fields that negatively affect crop growth and yields [21]. These state-of-the-art technologies for the development and implementation of site-specific management are integral part of PA [16].

Remote sensing systems, using information and communication technologies, usually generate a large volume of spectral data due to high spatial/spectral/radiometric/temporal resolutions needed for application in PA [26]. Emerging data processing techniques such as Big Data analysis, artificial intelligence, and machine learning have been utilized to draw useful information from the large volume of data [27]. Also, cloud computing systems have been used to store, process, and distribute/utilize such a large amount of data for applications in PA [28–30]. All these advanced data acquisitions and processing techniques have been applied globally, to aid the decision-making process for field crops, horticulture, viticulture, pasture, and livestock [27,31–36].

In the past, several studies have provided reviews of remote sensing techniques and applications in agriculture. While some studies focused on specific application areas such as soil properties estimation [37], evapotranspiration (ET) estimation [38,39], and disease and pest management [40], others included more than one area of applications [41–43]. Many of these studies reflected the state of the art of remote-sensing-based techniques along with their limitations and future challenges for application in agriculture. Some of these notable efforts include Mulla et al. [42], Weiss et al. [43], Maes and Steppe [44], and Angelopoulou et al. [45]. The primary purpose of this review is to complement these efforts to provide a comprehensive background and knowledge on applications of remotely sensed data and technologies in agriculture, focusing on precision agriculture. Specifically, we provide an overview of remote sensing systems, techniques, and applications in irrigation management, nutrients management, disease and pest management, and yield estimation along with a synthesis table of vegetation indices used for a variety of applications in PA. The rest of this paper is organized as follows.

Section 2 describes the types of remote sensing systems covering different sensors and platforms used for applications in PA. Section 3 provides a brief history of applications of remote sensing in agriculture with a focus on PA. Section 4 discusses some popularly used vegetation indices derived from remote sensing data and their applications in PA. Section 5 has five sub-sections discussing recent applications of remote sensing in PA for (i) irrigation water management, (ii) nutrient management, (iii) disease management, (iv) weed management, and (v) crop monitoring and yield. Particularly, in Section 5, we focus our review/discussion on studies published during 2015–2020. The final section in the paper discusses the progress made, needs, and challenges for applications of remote sensing in PA.

2. Remote Sensing Systems Used in Precision Agriculture

Remote sensing systems used for PA, and agriculture in general, can be classified based on (i) sensor platform and (ii) type of sensor. Sensors are typically mounted on satellites, aerial, and ground-based platforms (Figure 1). Since the 1970s, satellite products have been extensively used for PA. Recently, aerial platforms, which include aircraft and unmanned aerial vehicles (UAVs) are also used in PA. Ground-based platforms used for PA can be grouped into three categories: (i) hand-held, (ii) free standing in the field, and (iii) mounted on tractor or farm machinery. Ground-based systems are also referred to as proximal remote sensing systems because they are located in close proximity to the target surface (land surface or plant) as compared to aerial or satellite-based platforms.

Sensors used for remote sensing differ based on the spatial, spectral, radiometric, and temporal resolution they offer [46]. Spatial resolution of a sensor is defined by the size of the pixel that represents the area on the ground. Sensors with high spatial resolution tend to have small footprints, and sensors with large footprints tend to have a low spatial resolution. Temporal resolution can be considered to be associated with the sensor platform rather than the sensor itself. For example, temporal resolution for a satellite is the time it takes to complete an orbit and revisit the same observation area. Spectral resolution of a sensor is indicated by the number of bands captured in the given range of electromagnetic spectrum [47]. Hyperspectral images contain a large number (10s to 100s) of contagious bands of narrow width (<20 nm) separated by small increments in wavelength [48]. Numerous vegetation indices and statistical and machine learning approaches, such as deep convolutional neural network and random forest, have been applied to reduce the dimensionality of hyperspectral data to extract useful information on crop conditions [49–51]. More recently, quantification of solar-induced chlorophyll fluorescence (SIF) from hyperspectral images has increasingly been applied to estimate photosynthesis, plant nutrients, and biotic and abiotic stresses such as disease and water stress [49–54].

Although numerous recent satellites provide high spatial (<5 m) and temporal (daily) resolution images, most publicly available satellite products are rather coarse for many PA applications. Appropriate spatio-temporal resolution required for PA depends on several factors including management objectives, size of the field, and the ability of farm equipment to vary the inputs (irrigation, fertilizer, pesticide, etc.) application rates. Crop biomass and yield estimation typically require higher spatial resolution (1–3 m) compared to variable rate fertilizer and irrigation (5–10 m) applications [42]. Furthermore, weed mapping and variable herbicide application require spatial resolution that is finer than the weed patches (e.g., 5–50 cm) [55,56]. Aerial platforms such as UAVs generally provide higher spatial resolution (<5 m) images as compared to the satellites. Thus, UAVs and other ground-based platforms offer greater flexibility in providing images at fine spatial and temporal resolution (more frequent) or as needed.

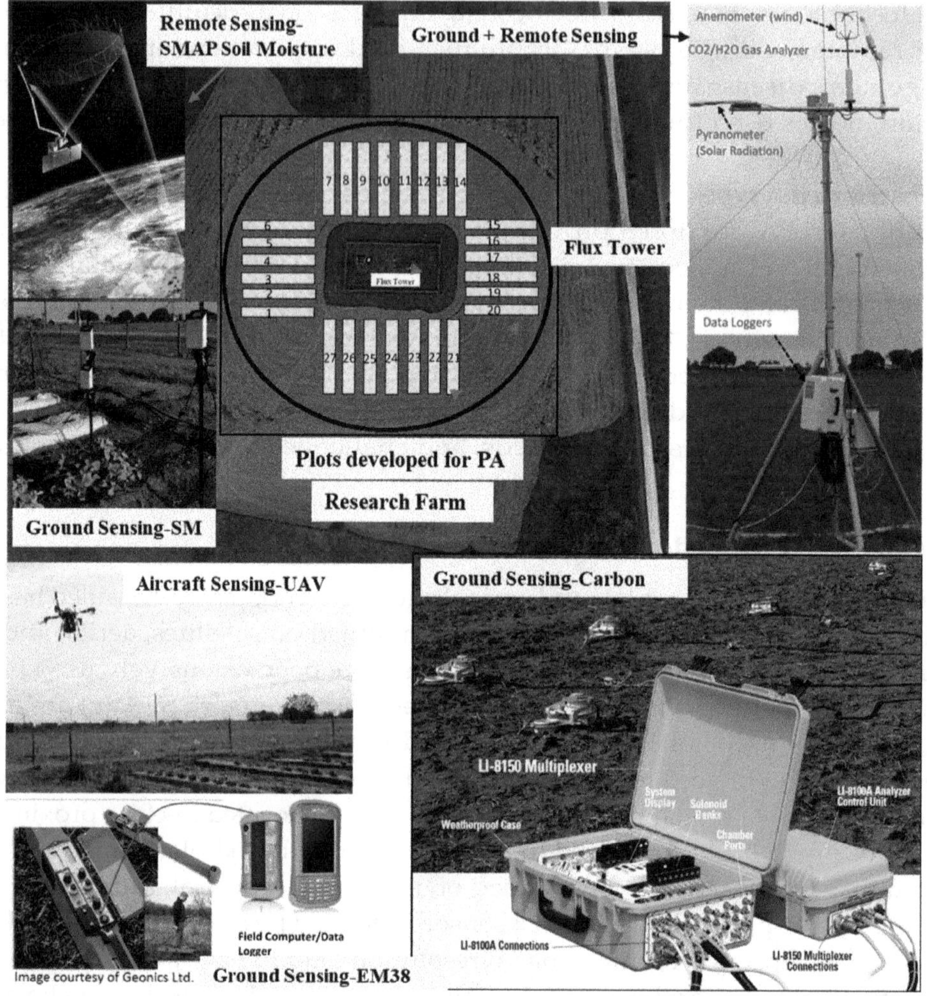

Figure 1. A typical layout for the remote, aircraft, and ground sensing systems deployed for precision agriculture. SMAP = Soil Moisture Active Passive; PA = Precision Agriculture; SM = Soil Moisture; UAV = Unmanned Aerial Vehicle; EM = Electromagnetic. (UAV photo credit—Dr. Ripendra Awal, PVAMU, Prairie View, TX, USA).

Sensors mounted on satellites, airplanes, and UAVs are generally passive sensors, i.e., they do not have their own light source. However, some of the satellites have active sensors onboard, such as the active microwave instrument (AMI) on the ERS-1/2. Many of the ground-based remote sensing platforms have active proximity sensors. For example, commercially available variable fertilizer rate application systems such as Green Seeker and Crop Circle have active proximity sensors. In such systems, variations in daylight have minimum effect on measured reflectance, thus providing more accurate and reproducible normalized difference vegetation index (NDVI) or other vegetation indices (VI) used for crop nutrient status assessment.

Two other sensors (thermal infrared and microwave) mounted on recent satellites are increasingly being used in agriculture. Thermal infrared sensors measure energy emitted from a target (e.g., crops) to estimate its temperature, which could be further used to estimate crop water stress, ET, and irrigation requirements [57]. Microwave sensors work in a similar way as the thermal sensors to measure the emitted energy (although in longer microwave wavelengths) from the land surface. Microwave sensors are mainly used to estimate soil moisture contents and crop water use over large areas [58]. Also, microwaves can penetrate through clouds, which is advantageous over other types of sensors that use visible and NIR wavelengths.

However, coarse spatial resolution (10s of km) of the microwave satellite sensors, especially passive sensors, limits their application in PA. Recently, many methods have been developed and used to downscale passive microwave data to a finer resolution for use in PA [59–62]. Active microwave sensors (e.g., synthetic aperture radar—SAR) generally offer a higher spatial resolution; however, they are also more sensitive to surface roughness (e.g., vegetation) that can introduce errors in soil moisture estimation. Overall, there are a variety of remote sensors and platforms available that can be used to generate high-resolution (spatial, spectral, radiometric, and temporal) images critical to develop and implement site-specific management.

3. Historical Applications of Remote Sensing in Agriculture

Researchers have long recognized the need to map soil and land use databases for sustainable management of natural resources at local, regional, and national scale [63,64]. Knowledge of soil physical, biological, and chemical properties is important to design and implement irrigation, drainage, nutrient, and other crop management strategies, which are essential components of PA. Similarly, land use mapping can help assess the impacts of existing management and policy at regional to national scale. A traditional approach of using remote sensing techniques in agriculture has been around even before 1958, when the term "remote sensing" was first introduced [65]. For example, aerial photography has been used to map soils, land use, and crop conditions in the United States during the 1930 and 1940s [66]. However, these conventional methods of soil mapping and land use classification (e.g., low altitude photography and ground crews) typically involve extensive fieldwork and laboratory analysis, which are expensive and time-consuming [67,68]. Advent of satellite remote sensing during later years facilitated more efficient and effective mapping of land use and land cover at regional, national, and global scales.

Launch of Vanguard 2 and TIROS 1 in 1959 and 1960, respectively, marked the start of satellite remote sensing for meteorological information [69]. However, the era of satellite remote sensing for agriculture started with the launch of Landsat 1 (formerly known as the Earth Resources Technology Satellite—ERTS) on 23 July 1972 by the National Aeronautics and Space Administration (NASA). NASA and the US Department of the Interior through the US Geological Survey (USGS) jointly manage the Landsat program. After Landsat 1, a series of Landsat satellites (Landsat 2–8) were launched to provide high quality images to researchers, land managers, and policy makers to help in the management of natural resources globally (Table 1). Images acquired from Landsat have been used for land use classification, crop classification, and monitoring and irrigation water requirement estimations in many parts of the world [70–75]. Later, in 1984, the Landsat 5 Thematic Mapper was launched to collect higher resolution (30 m) images in more bands in visible and NIR region. Currently, the USGS-NASA is planning to launch Landsat 9 (resolution 30 m, 100 m) by mid-2021. In 1986 and 1988, France and India also launched the SPOT 1 and IRS-1A satellites, respectively (Table 1).

Table 1. Spatiotemporal resolutions of the satellite sensors used for precision agricultural (PA) applications. Satellites that provide high spatial (<30 m) and temporal resolutions (e.g., daily) are more suitable for PA.

Satellite (Years Active)	Sensor (Spatial Resolution)	Temporal Resolution	Application in Precision Agriculture
Landsat 1 (1972–1978)	MS (80 m)	18 days	Crop growth [76]
AVHRR (1979–present)	MS (1.1 km)	1 day	Nutrient management [77]
Landsat 5 TM (1984–2013) Landsat 7 (1999–present) Landsat 8 (2013–present)	MS and Thermal (60 m–Landsat 7, 100 m–Landsat 8, 120 m–Landsat 5)	16 days	Biomass [78]; crop yield [79]; crop growth [80]

Table 1. *Cont.*

Satellite (Years Active)	Sensor (Spatial Resolution)	Temporal Resolution	Application in Precision Agriculture
SPOT 1 (1986–1990) SPOT-2 (1990–2009)	MS (20 m)	2–6 days	Water management [81]
IRS 1A (1988–1996)	MS (72 m)	22 days	Water management, nutrient management [82]
LiDAR (1995)	VIS (10 cm)	N/A	Topography, nutrient management [83]
RadarSAT (1995–2013)	C-band SAR (30 m)	1–6 days	Crop growth [84]
IKONOS (1999–2015)	MS (3.2 m)	3 days	Crop yield [85]; soil properties [86]; nutrient management [77]; ET estimation [87]
EO-1 Hyperion (2000–2017)	HS (30 m)	16 days	Disease [88,89]
Terra/Aqua MODIS (Terra-1999–present, Aqua-2002–present)	MS (SpectroRadiometer; 250–1000 m)	1–2 days	Crop yield [90]; crop growth [91]
Terra-ASTER (2000–present)	MS and Thermal (15 m–V, NIR, 30 m–SWIR, 90 m–TIR)	16 days	Water management [92]
QuickBird (2001–2014)	MS (2.44 m)	1–3.5 days	Disease [93]
AQUA AMSR-E (2002–2016)	MS (Microwave Radiometer; 5.4 km–56 km)	1–2 days	Water management [94]
Spot-5 (2002–2015)	MS (V, NIR–10 m, SWIR–20 m)	2–3 days	Crop yield [95]
ResourceSat-1 (2003–2013)	MS (5.6m–V, 23.5 m–SWIR)	5 days	Nutrient management [96]
KOMPSAT-2 (2006–present)	MS (4 m)	5.5 days	Crop yield [97]
Radarsat-2	C-band SAR (1–100 m)	3 days	LAI and biomass [98]
RapidEye (2008–present)	MS (6.5 m)	1–5.5 days	Water management [99]; crop yield [100]; crop growth and chlorophyll [101]
GeoEye-1 (2008–present)	MS (1.65 m)	2.1–8.3 days	Nutrient management [102]
WorldView-2 (2009–present)	MS (1.4 m)	1.1 days	Crop growth [103]
Pleiades-1A (2011–present) Pleiades-1B (2012–present)	MS (2 m)	1 day	Crop growth [104,105]
VIIRS Suomi-NPP (2011–present) VIIRS-JPSS-1 (2017–present)	MS (IR Radiometer, 375 m and 750 m)	16 day (repeat)	Crop management (NDVI [106])
KOMPSAT-3 (2012–present)	MS (2.8 m)	1.4 days	Crop growth [107]
Spot-6 (2012–present), Spot-7 (2014–present)	MS (6 m)	1-day	Disease [108]
SkySat-1 (2013–present) SkySat-2 (2014–present)	MS (1 m)	sub-daily	Crop growth [109]
Worldview-3 (2014–present)	SS (1.24 m)	<1 day	Crop growth [110]; weed management [102]
Sentinel-1 (2014–present)	C-band SAR (5–40 m)	1–3 days	Crop growth
Sentinel-2 (2015–present)	MS (10 m–V and NIR, 20 m–Red edge and SWIR, 60 m–2 NIR)	2–5 days	Yield [111]; N management [112]
KOMPSAT-3A (2015–present)	MS (V NIR–2.2 m, SWIR–5.5 m)	1.4 days	Disease [113]
SMAP (2015–present)	L-band SAR (1–3 km) and radiometer (40 km)	2–3 days	Crop yield [114]; water management [115]
TripleSat (2015–present)	MS (3.2 m)	1 day	Crop growth [116]
ECOSTRESS-PHyTIR (2018–present)	Thermal (38 × 69 m)	1–5 days	ET [117]

Satellite products from these missions were used for land use and crop classification in many large regions of the world. In addition, satellite products are used to monitor soil health, vegetation health, and hydrologic and climatic parameters, which are important for PA (e.g., soil organic carbon, soil moisture, NDVI, leaf area index (LAI), groundwater, and rainfall). Use of the satellite images proved to be cost-effective compared to aerial photography previously used for land use classification over large regions. However, coarse spatiotemporal resolution satellite products are not quite adequate for many PA applications.

Satellites adequate for PA, such as IKONOS, were launched in the late 1990s. IKONOS, launched in 1999, collected imageries at 4-m spatial resolution in visible and NIR bands with a revisit period of up to five days [42]. Imageries collected from IKONOS have been used for multiple purposes in PA, including soil mapping, crop growth and yield prediction, nutrient management, and ET estimation [77,85–87]. Launching of numerous nanosatellite constellations during later years addressed further limitations associated with spatial, spectral, and temporal resolution of the satellite imagery [110]. Nanosatellite constellations consist of a large number of small satellites with compact sensors that are cheaper and replaceable [91].

Nanosatellites and other satellites launched after 2000, such as GeoEye-1 (2008), Pleiades-1A (2011), Worldview-3 (2014), SkySat-2 (2014), and Superview-1 (2018), collect multispectral images at a high spatial resolution of ≤2 m with a daily or sub-daily revisit period. Pleiades-1A and Worldview-3 have been used for many PA applications requiring high spatial resolution imagery, including disease and crop water stress detection [118–120]. To take advantage of a wide variety of publicly available satellite data, several data fusion approaches have been proposed to combine high/moderate spatial resolution data with high temporal resolution data (and vice-versa) to generate high spatial–temporal resolution data products [121,122]. Satellite data with moderate spatial resolution but high temporal resolution (e.g., Sentinel 2) can also be used with reference ground truth data to help develop PA decision support systems [112].

Despite significant advances in spatial, spectral, and temporal resolution of satellite sensors, the use of satellite images is still limited in commercial agriculture production. Limited flexibility in on-demand imaging solutions, high costs, cloud cover restriction, and lack of automated or established frameworks for image analysis and application are factors affecting large-scale adoption of satellite imageries in PA [123]. These limitations have promoted interest in low-cost proximal remote sensing techniques, including UAVs. Use of UAVs and hand-held, tractor-mounted, and other sensors mounted on farm machinery (e.g., spray boom, fertilizer applicator) has increased tremendously during the last two decades. UAVs with multispectral, hyperspectral, and thermal sensors can provide on-demand information at a spatial scale necessary for PA operations. Getting continuous or frequent satellite scanning during a crop growing season can be problematic due to cloud cover and/or other limitations/uncertainties associated with the sensor platform (e.g., revisit period) [124]. However, UAVs can be flown multiple times during a growing season to acquire information on a cm-scale as required. Most satellites do not offer the data at cm-scale needed for many field-scale PA applications such as weed mapping and disease detection [125,126].

Unprecedented availability of low-cost UAVs is likely to change the face of PA in the future. The average size of a farm in the United States was 179 ha in 2018 [127] and is even smaller in other countries [126]. Acquiring high spatial resolution images obtained from commercial satellites can be expensive, especially for small farms, as many of these images are not available for free. In addition, flying airplanes to obtain such images may be cost-prohibitive for small farms. Images acquired from UAVs offer a low cost alternative to expensive airplane and satellite products [126]. Although the cost of utilizing UAVs (including equipment cost, data processing, and software) at a commercial scale is likely to be a sizable investment for limited resources farmers, continuous development of low-cost sensor technologies and input-production cost savings and/or benefits are likely to outweigh these costs in the future [128,129]. Growth in the development and use of UAVs has enabled the acquisition of high spatial, spectral, and temporal resolution data needed to implement PA management at a crop field or farm scale. Multispectral, high spatial resolution data acquired with UAVs could also be used with available satellite data for scale-up applications over large areas [130].

4. Vegetation Indices

Solar radiation reflected by plants depends on the chemical and morphological characteristics of the plant. Plant type, water content, and canopy characteristics affect the light reflected in each spectral band differently. Measured reflected light in ultraviolet, visible (blue, green, red), and near- and mid-infrared portions of the spectrum has commonly been used to develop various vegetation indices that provide useful information on plant structure and conditions [131] (Table 2). Vegetation indices are mathematical expressions that combine measured reflectance in many spectral bands to produce a value that helps assess crop growth, vigor, and several other vegetation properties such as biomass and chlorophyll content [132]. Mapping of these indices can help understand spatio-temporal variability in crop conditions, which is crucial for PA applications.

Popularly used vegetation indices such as normalized difference vegetation index (NDVI), green NDVI (GNDVI), and soil adjusted vegetation index (SAVI) (Table 2) utilize the fact that within the visible range of spectrum, plant reflectance is low in blue and red regions, while it peaks in the green region (Figure 2). Plant pigments, mainly chlorophyll and carotenoids, adsorb strongly in the visible part of the spectrum except for the green region. However, such strong adsorption does not occur in the NIR part of the spectrum, thus causing high reflectance in NIR region from green and healthy plants (Figure 2). The NDVI uses measured reflectance values in red and NIR regions to provide valuable information on crop growth (LAI, biomass), vigor, and photosynthesis (Table 2). The value of NDVI ranges from −1 to 1, where positive values indicate increasing greenness (LAI and vigor), and negative values indicate non-vegetated surfaces such as urban areas, bare soil/land, water, and ice. External factors to the vegetation conditions such as solar and viewing geometry, soil and crop residue on the land surface, and atmospheric effects may cause interferences in spectral signals [133]. NDVI is sensitive to confounding effects caused due to soils, atmosphere, cloud, and leaf canopy shadow that may result in erroneous information on crop or plant conditions [134,135]. In addition, NDVI is also known to be insensitive to changes in LAI and biomass after reaching a threshold (saturation), especially in dense vegetative conditions [136,137]. A large number of alternative indices have been developed to address these shortcomings in NDVI [135,138,139]. Some of the indices that address these limitations are the soil adjusted vegetation index (SAVI), atmospherically resistant vegetation index (ARVI), and wide dynamic range vegetation index (WDRVI). Red edge based vegetation indices such as red-edge NDVI (RNDVI), normalized difference red edge (NDRE), and red edge difference vegetation index (REDVI) have been shown to perform better than NDVI in estimating plant nutrient status, LAI, and biomass in dense vegetation conditions such as those present during the later growth stages of corn [139–142].

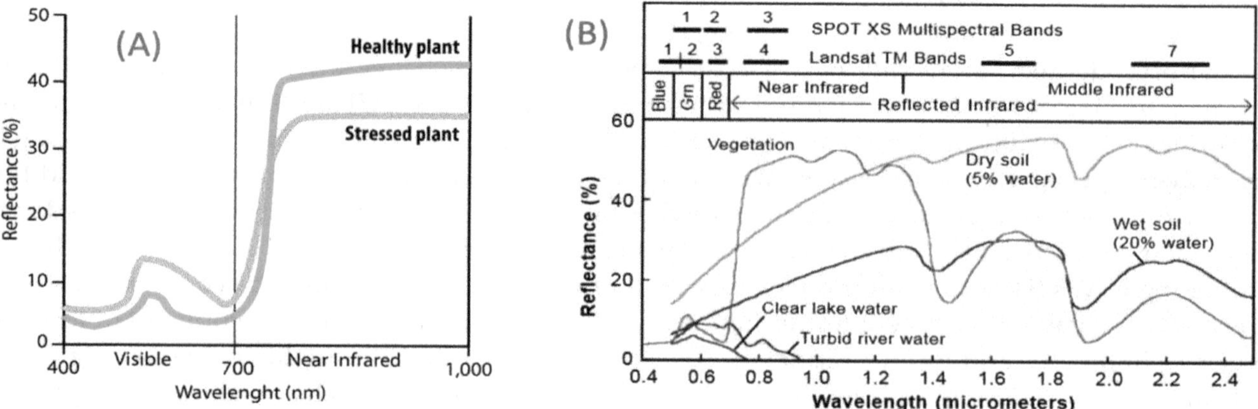

Figure 2. Typical reflectance spectrum of (**A**) a healthy and a stressed plant (taken from Govaerts and Verhulst [143]) and (**B**) soil, water, and vegetation (taken from Mondal [82]).

Table 2. Some recently used vegetation indices for remote sensing applications in precision agriculture *.

Index	Definition/Equation	Applications (References)
Normalized difference vegetation index (NDVI)	$\frac{R_{NIR}-R_{red}}{R_{NIR}+R_{red}}$	Biomass [144]; breeding, phenotyping [145]; yield [146]; disease [108]; n-management [147]; soil moisture [148]; water stress [149]
Green NDVI (GNDVI)	$\frac{R_{NIR}-R_{green}}{R_{NIR}+R_{green}}$	Water stress [150]; yield [151]; biomass [28,152,153]; disease [154]
Normalized difference red edge (NDRE)	$\frac{R_{NIR}-R_{red\ edge}}{R_{NIR}+R_{red\ edge}}$	Crop yield and biomass [155]; N-management [147]; disease [154,156]
Red edge normalized difference vegetation index (RENDVI)	$\frac{R_{NIR}-R_{red\ edge}}{R_{NIR}+R_{red\ edge}}$	Yield [100,111]; irrigation management [99]; N-status/application [140]; disease [156]
Soil adjusted vegetation index (SAVI)	$\frac{(R_{NIR}-R_{red})(1+L)}{R_{NIR}+R_{red}+L}$ L-soil conditioning index	Yield [79]; biomass [28,153]; disease [157]; N-concentration and uptake [142]; water stress [158]
Modified soil adjusted vegetation index (MSAVI)	$\frac{2R_{NIR}+1-\sqrt{(2R_{NIR}+1)^2-8(R_{NIR}-R_{red})}}{2}$	Biomass [153]; crop yield [159]; N-uptake [142]; chlorophyll content [112,160]
Renormalized difference vegetation index (RDVI)	$\frac{R_{NIR}-R_{red}}{\sqrt{R_{NIR}+R_{red}}}$	Crop yield [159]; N-uptake [142]; soil moisture [148]; biomass [28]
Wide dynamic range vegetation index (WDRVI)	$\frac{\alpha R_{NIR}-R_{red}}{\alpha R_{NIR}+R_{red}}$	N-Application, yield [161]; crop growth (LAI) [162]; disease [113]
Atmospherically resistant vegetation index (ARVI)	$\frac{R_{NIR}-R_{RedBlue}}{R_{NIR}+R_{RedBlue}}$	Disease [108]; weed mapping [163]
Atmospherically effect resistant vegetation index (IAVI)	$\frac{R_{NIR}-(R_{red}-\lambda(R_{blue}-R_{red}))}{R_{NIR}+(R_{red}-\lambda(R_{blue}-R_{red}))}$	Crop yield [164]
Ratio vegetation index (RVI)	$\frac{R_{NIR}}{R_{red}}$	Crop yield [159]; biomass [28]
Difference vegetation index (DVI)	$R_{NIR}-R_{Red}$	Disease [154]; crop yield [159]; LAI [142]
Red edge DVI (REDVI)	$R_{NIR}-R_{RedEdge}$	Crop yield and biomass [155]; biomass, N-uptake, and concentration [142]
Transformed soil adjusted vegetative index (TSAVI)	$\frac{a(R_{NIR}-aR_{Red}-b)}{R_{Red}+aR_{NIR}-ab}$	Water stress [158]; crop yield [165]
Plant senescence reflectance index PSRI)	$\frac{R_{680}-R_{550}}{R_{750}}$	Disease [166]; yield [167]; biomass [28]
Normalized pigment chlorophyll ratio index (NPCI)	$\frac{R_{680}-R_{430}}{R_{680}+R_{430}}$	Water stress [168]
Chlorophyll absorption ratio index (CARI)	$\frac{R_{700}}{R_{670}}\times\frac{aR_{670}+bR_{670}}{\sqrt{a^2+1}}$	Chlorophyll content [169]
Modified chlorophyll and reflectance index (MCARI)	$\frac{R_{RedEdge}}{R_{Red}}\times$ $\{(R_{RedEdge}-R_{Red})-0.2(R_{ReEdge}-R_{Green})\}$	Crop growth–chlorophyll content [101]
Chlorophyll vegetation index (CVI)	$\frac{R_{NIR}}{R_{Green}}*\frac{R_{Red}}{R_{Green}}$	Crop yield [170]; crop growth-chlorophyll content [101]; yield [111]
Chlorophyll index (CI)	$\frac{R_{NIR}}{R_{RedEdge}}-1$	Chlorophyll and N-content [171]
Optimized soil adjusted vegetation index (OSAVI)	$\frac{1.16(R_{NIR}-R_{Red})}{R_{NIR}+R_{Red}+0.16}$	Disease [153]; crop yield [159]; biomass, N-uptake [28,142]; soil moisture [148]; water stress [158]
Photochemical reflectance index (PRI)	$\frac{R_{531}-R_{570}}{R_{531}+R_{570}}$	Disease [172]; leaf water stress (PRI_{norm}), canopy temperature and yield (PRI_{550}) [148]; water stress (PRI, $PRI_{550-515}$, PRI_{norm} [149]
Water balance index	$\frac{R_{1500}-R_{531}}{R_{1500}+R_{531}}$	Irrigation scheduling [173]
Normalized difference water content (NDWI)		Vegetation water content [174]
Shortwave infrared water stress index (SIWSI)	$\frac{R_{858.5}-R_{1640}}{R_{858.5}+R_{1640}}$	Leaf water content (water stress; [175])

Table 2. *Cont.*

Index	Definition/Equation	Applications (References)
Degrees above non-stressed canopy (DANS)	$\min(0, T_c - T_{cNonStressed})$	Water stress [176]; ET [177]
Degrees above canopy threshold (DACT)	$\max(0, T_c - T_{critical})$	ET [177]; water stress [176]
Triangular vegetation index (TVI)	$0.5[120(R_{NIR} - R_{Green}) - 200(R_{Red} - R_{Green})]$	Disease [108,156,172]
Red-edge inflection point (REIP)		Yield and biomass [155]
Enhanced vegetation index (EVI)	$\frac{2.5(R_{NIR}-R_{Red})}{(R_{NIR}+6R_{Red}-7.5R_{Blue}+1)}$	Disease [157]; biomass [28]
Normalized water index (NWI)	$\frac{R_{970}-R_{900}}{R_{970}+R_{900}}$	Soil moisture and crop yield [148]

* This list is an effort to compile some recently used vegetation indices, it is not meant to be a comprehensive list as there are many more indices that have been used in PA applications.

Measurement of reflectance or emissivity in near- and mid-infrared bands is particularly useful in developing indices that help to understand several intrinsic plant characteristics such as water content, pigments, sugar, carbohydrate, and protein contents [131]. Reflected or emitted radiation in thermal infrared bands is directly related to the plant temperature. Since plant temperature is related to plant transpiration rate, indices obtained from thermal/infrared reflectance data can be used to understand plant water status and other biotic and abiotic stresses such as disease [178,179]. In the past, many indices have been developed based on infrared and thermal reflectance or emission such as crop water stress index (CWSI) and shortwave infrared water stress index (SIWSI; Table 2). These indices have been used for a variety of applications in PA, including water stress and drought monitoring, soil moisture, plant disease, and crop yield estimations [57] (Table 2).

5. Applications

5.1. *Irrigation Water Management*

Application time and rate of irrigation play an important role in mitigating crop water stress and achieving optimum crop growth and yield. A variety of irrigation management practices are used by farmers depending on many factors including water availability, existing water management infrastructure at the farm (e.g., storage and conveyance system, type of irrigation system), local/regional water laws, economic status, size of the farm, knowledge of farmer, and others [180,181].

Many farmers apply uniform irrigation at regular intervals based on their prior knowledge or experience of farming, soils, and climate at the location [182]. Large commercial farmers deploy soil moisture monitoring systems (wired or wireless moisture sensors) to irrigate (automatically or manually operation mode) based on the measured soil moisture data and crop/plant water requirements [183,184]. Local and regional agricultural agencies may provide irrigation advisory services based on the observed climate and weather conditions in the area [185,186].

Almost all of these conventional farming practices do not consider the variability within a field and use a uniform irrigation rate for the entire field. Remote sensing data can help discern the variability within the field and apply variable rate irrigation with commonly used irrigation systems such as a center pivot. Variable rate application can help mitigate water stress arising from extreme wet and dry conditions to achieve uniformly high yields in all parts in the field while reducing water and nutrient losses [187,188]. Remote sensing images, collected multiple times during a growing season, are used to determine various indicators of crop water demand such as ET, soil moisture, and crop water stress. These indicators are used to estimate crop water requirement and schedule irrigation precisely.

5.1.1. Water Stress

Remote sensing products, either in optical, thermal, and microwave bands, have been used to develop and test multiple indices and techniques for precision water management [189] (Table 2).

For example, normalized difference vegetation index (NDVI) and soil adjusted vegetation index (SAVI), developed from optical images, can be used to diagnose water stress and soil moisture conditions for many crops (Table 2). As shown in Table 2, these indices, combined with forecasted weather data, could be used for irrigation scheduling. Thermal remote sensing–based crop water stress index (CWSI) is a popular indicator used to estimate irrigation water demand and scheduling [57].

$$CWSI = \frac{(T_c - T_a) - (T_c - T_a)_{LL}}{(T_c - T_a)_{UL} - (T_c - T_a)_{LL}}$$

where T_c is the canopy temperature extracted from thermal images, and T_a is the air temperature. *LL* and *UL* indicate the upper and lower limit of the difference in canopy and air temperature. Conceptually, the lower limit (*LL*) corresponds to the condition when the canopy is transpiring at the potential rate, and the upper limit (*UL*) corresponds to the condition when transpiration from the canopy is ceased. Multiple methods have been used to calculate the *UL* and *LL* of difference in canopy and air temperature, each having their own set of strengths and weaknesses [57]. *CWSI* has been extensively used for precision irrigation management in orchards [44,190]. For example, Katsigiannis et al. [123] used an autonomous multi-sensor (multi-spectral and thermal sensor) UAV system to develop *CWSIs* maps for irrigation scheduling and management in kiwi, pomegranate, and vine fields. However, some studies have indicated that more research is needed to establish climate-soil-crop specific trigger/threshold values to enable the use of *CWSI* for irrigation scheduling [191].

5.1.2. Evapotranspiration (ET)

Evapotranspiration (ET), the largest water flux from the Earth's surface to the atmosphere, is a critical component of the hydrologic cycle and water balance. Conventional methods of ET measurement (e.g., weighing lysimeter and eddy covariance) are generally expensive and do not provide spatially variable ET estimates resulting from differences in land use, soils, topography and other hydrologic processes [192,193]. Remote sensing data is widely used to estimate ET, which is needed to determine crop water requirements to schedule irrigation [193–195]. ET estimation approaches, based on the remote sensing data, can be grouped into three categories: (i) surface energy balance, (ii) crop coefficient, and (iii) the Penman–Monteith method [193,194,196]. Many studies in the past have provided a review of remote-sensing-based ET estimation techniques [39,192,197], including a recent review from Zhang et al. [198] that discussed the theories of development of several ET estimation approaches/methods along with their advantages and limitations. The surface energy balance approach has been extensively used for ET estimation in the past five years. Some of the studies have used hybrid methods to combine crop coefficient and energy balance approaches for ET estimation [195]. In a surface energy balance approach, net radiation flux (R_n), soil heat flux (G), and sensible heat flux (H) are typically calculated from remotely sensed data in visible, near-infrared and thermal infrared bands, while the latent heat flux (λET) is calculated as a reminder of the term in the energy balance equation [195,196].

$$\lambda ET = R_n - G - H$$

Recently, Liou and Kar [192] and McShane et al. [197] provided a detailed review of various surface energy balance algorithms used for estimating landscape-scale ET at high spatial resolution and discussed their physical basis, assumptions, uncertainties, and limitations. Surface energy balance techniques use a variety of empirical and physically based models, input data, and assumptions to fully or partially solve the energy balance equation for ET estimation. Popularly used methods such as surface energy balance algorithm for land (SEBAL) and mapping ET at high resolution with internalized calibration (METRIC) fully solve the energy balance equation by determining the sensible heat flux (H) based on the identification of dry and wet pixels in the image [199]. Empirical constants derived from dry and wet pixels and measured radiometric surface temperature are then used to determine pixel-wise surface-air temperature difference and sensible heat-flux. In METRIC, a variant of SEBAL, net radiation is calculated using measured narrowband reflectance and radiometric surface temperature data, while soil heat flux is estimated using net radiation, surface temperature and vegetation indices [197]. Most of the surface energy balance methods typically use similar methods to calculate net radiation and soil heat flux; the major difference lies in the way sensible heat flux is calculated [200]. The primary difference between METRIC and SEBAL is that the METRIC method uses reference ET data calculated from ground-based weather data to calibrate sensible heat and ET estimations obtained from the surface energy balance [199]. Surface energy balance methods are also classified based on the sensible heat sources; two-source energy balance methods account for the individual contributions of soil and vegetation to the total heat flux, while one-source methods do not distinguish between soil and vegetation [195]. Simpler surface energy balance methods, such as simplified surface energy balance index (S-SEBI) and the operational simplified energy balance model (SSEBop) does not require pixel-wise calculation of sensible heat flux and can thus be considered as partial surface energy balance models [199]. Recently, several studies have evaluated these energy balance approaches with satellite data (e.g., Landsat 7 and 8) to estimate ET for irrigation management in PA [201–203]. However, coarse spatio-temporal resolution of thermal satellite data limits the application of surface energy balance approaches in PA. More studies are needed to develop and evaluate downscaling techniques that can provide ET estimates at a spatio-temporal resolution necessary for irrigation scheduling in PA.

Crop coefficient based approach for ET estimation involves developing a statistical relationship between a vegetation index (e.g., NDVI, SAVI) and basal crop coefficient (or crop coefficient). A product (ET_{ref}*K_c) of reference ET (ET_{ref}), estimated from weather data, and crop coefficient (K_c) provides an estimation of potential crop water requirements which can be used in a water balance model to determine irrigation water requirements [122,204–207]. In the third approach, crop biophysical parameters such as LAI, canopy height, and albedo, estimated from remotely sensed data, are used to determine unknown parameters in Penman-Monteith equation [196,208] for ET estimation. Calculated ET from these approaches is generally used in a soil water balance model to estimate crop irrigation requirements. Bhatti et al. [205] used Landsat 7 and 8 and UAV data to estimate crop coefficient based ET and variable rate irrigation (VRI) requirements for corn and soybean grown in Nebraska, USA. Similarly, Stone et al. [209] demonstrated that variable rate irrigation scheduling based on the NDVI derived crop coefficient and ET can provide corn yields similar to the other precision irrigation techniques relying on measured in-situ soil moisture data. Using crop parameters (LAI and albedo) derived from Sentinel 2A/B images, Bonfante et al. [210] calculated Penman–Monteith ET and time-variable irrigation requirements to demonstrate the application of a publicly available decision support system for maximizing corn yields in Italy.

A large number of approaches exist for ET estimation based on the remote sensing data, each having its own set of advantages and limitations. Some surface energy balance approaches such as S-SEBI do not require any ground-based measurements, and thus, ET can be estimated solely based on the remote sensing data [192]. However, surface energy balance approaches work for only clear sky conditions and suffer from various uncertainties related to the retrieval/estimation of surface temperature, solar parameters, land surface variables (e.g., LAI, vegetation coverage, plant height) and other parameters due to uncertainties in surface emissivity, atmospheric corrections, diurnal variation and aerodynamic surface characteristics [192,198]. In addition, ET estimates derived by solving the surface energy balance equations are instantaneous and require temporal extrapolation to generate daily or larger time step estimates using further assumptions and methods [198]. The Penman–Monteith approach, which uses a process-based model, can provide temporally continuous ET estimations. However, it requires several meteorological parameters that are not easily available at the needed spatio-temporal resolution. Studies have shown that surface energy balance methods such as METRIC and SSEBop provide reasonably accurate (80–95% accuracy) ET estimations on daily to seasonal/annual scales [197]. Generally, ET estimation error of less than 1 mm/day can be considered to be reasonable [211]. However, it also depends on the crop and growth stage; higher accuracies (mm/day) are desirable during initial crop growth stages when ET is lower. However, the accuracy of many of these models (e.g., SEBAL) varies greatly from one spatial and temporal scale to another, making it difficult to quantify the uncertainty in ET estimation [194]. In order to advance the development of remote sensing based ET estimation approaches, further studies are needed to identify and characterize the spatio-temporal structure of uncertainties in the ET estimation due to forcing errors, process errors, and parametrization errors, among others [198]. Advanced approaches that use the knowledge developed from process-based physical models to complement the machine learning based surface energy balance methods need to be explored to develop more accurate and reliable ET estimates at a scale necessary for PA management [212].

5.1.3. Soil Moisture

Remote sensing data acquired in multiple bands, including optical, thermal, and microwave, have been used to estimate soil moisture globally [28,193,213]. Optical and thermal remote sensing data has been extensively used for soil moisture and ET estimations in an approach referred to as "triangle" or "trapezoid" or land surface temperature-vegetation index (LST-VI) method [198,214–216]. The triangle or LST-VI method is based on the physical relationship between the land surface temperature (and thus soil moisture and latent heat fluxes) and vegetative cover characteristics. Soil moisture estimation in this method is based on the interpretation of the pixel distribution in the LST-VI plot-space. If a sufficiently large number of pixels are present in an image covering a full range of soil moisture and vegetation density and when cloud, surface water, and other outliers are removed, the LST-VI space resembles a triangle or trapezoid [214]. One edge of the LST-VI triangle or trapezoid falling toward higher temperatures represent dry edge (low soil moisture), while the opposite side represents the wet edge (high soil moisture) [217]. Triangular or trapezoidal shape of the LST-VI space is formed due to the low sensitivity of LST to soil moisture under dense vegetative conditions, which is contrary to the high sensitivity of LST to soil moisture under bare soil or sparse vegetation conditions. Once upper and lower limit moisture content for wet and dry edges is determined, theoretically, soil moisture for remaining pixels can be estimated using interpolation techniques. The triangle method uses a simple parametrization approach and does not require ancillary atmospheric or surface data for soil moisture estimation [214,218]. However, a subjective determination of wet and dry edges in the triangle method can introduce significant uncertainties in soil moisture estimation especially over relatively homogenous land-surface areas (e.g., rainfed agriculture during dry season) and after rainfall events when there is a general lack of variability in soil moisture to form an LST-VI triangle [215]. In addition, the traditional triangle method requires individual parametrization for each observation date, which is a time-consuming and computationally demanding process [60]. The traditional triangle method also needs both optical and thermal data that may not be available in certain instances (e.g., Sentinel-2).

Recently, a newer generation of triangular models has been developed and tested for high spatial resolution mapping of soil moisture in PA applications [59,60]. One such technique, called the optical trapezoid model (OPTRAM), replaces the LST in the traditional triangle model with short-wave-infra-red transformed reflectance (STR). Analogous to the traditional triangle model, soil moisture in OPTRAM is estimated based on the interpretation of STR-VI space [60]. Using Sentinel-2 and Landsat-8 data, Sadeghi et al. [53] showed reasonably accurate (<0.04 cm^3/cm^3) soil moisture estimations with OPTRAM model for grass and cropland dominated watersheds in Arizona and Oklahoma, USA. The OPTRAM model does not require thermal remote sensing data, thus making it more suitable for a wide range of remote sensing data. Unlike LST, surface reflectance (STR) is a function of surface properties and does not vary significantly with ambient atmospheric conditions, which eliminates the need to parametrize/calibrate the model for individual dates [216]. However, compared to the traditional triangle model, the OPTRAM model is more sensitive to saturated land areas, which may result in increased uncertainty in soil moisture estimates [60]. More studies are needed to further evaluate these newer models for their applicability in a diverse range of climatic, hydrologic, and environmental conditions.

Compared to the data acquired in visible, NIR, and SWIR bands, microwave remote sensing data have a greater potential to provide accurate soil moisture estimations [216]. Signals in visible and NIR regions have weaker penetration ability, compared to microwave, and is more likely to get affected by interferences caused due to atmospheric and cloud conditions [135]. Microwave sensors measure dielectric properties of soil based on land surface emissivity or scattering for soil moisture estimation. Several satellites with active and passive microwave sensors have been launched for soil moisture monitoring such as the advanced microwave scanning radiometer-earth observing system (AMSR-E), soil moisture and ocean salinity (SMOS), soil moisture active passive (SMAP), and Sentinel-1 [216].

Active microwave sensors provide higher spatial resolution as compared to the passive sensors. However, active sensors also suffer from measurement uncertainties caused due to land surface roughness and vegetative cover or canopy [219]. On the other hand, although passive sensors are more accurate and provide better temporal resolution, they also provide a coarser spatial resolution (e.g., 10s of km) [215]. Typically, watershed and regional scale hydrologic and agricultural applications, especially PA, require finer resolution data [220].

Several spatial downscaling techniques have been developed and used in the past to address the coarse spatial resolution limitation of passive microwave sensors. These techniques can be broadly categorized into two main groups: (i) satellite-based methods and (ii) modeling methods [221]. Typically, satellite-based methods use vegetation, surface temperature, and other geophysical and environmental information obtained from high spatial resolution satellite data (in optical and thermal bands) to downscale coarse resolution soil moisture data [61,62].

Machine learning and data mining techniques have been used with topography, land surface temperature, albedo, land cover, NDVI, and ET obtained from high spatial resolution satellite data to downscale coarse resolution soil moisture data from passive microwave sensors [135,222]. Higher resolution (1 km) land surface temperature and vegetation data derived from moderate resolution imaging spectroradiometer (MODIS) have popularly been used in many studies to downscale coarse resolution soil moisture data [135,223,224]. Many others have used higher spatial resolution backscatter data from active microwave sensors to disaggregate and downscale brightness and soil moisture data from passive sensors [225]. Some studies have used a combination of SAR (Sentinel 1/2) and optical remote sensing data with machine learning techniques to downscale passive microwave (e.g., SMAP) data [226,227]. Modeling-based methods use high spatial resolution data derived from physically-based and statistical models with data assimilation approaches to downscale coarse-resolution soil moisture data [221,228]. Despite the development of these downscaling techniques, spatio-temporal resolution of soil moisture estimates generated from microwave data is still coarse and/or lacks the accuracy needed for PA, limiting their application for irrigation scheduling in PA. Recent launches of multi-polarization sensors (e.g., Radarsat-2) along with the development of new polarization parameters and polarization decomposition methods can potentially help generating soil moisture and crop information at higher spatial resolution [229]. Multi-polarization SAR data can be combined with optical satellite data to generate soil moisture estimates at high spatial resolution [230]. Further efforts are needed to develop advanced downscaling techniques that can combine data from multiple sources (e.g., optical and microwave remote sensing, biophysical and hydrologic models) and provide temporally continuous (e.g., daily) soil moisture estimates at the finer spatial resolution necessary for PA [221]. Overall, in addition to satellites, high spatial resolution data acquired from UAVs in optical, thermal, infrared, and microwave bands carry great promise for soil moisture mapping and other PA applications [231–233].

5.2. Nutrient Management

Timely and appropriate application of fertilizers is essential to optimize crop growth and yields while minimizing environmental damage through nutrient losses to groundwater and surface water. Typically, a recommended rate of fertilizer is uniformly applied during planting and later crop growth stages. However, the fertilizer requirement of crops varies spatially and temporally (during and among seasons) due to differences in soils, management, topography, weather, and hydrology [12,234]. Mapping of such variability in crop nutrient status/requirement for PA applications could be challenging with conventionally used tools such as chlorophyll meters [165].

Several vegetation indices (e.g., NDVI, SAVI), derived from remote sensing data, have been shown to be significantly correlated with plant chlorophyll content, photosynthetic activity, and plant productivity (Table 2). Mapping of these indices can thus help understand the spatial variability in crop nutrient status, which is important for PA. Recently, several tractor mounted remote sensors have become available that can measure plant nutrient status for real-time application of spatially-variable fertilizer rates. Green Seeker, Yara N-sensor, and Crop Circle are some examples of commercially available hand-held and tractor mounted remote sensors that use crop reflectance data to determine and apply spatially variable fertilizer rates in real-time [235].

In tractor mounted systems, remote sensors are usually mounted ahead of the spray boom. Nitrogen (N) application rates in these systems are determined based on the calculated vegetation indices (e.g., NDVI), which are further communicated to nutrient applicator/spreader for real-time fertilizer application. Different algorithms are used to convert the measured vegetation indices into recommended N-application rates. In general, the N-application rates are calculated by comparing measured vegetation indices in the target field with a reference vegetation index measured in a well fertilized (N-rich) plot/strip that is representative of the target field. Several fertilizer rate calculation algorithms (e.g., the nitrogen fertilizer optimization algorithm) [236,237] have been developed and successfully implemented in these commercially available sensors to determine vegetation-indices based in-season N-requirements for many crops [238,239].

Despite the commercialization of variable rate N-management technologies based on proximal remote sensing, farmer's adoption remains low in many agricultural enterprises [240]. A lack of clear evidence on significant economic benefits (crop yield and/or profits), especially in commercial farm settings (i.e., large fields) is a limiting factor for the large scale adoption of these technologies [241]. To further refine these remote sensing based technologies and enhance its benefits, research is being conducted with UAVs and other remote sensors for a variety of crops in different climatic regions. Maresma et al. [165] used images captured from a UAV to determine the suitability of several vegetation indices and crop height in determining in-season fertilizer application rates for corn grown in Spain. Cao et al. [242] showed that Green Seeker and Crop Circle sensors resulted in reduced N-fertilizer use and increased N-productivity for winter wheat production in China. Overall, remote sensing based mapping of crop nutrient status in PA can help increase crop nutrient use efficiency while maintaining/increasing crop yields and reducing off-site nutrient losses that are damaging to the environment.

Remote sensing has also been used to determine soil organic matter and phosphorus content to develop spatial maps that can aid in site-specific management [243–246]. Blasch et al. [243] used multi-temporal satellite images from RapidEye in conjunction with principal component analysis to develop field-scale soil organic matter maps to aid in site-specific precision management. Castaldi et al. [245] used images obtained from Sentinel-2 to develop soil organic carbon maps at both field and regional scales in Germany, Luxembourg, and Belgium and showed that the satellite data have an adequate spatial resolution for field scale planning in PA management. Crop growth is a function of several biotic and abiotic factors related to soils, management, topography, hydrology, and other environmental variables. Therefore, vegetation indices derived from remote sensing of crop growth/status are reflective of the combined variability in these factors or stressors. While using the vegetation indices to determine the plant nutrient status or N-application rates, one should also consider potentially confounding effects caused due to other stressors such as moisture stress and disease.

5.3. Disease Management

Diseases can cause a significant loss of crop production and farm profits. Early detection of plant disease and its spatial extent can help contain the disease spread and reduce production losses.

Field scouting, a conventional method of disease detection, is time consuming, labor intensive, and prone to human error [126]. In addition, with field scouting, it may be difficult to detect the disease during the early stages when the symptoms are not fully visible. Furthermore, some diseases do not show any visible symptoms, or the effect may not be noticeable until it is too late to act [247]. It is also difficult to map the spatial extent and severity of the disease spread with the traditional method of field scouting.

Remote sensing could be used to monitor the disease efficiently, especially in the early stages of disease development, when it may be difficult to discern the signs of disease with field scouting. Multiple techniques using RGB, multi-spectral, hyperspectral, thermal, and fluorescence imaging have been used to identify diseases in a variety of crops [178]. Di Gennaro et al. [248] found a good correlation between grapevine leaf stripe disease and NDVI generated from UAV imageries in Italy. Abdulridha et al. [172] used a machine learning approach with vegetation indices derived from hyperspectral UAV images to detect citrus canker with 96% accuracy even during the early stage of disease development. As the differences in spectral signatures observed in a field can occur due to a combination of biotic and abiotic stresses, it could be challenging to discern the effects caused due to individual stressors (e.g., disease, water or nutrient stress). As compared to the typically used vegetation indices (e.g., NDVI), the development of disease specific spectral disease indices (SDI) can increase the accuracy of disease detection and differentiation under real world field conditions [249,250]. Higher disease detection accuracies during critical crop growth stages such as flowering can help implement effective management plans in a timely manner to mitigate or avoid large yield and fruit quality losses. Use of SDIs, in place of typical VIs, can also reduce the complexity of disease detection methods and increase the system efficacy by reducing the computational demand [250]. Although studies have been conducted for plant disease classification [251], further efforts are needed to develop more accurate, automated and reproducible methodologies for disease detection under diverse climatic and real-world field conditions [40].

5.4. *Weed Management*

Conventional weed management approaches involving a uniform application of herbicide is an inefficient practice that increases the risk for off-site pesticide losses [252]. Application of herbicide at a variable rate as per need can help enhance the treatment efficiency and reduce input costs and environmental pollution [253]. Remote sensing has widely been used for mapping weed patches in crop fields for site-specific weed management [43]. Weeds can be identified or differentiated from crop plants based on their peculiar spectral signature related to their phenological or morphological attributes that are different from the crop. Over the last few years, machine learning approaches have emerged as a highly accurate and efficient method of image classification for weed mapping [254–256]. Two general types of image classification approaches are commonly used for weed mapping; supervised and unsupervised classification. Although each method has its own strengths and weaknesses, supervised classification is time-intensive and requires more manual work [26].

UAVs have been the most popular remote sensing platform for weed mapping and management primarily due to its capability to produce cm-scale resolution (5 cm) [257]) images needed for weed detection and mapping. Using the fully convolutional network approach, Huang et al. [254] mapped weeds in a rice field in China with up to 90% accuracy. Partel et al. [255] developed a target weed sprayer using deep learning neural network approach for ground-sensor based weed detection that yielded in 71% application accuracy in experimental fields in Florida, USA. However, the commercial adoption of these technologies remains challenging, given the expertise required to use advanced software and technical processes involved in their applications [258].

5.5. *Crop Monitoring and Yield*

Monitoring crop growth and yield are necessary to understand the crop response to the environment and agronomic practices and develop effective management plans for fieldwork and/or remedies [259]. LAI and biomass are two essential indicators of crop health and development [28]. LAI is also used as an input in many crop growth and yield forecasting models [260]. In-situ methods of LAI estimation (physical and optical) are time consuming and labor intensive, similar to the destructive field methods used for biomass estimation. Also, these methods do not provide a spatial variability map of crop growth and biomass [261,262]. Remote sensing data on crop growth (e.g., LAI) and biomass can help obtain valuable information on site-specific properties (e.g., soils, topography), management (e.g., water, nutrient, and other inputs), and various biotic and abiotic stressors (e.g., diseases, weeds, water, and nutrient stress) [263]. Similarly, remote sensing data can also be used to map differences in tillage and residue management and its effects on crop growth [264]. Several studies have used hyperspectral images with various machine learning and classification techniques to map tillage and crop residue in agricultural fields [265,266]. Such information on crop conditions and tillage practices can help develop site specific management plans, including variable water, nutrient, and pesticide application to increase production and management efficiency.

Remote sensing data have been used to estimate LAI and biomass for a variety of crops, including row crops, orchards, and vine crops [267–269]. Typically, such studies use a set of reference data (e.g., measured LAI and corresponding vegetation indices) to develop a regression or machine learning based approach to estimate LAI and/or biomass for a target field. Yue et al. [262] used multiple spectral indices in conjunction with measured plant height to estimate biomass (R^2 = 0.74) in multiple irrigations and nutrient treatment plots for winter wheat grown in China. Ali et al. [269] used red-edge position (REP) extracted from hyperspectral images to predict LAI (R^2 = 0.93) and chlorophyll content (R^2 = 0.90) for Kinnow mandarins grown in Pakistan. REP is the position of the main inflection point of the red-NIR slope created due to strong chlorophyll absorption in the red spectrum and canopy scattering in the NIR region [269]. Reliable LAI estimation from reflectance data could be difficult, especially during early crop growth stages due to interference from the bare soil surface. To overcome this limitation, modified vegetation indices adjusted for soil and other interferences have been proposed and used to estimate LAI [270]. Recently, red-edge based vegetation indices have been shown to be promising for estimating LAI in multiple crops [271].

For PA, knowledge of spatial variability in crop yield is important to understand crop response to management practices and environmental stressors. Remote sensing derived crop biophysical parameters, or vegetation indices have a strong correlation with observed crop yield and biomass, indicating their potential for use in yield estimation [259]. Crop yield estimation from remotely sensed data has generally been conducted in two ways.

First, biophysical parameters (e.g., LAI) derived from remotely sensed data are used in a crop model to estimate the crop yield and biomass. Second, statistical (e.g., regression) or empirical relationships are developed between remote sensing derived crop parameters/indices (e.g., NDVI, LAI) and observed crop yield and biomass in a representative crop field. The developed regression model or empirical relationship could then be used to map crop yield at a target crop field. Crop modeling is a data-intensive approach, which requires a large amount of information as model input parameters, meteorological data, and observed yield and biomass data.

Maresma et al. [161] used a regression based approach to evaluate the relationship between corn yield and biomass and spectral indices measured at the V12 stage. Similar to other studies, they also found that the red-based indices NDVI and wide dynamic range vegetation index (WDRVI) had the highest correlation with grain yields for a range of fertilizer application rates. Spatial mapping of crop biophysical parameters or indices at multiple times during a growing season is likely to provide a better estimation of crop biomass and yield as compared to a single snapshot during the season [263].

6. Progress Made, Needs, and Challenges

Remote sensing has potential applications in almost every aspect of PA, from land preparation to harvesting. The abundance of high spatial resolution multi-temporal satellite data along with low-cost UAVs and commercially available ground-based proximity sensors have changed the face of PA. A large number of advanced techniques, including empirical, regression, and various forms of machine learning approaches have been used to explore the potential applications of remote sensing in PA. Similarly, many vegetation indices have been developed and tested for their ability to help PA operations, including variable fertilizer management, irrigation scheduling, disease control, weed mapping, and yield forecasting. However, many challenges need to be addressed before the remote sensing technologies can potentially see a large-scale adoption in commercial and non-commercial agriculture.

Although most of the satellite data are available for free, it may require a significant amount of technical knowledge and expertise to process them for real-world applications. For example, image pre-processing and post-processing require expert knowledge and software. In addition, many PA operations such as disease and weed management require fine spatial resolution (cm-scale) data with high spectral and temporal (e.g., daily) resolution. Most of the publicly available satellite data do not meet such requirements. Furthermore, cloudy days and variable or inconsistent irradiance or sunlight may render many satellite images unsuitable for use.

Users/farmers may need to purchase high resolution (spatial, temporal and spectral) satellite data, which can be cost-prohibitive, especially for small farms. However, images acquired from UAVs are likely to offer a low-cost alternative for small farm operations [272]. Use of UAVs and tractor-mounted sensors also involve the use of special software for data analysis and need professional operators (e.g., drone licensing) [235]. Hyperspectral images acquired from the state of the art sensors mounted on some of the recently launched satellites and UAVs provide a large amount of information on crop biophysical parameters. However, these sensors are expensive (UAVs), and the processing of imageries is complex [246]. There is a need to explore and develop advanced information and communication technologies as well as chemometric and spectral decomposition methods to synthesize and generate practical information needed for PA applications. Artificial intelligence techniques, including machine learning, has great potential to generate spatially and temporally continuous information from instantaneous satellite data at a scale necessary for many PA applications [212]. Hybrid methods, combining the knowledge obtained from physically based models, can complement such AI techniques to help develop techniques useful in PA decision making [43,212].

Despite a large number of studies on remote sensing applications in PA, there is a general lack of established techniques and/or framework that are accurate, reproducible, and applicable under a wide variety of climatic, soil, crop, and management conditions. Accuracies of methods using remote sensing (satellite, aerial, and UAV) data depends on a variety of factors including image resolution (spatial, spectral, and temporal); atmospheric, climatic, and weather conditions; crop and field conditions (e.g., growth stage, land cover); and the analyses technique (e.g., regression-based, machine learning, physically based modeling). For example, the accuracy of surface energy balance techniques for ET estimation varies significantly in space and time, causing a large uncertainty in PA decision making. More studies are needed to understand the spatio-temporal structure of uncertainty in estimating ET, soil moisture, disease stress, and other crop parameters. Spectral signature from a crop is a reflection of crop status/response to site characteristics (e.g., soil, topography), management, and simultaneously acting multiple biotic and abiotic stressors (e.g., diseases, weeds, nutrient and water stress, etc.). A disease detection method found suitable under controlled experimental conditions may not perform similarly well in real-world conditions where multiple biotic and biotic stressors govern crop response or conditions. Given the complexity of image processing methods and the amount of technical knowledge and expertise it requires for application, there is a need to explore and develop a simple and reliable workflow for image pre-processing, analysis and application in real time. Major challenges and gaps remain in the development of tools and frameworks that can facilitate the use of satellite data for real-time applications by the end-users. Development of accurate, user-friendly systems is likely to result in wider adoption of remote sensing data in commercial and non-commercial PA operations.

Author Contributions: R.P.S., R.L.R., and S.K.S. developed this concept, including method and approach to be used; R.P.S. and R.L.R. outlined the manuscript. R.P.S. and R.L.R. reviewed remotely sensed data; S.K.S. contributed to the discussion of this manuscript. All authors have read and agreed to the published version of the manuscript.

Acknowledgments: This work was supported by the Evans–Allen project of the United States Department of Agriculture (USDA), National Institute of Food and Agriculture. We thank Richard McWhorter, Samiksha Ray, and Tosin Kayode for their time to edit this paper.

References

1. Awokuse, T.O.; Xie, R. Does agriculture really matter for economic growth in developing countries? *Can. J. Agric. Econ.* **2015**, *63*, 77–99.
2. Gillespie, S.; Van den Bold, M. Agriculture, food systems, and nutrition: Meeting the challenge. *Glob. Chall.* **2017**, *1*, 1600002.
3. Patel, R. The long green revolution. *J. Peasant Stud.* **2013**, *40*, 1–63.
4. Pingali, P.L. Green revolution: Impacts, limits, and the path ahead. *Proc. Natl. Acad. Sci. USA* **2012**, *109*, 12302–12308.
5. Wik, M.; Pingali, P.; Broca, S. *Background Paper for the World Development Report 2008: Global Agricultural Performance: Past Trends and Future Prospects*; World Bank: Washington, DC, USA, 2008.
6. World Bank Group. Available online: https://openknowledge.worldbank.org/handle/10986/9122 (accessed on 21 May 2020).
7. Konikow, L.F. Long-term groundwater depletion in the United States. *Groundwater* **2015**, *53*, 2–9.
8. Kleinman, P.J.; Sharpley, A.N.; McDowell, R.W.; Flaten, D.N.; Buda, A.R.; Tao, L.; Bergstrom, L.; Zhu, Q. Managing agricultural phosphorus for water quality protection: Principles for progress. *Plant Soil* **2011**, *349*, 169–182.
9. Wen, F.; Chen, X. Evaluation of the impact of groundwater irrigation on streamflow in Nebraska. *J. Hydrol.* **2006**, *327*, 603–617.
10. Konikow, L.F.; Kendy, E. Groundwater depletion: A global problem. *Hydrogeol. J.* **2005**, *13*, 317–320.
11. Sishodia, R.P.; Shukla, S.; Graham, W.D.; Wani, S.P.; Jones, J.W.; Heaney, J. Current, and future groundwater withdrawals: Effects, management and energy policy options for a semi-arid Indian watershed. *Adv. Water Resour.* **2017**, *110*, 459–475.
12. Hendricks, G.S.; Shukla, S.; Roka, F.M.; Sishodia, R.P.; Obreza, T.A.; Hochmuth, G.J.; Colee, J. Economic and environmental consequences of overfertilization under extreme weather conditions. *J. Soil Water Conserv.* **2019**, *74*, 160–171.
13. Delgado, J.; Short, N.M.; Roberts, D.P.; Vandenberg, B. Big data analysis for sustainable agriculture. *FSUFS* **2019**, *3*, 54.
14. Berry, J.K.; Delgado, J.A.; Khosla, R.; Pierce, F.J. Precision conservation for environmental sustainability. *J. Soil Water Conserv.* **2003**, *58*, 332–339.
15. Srinivasan, A. (Ed.) *Handbook of Precision Agriculture: Principles and Applications*; Food Products Press, Haworth Press Inc.: New York, NY, USA, 2006; ISBN 978-1-56022-955-1.
16. Aubert, B.A.; Schroeder, A.; Grimaudo, J. IT as enabler of sustainable farming: An empirical analysis of farmers' adoption decision of precision agriculture technology. *Decis. Support Syst.* **2012**, *54*, 510–520. [CrossRef]
17. Pierpaolia, E.; Carlia, G.; Pignattia, E.; Canavaria, M. Drivers of precision agriculture technologies adoption: A literature review. *Proc. Technol.* **2013**, *8*, 61–69. [CrossRef]
18. Gebbers, R.; Adamchuk, V. Precision agriculture and food security. *Science* **2010**, *327*, 828–831. [CrossRef]
19. Zhang, N.; Wang, M.; Wang, N. Precision agriculture—A worldwide overview. *Comput. Electron. Agric.* **2002**, *36*, 113–132. [CrossRef]
20. Bongiovanni, R.; Lowenberg-DeBoer, J. Precision agriculture and sustainability. *Precis. Agric.* **2004**, *5*, 359–387. CrossRef]
21. Koch, B.; Khosla, R.; Frasier, W.M.; Westfall, D.G.; Inman, D. Economic feasibility of variable-rate nitrogen application utilizing site-specific management zones. *Agron. J.* **2004**, *96*, 1572–1580. [CrossRef]
22. Hedley, C. The role of precision agriculture for improved nutrient management on farms. *J. Sci. Food Agric.* **2014**, *95*, 12–19. [CrossRef]
23. Boursianis, A.D.; Papadopoulou, M.S.; Diamantoulakis, P.; Liopa-Tsakalidi, A.; Barouchas, P.; Salahas, G.; Karagiannidis, G.; Wan, S.; Goudos, S.K. Internet of Things (IoT) and Agricultural Unmanned Aerial Vehicles (UAVs) in smart farming: A comprehensive review. *IEEE Internet Things* **2020**. [CrossRef]

24. Jha, K.; Doshi, A.; Patel, P.; Shah, M. A comprehensive review on automation in agriculture using artificial intelligence. *Artif. Intell. Agric.* **2019**, *2*, 1–12. [CrossRef]
25. Elijah, O.; Rahman, T.A.; Orikumhi, I.; Leow, C.Y.; Hindia, M.N. An overview of Internet of Things (IoT) and data analytics in agriculture: Benefits and challenges. *IEEE Internet Things* **2018**, *5*, 3758–3773. [CrossRef]
26. Huang, Y.; Chen, Z.; Yu, T.; Huang, X.; Gu, X. Agricultural remote sensing big data: Management and applications. *J. Integr. Agric.* **2018**, *7*, 1915–1931. [CrossRef]
27. Kamilaris, A.; Kartakoullis, A.; Prenafeta-Boldú, F.X. A review on the practice of big data analysis in agriculture. *Comput. Electron. Agric.* **2017**, *143*, 23–37. [CrossRef]
28. Zhou, L.; Chen, N.; Chen, Z.; Xing, C. ROSCC: An efficient remote sensing observation-sharing method based on cloud computing for soil moisture mapping in precision agriculture. *IEEE J. Sel. Top. Appl. Earth Obs. Remote Sens.* **2016**, *9*, 5588–5598. [CrossRef]
29. Khattab, A.; Abdelgawad, A.; Yelmarthi, K. Design and implementation of a cloud-based IoT scheme for precision agriculture. In Proceedings of the 2016 28th International Conference on Microelectronics (ICM), Giza, Egypt, 17 December 2016; Volume 4, pp. 201–204.
30. Pavo'n-Pulido, N.; Lo'pez-Riquelme, J.A.; Torres, R.; Morais, R.; Pastor, J.A. New trends in precision agriculture: A novel cloud-based system for enabling data storage and agricultural task planning and automation. *Precis. Agric.* **2017**, *18*, 1038–1068. [CrossRef]
31. Say, M.S.; Keskin, M.; Sehri, M.; Sekerli, Y.E. Adoption of precision agriculture technologies in developed and developing countries. *TOJSAT* **2018**, *8*, 7–15.
32. Rokhmana, C.A. The potential of UAV-based remote sensing for supporting precision agriculture in Indonesia. *Proc. Environ. Sci.* **2015**, *24*, 245–253. [CrossRef]
33. Chivasa, W.; Mutanga, O.; Biradar, C. Application of remote sensing in estimating maize grain yield in heterogeneous African agricultural landscapes: A review. *Int. J. Remote Sens. Appl.* **2017**, *38*, 6816–6845. [CrossRef]
34. Schellberg, J.; Hill, M.J.; Gerhards, R.; Rothmund, M.; Braun, M. Precision agriculture on grassland: Applications, perspectives and constraints. *Eur. J. Agron.* **2008**, *29*, 59–71. [CrossRef]
35. Maia, R.F.; Netto, I.; Tran, A.L.H. Precision agriculture using remote monitoring systems in Brazil. In Proceedings of the 2017 IEEE Global Humanitarian Technology Conference (GHTC), San Jose, CA, USA, 19 October 2017; pp. 1–6.
36. Borgogno-Mondino, E.; Lessio, A.; Tarricone, L.; Novello, V.; Palma, D.L. A comparison between multispectral aerial and satellite imagery in precision viticulture. *Precis. Agric.* **2018**, *19*, 195–217. [CrossRef]
37. Ge, Y.; Thomasson, J.A.; Sui, R. Remote sensing of soil properties in precision agriculture: A review. *Front. Earth Sci.* **2011**, *5*, 229–238. [CrossRef]
38. Courault, D.; Seguin, B.; Olioso, A. Review on estimation of evapotranspiration from remote sensing data: From empirical to numerical modeling approaches. *Irrig. Drain. Syst.* **2005**, *19*, 223–249. [CrossRef]
39. Maes, W.H.; Steppe, K. Estimating evapotranspiration and drought stress with ground-based thermal remote sensing in agriculture: A review. *J. Exp. Bot.* **2012**, *63*, 4671–4712. [CrossRef]
40. Zhang, J.; Huang, Y.; Pu, R.; Gonzalez-Moreno, P.; Yuan, L.; Wu, K.; Huang, W. Monitoring plant diseases and pests through remote sensing technology: A review. *Comput. Electron. Agric.* **2019**, *165*, 104943. [CrossRef]
41. Atzberger, C. Advances in remote sensing of agriculture: Context description, existing operational monitoring systems and major information needs. *Remote Sens. Environ.* **2013**, *5*, 949–981. [CrossRef]
42. Mulla, D.J. Twenty-five years of remote sensing in precision agriculture: Key advances and remaining knowledge gaps. *Biosyst. Eng.* **2013**, *114*, 358–371. [CrossRef]
43. Weiss, M.; Jacob, F.; Duveillerc, G. Remote sensing for agricultural applications: A meta-review. *Remote Sens. Environ.* **2020**, *236*, 111402. [CrossRef]
44. Maes, W.H.; Steppe, K. Perspectives for remote sensing with unmanned aerial vehicles in precision agriculture. *Trends Plant Sci.* **2019**, *24*, 152–154. [CrossRef]
45. Angelopoulou, T.; Tziolas, N.; Balafoutis, A.; Zalidis, G.; Bochtis, D. Remote sensing techniques for soil organic carbon estimation: A review. *Remote Sens.* **2019**, *11*, 676. [CrossRef]
46. Santosh, K.M.; Sundaresan, J.; Roggem, R.; Déri, A.; Singh, R.P. *Geospatial Technologies and Climate Change*; Springer International Publishing: Dordrecht, The Netherlands, 2014.
47. Nowatzki, J.; Andres, R.; Kyllo, K. Agricultural Remote Sensing Basics. NDSU Extension Service Publication. 2004. Available online: www.ag.ndsu.nodak.edu (accessed on 23 September 2020).
48. Teke, M.; Deveci, H.S.; Haliloğlu, O.; Gürbüz, S.Z.; Sakarya, U. A short survey of hyperspectral remote

sensing applications in agriculture. In Proceedings of the 2013 6th International Conference on Recent Advances in Space Technologies (RAST), Istanbul, Turkey, 12 June 2013; pp. 171–176.

49. Chang, C.Y.; Zhou, R.; Kira, O.; Marri, S.; Skovira, J.; Gu, L.; Sun, Y. An Unmanned Aerial System (UAS) for concurrent measurements of solar induced chlorophyll fluorescence and hyperspectral reflectance toward improving crop monitoring. *Agric. For. Meteorol.* **2020**, *294*, 1–15. [CrossRef]
50. Nagasubramanian, K.; Jones, S.; Singh, A.K.; Sarkar, S.; Singh, A.; Ganapathysubramanian, B. Plant disease identifcation using explainable 3D deep learning on hyperspectral images. *Plant Methods* **2019**, *15*, 1–10. [CrossRef]
51. Chlingaryan, A.; Sukkarieh, S.; Whelan, B. Machine learning approaches for crop yield prediction and nitrogen status estimation in precision agriculture: A review. *Comput. Electron. Agric.* **2018**, *151*, 61–69. [CrossRef]
52. Camino, C.; González-Dugo, V.; Hernández, P.; Sillero, J.C.; Zarco-Tejada, P.J. Improved nitrogen retrievals with airborne-derived fluorescence and plant traits quantified from VNIR-SWIR hyperspectral imagery in the context of precision agriculture. *Int. J. Appl. Earth Obs. Geoinf.* **2018**, *70*, 105–117. [CrossRef]
53. Zarco-Tejada, P.J.; González-Dugo, M.V.; Fereres, E. Seasonal stability of chlorophyll fluorescence quantified from airborne hyperspectral imagery as an indicator of net photosynthesis in the context of precision agriculture. *Remote Sens. Environ.* **2016**, *179*, 89–103. [CrossRef]
54. Mohammed, G.H.; Colombo, R.; Middleton, E.M.; Rascher, U.; van der Tole, C.; Nedbald, L.; Goulas, Y.; Pérez-Priego, O.; Damm, A.; Meroni, M.; et al. Remote sensing of solar-induced chlorophyll fluorescence (SIF) in vegetation: 50 years of progress. *Remote Sens. Environ.* **2019**, *231*, 1–39.
55. Fernández-Quintanilla, C.; Peña, J.M.; Andújar, D.; Dorado, J.; Ribeiro, A.; López-Granados, F. Is the current state of the art of weed monitoring suitable for site-specific weed management in arable crops? *Weed Res.* **2018**, *58*, 259–272.
56. Castaldi, F.F.; Pelosi, F.; Pascucci, S.; Casa, R. Assessing the potential of images from unmanned aerial vehicles (UAV) to support herbicide patch spraying in maize. *Precis. Agric.* **2017**, *18*, 76–94.
57. Khanal, S.; Fulton, J.; Shearer, S. An overview of current and potential applications of thermal remote sensing in precision agriculture. *Comput. Electron. Agric.* **2017**, *139*, 22–32.
58. Palazzi, V.; Bonafoni, S.; Alimenti, F.; Mezzanotte, P.; Roselli, L. Feeding the world with microwaves: How remote and wireless sensing can help precision agriculture. *IEEE Microw. Mag.* **2019**, *20*, 72–86.
59. Babaeian, E.; Sidike, P.; Newcomb, M.S.; Maimaitijiang, M.; White, S.A.; Demieville, J.; Ward, R.W.; Sadeghi, M.; LeBauer, D.S.; Jones, S.B.; et al. A new optical remote sensing technique for high resolution mapping of soil moisture. *Front. Big Data* **2019**, *2*, 37. [CrossRef]
60. Sadeghi, M.; Babaeian, E.; Tuller, M.; Jones, S.B. The optical trapezoid model: A novel approach to remote sensing of soil moisture applied to Sentinel-2 and Landsat-8 observations. *Remote Sens. Environ.* **2017**, *198*, 52–68. [CrossRef]
61. Fang, B.; Lakshmi, V.; Bindlish, R.; Jackson, T.J. AMSR2 soil moisture downscaling using temperature and vegetation data. *Remote Sens.* **2018**, *10*, 1575. [CrossRef]
62. Im, J.; Park, S.; Rhee, J.; Baik, J.; Choi, M. Downscaling of AMSR-E soil moisture with MODIS products using machine learning approaches. *Environ. Earth Sci.* **2016**, *75*, 1–19.
63. Pereira, P.; Brevik, E.; Muñoz-Rojas, M.; Miller, B. *Soil Mapping and Process Modeling for Sustainable Land Use Management*; Elsevier: Amsterdam, The Netherlands, 2017.
64. Metternicht, G. *Land Use and Spatial Planning: Enabling Sustainable Management of Land Resources*; Springer: New York, NY, USA, 2018.
65. Nellis, M.D.; Price, K.P.; Rundquist, D. Remote sensing of cropland agriculture. In *The SAGE Handbook of Remote Sensing*; Sage: London, UK, 2009; Volume 1, pp. 368–380.
66. With, K.A. *Essential of Landscape Ecology*; Oxford University Press: Oxford, UK, 2019.
67. Forkuor, G.; Hounkpatin, O.K.L.; Welp, G.; Thiel, M. High resolution mapping of soil properties using remote sensing variables in southwestern burkina faso: A comparison of machine learning and multiple linear regression models. *PLoS ONE* **2017**, *12*, e0170478. [CrossRef]
68. Still, D.A.; Shih, S.F. Using Landsat data to classify land use for assessing the basinwide runoff index 1. *J. Am. Water Resour. Assoc.* **1985**, *21*, 931–940. [CrossRef]
69. Kidder, S.Q.; Kidder, R.M.; Haar, T.H.V. *Satellite Meteorology: An Introduction*; Academic Press: San Diago, CA, USA, 1995; p. 466.
70. Odenyo, V.A.O.; Pettry, D.E. Land-use mapping by machine processing of Landsat-1 data. *PERS* **1977**, *43*, 515–523.
71. Welch, R.; Pannell, C.W.; Lo, C.P. Land use in Northeast China, 1973: A view from Landsat-1. *AAAG* **1975**, *65*, 595–596. [CrossRef]

72. Kirchhof, W.; Haberäcker, P.; Krauth, E.; Kritikos, G.; Winter, R. Evaluation of Landsat image data for land-use mapping. *Acta Astronaut.* **1980**, *7*, 243–253. [CrossRef]
73. Blair, B.; Baumgardner, M.F. Detection of the green and brown wave in hardwood canopy covers using multidate, multispectral data from Landsat-11. *Agron J.* **1977**, *69*, 808–811. [CrossRef]
74. Bauer, M.E.; Cipra, J.E.; Anuta, P.E.; Etheridge, J.B. Identification and area estimation of agricultural crops by computer classification of Landsat MSS data. *Remote Sens. Environ.* **1979**, *8*, 77–92. [CrossRef]
75. Estes, J.E.; Jensen, J.R.; Tinney, L.R. Remote sensing of agricultural water demand information: A California study. *Water Resour. Res.* **1978**, *14*, 170–176. [CrossRef]
76. Leslie, C.R.; Serbina, L.O.; Miller, H.M. *Landsat and Agriculture—Case Studies on the Uses and Benefits of Landsat Imagery in Agricultural Monitoring and Production*; US Geological Survey Open-File Report; US Geological Survey: Reston, VA, USA, 2017; Volume 1034, p. 27.
77. Seelan, S.K.; Laguette, S.; Casady, G.M.; Seielstad, G.A. Remote sensing applications for precision agriculture: A learning community approach. *Remote Sens. Environ.* **2003**, *88*, 157–169. [CrossRef]
78. Scudiero, E.; Corwin, D.L.; Wienhold, B.J.; Bosley, B.; Shanahan, J.F.; Johnson, C.K. Downscaling Landsat 7 canopy reflectance employing a multi-soil sensor platform. *Precis. Agric.* **2016**, *17*, 53–73. [CrossRef]
79. Venancio, L.P.; Mantovani, E.C.; do Amaral, C.H.; Neale, C.M.U.; Gonçalves, I.Z.; Filgueiras, R.; Campos, I. Forecasting corn yield at the farm level in Brazil based on the FAO-66 approach and soil-adjusted vegetation index (SAVI). *Agric. Water Manag.* **2019**, *225*, 105779. [CrossRef]
80. Dong, T.; Liu, J.; Qian, B.; Zhao, T.; Jing, Q.; Geng, X.; Wang, J.; Huffman, T.; Shang, J. Estimating winter wheat biomass by assimilating leaf area index derived from fusion of Landsat-8 and MODIS data. *Int. J. Appl. Earth Obs. Geoinf.* **2016**, *49*, 63–74. [CrossRef]
81. Worsley, P.; Bowler, J. Assessing flood damage using SPOT and NOAA AVHRR data. *Geospat. Inf. Agric.* **2001**, 2–7. Available online: http://www.regional.org.au/au/gia/12/397worsley.htm#TopOfPage (accessed on 23 September 2020).
82. Mondal, P.; Basu, M. Adoption of precision agriculture technologies in India and in some developing countries: Scope, present status and strategies. *Prog. Nat. Sci.* **2009**, *19*, 659–666. [CrossRef]
83. Koenig, K.; Höfle, B.; Hämmerle, M.; Jarmer, T.; Siegmann, B.; Lilienthal, H. Comparative classification analysis of post-harvest growth detection from terrestrial LiDAR point clouds in precision agriculture. *ISPRS J. Photogramm. Remote Sens.* **2015**, *104*, 112–125. [CrossRef]
84. McNairn, H.; Ellis, J.; Van Der Sanden, J.J.; Hirose, T.; Brown, R.J. Providing crop information using RADARSAT-1 and satellite optical imagery. *ISPRS J. Photogramm. Remote Sens.* **2002**, *23*, 851–870. [CrossRef]
85. Enclona, E.A.; Thenkabail, P.S.; Celis, D.; Diekmann, J. Within-field wheat yield prediction from IKONOS data: A new matrix approach. *ISPRS J. Photogramm. Remote Sens.* **2004**, *25*, 377–388. [CrossRef]
86. Sullivan, D.G.; Shaw, J.N.; Rickman, D. IKONOS imagery to estimate surface soil property variability in two alabama physiographies. *Soil Sci. Soc. Am. J.* **2005**, *69*, 1789–1798.
87. Yang, G.; Pu, R.; Zhao, C.; Xue, X. Estimating high spatiotemporal resolution evapotranspiration over a winter wheat field using an IKONOS image based complementary relationship and Lysimeter observation. *Agric. Water Manag.* **2014**, *133*, 34–43. [CrossRef]
88. Omran, E.E. Remote estimation of vegetation parameters using narrow band sensor for precision agriculture in arid environment. *Egypt. J. Soil Sci.* **2018**, *58*, 73–92. [CrossRef]
89. Apan, A.; Held, A.; Phinn, S.; Markley, J. Detecting sugarcane 'orange rust' disease using EO-1 Hyperion hyperspectral imagery. *Int. J. Remote Sens.* **2004**, *25*, 489–498.
90. Filippi, P.; Jones, J.E.; Niranjan, S.; Wimalathunge, N.S.; Somarathna, D.S.N.P.; Liana, E.; Pozza, L.E.; Ugbaje, S.U.; Jephcott, T.G.; Paterson, S.E.; et al. An approach to forecast grain crop yield using multi-layered, multi-farm data sets and machine learning. *Precis. Agric.* **2019**, *20*, 1–16.
91. Houborg, R.; McCabe, M.F. High-resolution NDVI from planet's constellation of Earth observing nanosatellites: A new data source for precision agriculture. *Remote Sens.* **2016**, *8*, 768.
92. Mobasheri, M.R.; Jokar, J.; Ziaeian, P.; Chahardoli, M. On the methods of sugarcane water stress detection using Terra/ASTER images. *Am. Eurasian J. Agric. Environ. Sci.* **2007**, *2*, 619–627.
93. Santoso, H.; Gunawan, T.; Jatmiko, R.H.; Darmosarkoro, W.; Minasny, B. Mapping and identifying basal stem rot disease in oil palms in North Sumatra with QuickBird imagery. *Precis. Agric.* **2011**, *12*, 233–248.
94. Jackson, T.J.; Bindlish, R.; Klein, M.; Gasiewski, A.J.; Njoku, E.G. Soil moisture retrieval and AMSR-E validation using an airborne microwave radiometer in SMEX02. In Proceedings of the 2003 IEEE International Geoscience and Remote Sensing Symposium, Toulouse, France, 21–25 July 2003; Volume 1, pp. 401–403.

95. Yang, C.; Everitt, J.H.; Bradford, J.M. Evaluating high resolution SPOT 5 satellite imagery to estimate crop yield. *Precis. Agric.* **2009**, *10*, 292–303.
96. Sai, M.S.; Rao, P.N. Utilization of resourcesat-1 data for improved crop discrimination. *Int. J. Appl. Earth Obs. Geoinf.* **2008**, *10*, 206–210.
97. Lee, J.W.; Park, G.; Joh, H.K.; Lee, K.H.; Na, S.I.; Park, J.H.; Kim, S.J. Analysis of relationship between vegetation indices and crop yield using KOMPSAT (KoreaMulti-Purpose SATellite)-2 imagery and field investigation data. *JKSAE* **2011**, *53*, 75–82.
98. Gao, S.; Niu, Z.; Huang, N.; Hou, X. Estimating the Leaf Area Index, height and biomass of maize using HJ-1 and RADARSAT-2. *Int. J. Appl. Earth Obs. Geoinf.* **2013**, *24*, 1–18.
99. Siegfried, J.; Longchamps, L.; Khosla, R. Multisectral satellite imagery to quantify in-field soil moisture variability. *J. Soil Water Conserv.* **2019**, *74*, 33–40.
100. De Lara, A.; Longchamps, L.; Khosla, R. Soil water content and high-resolution imagery for precision irrigation: Maize yield. *Agron. J.* **2019**, *9*, 174.
101. Shang, J.; Liu, J.; Ma, B.; Zhao, T.; Jiao, X.; Geng, X.; Huffman, T.; Kovacs, J.M.; Walters, D. Mapping spatial variability of crop growth conditions using RapidEye data in Northern Ontario, Canada. *Remote Sens. Environ.* **2015**, *168*, 113–125.
102. Caturegli, L.; Casucci, M.; Lulli, F.; Grossi, N.; Gaetani, M.; Magni, S.; Bonari, E.; Volterrani, M. GeoEye-1 satellite versus ground-based multispectral data for estimating nitrogen status of turfgrasses. *Int. J. Remote Sens.* **2015**, *36*, 2238–2251.
103. Tian, J.; Wang, L.; Li, X.; Gong, H.; Shi, C.; Zhong, R.; Liu, X. Comparison of UAV and WorldView-2 imagery for mapping leaf area index of mangrove forest. *Int. J. Appl. Earth Obs. Geoinf.* **2017**, *61*, 22–31.
104. Kokhan, S.; Vostokov, A. Using vegetative indices to quantify agricultural crop characteristics. *Ecol. Eng.* **2020**, *21*, 122–129. [CrossRef]
105. Romanko, M. Remote Sensing in Precision Agriculture: Monitoring Plant Chlorophyll, and Soil Ammonia, Nitrate, and Phosphate in Corn and Soybean Fields. Ph.D. Thesis, Bowling Green State University, Bowling Green, OH, USA, 2017.
106. Skakun, S.; Justice, C.O.; Vermote, E.; Roger, J.C. Transitioning from MODIS to VIIRS: An analysis of inter-consistency of NDVI data sets for agricultural monitoring. *Int. J. Remote Sens.* **2018**, *39*, 971–992. [PubMed]
107. Kim, S.J.; Lee, M.S.; Kim, S.H.; Park, G. Potential application topics of kompsat-3 image in the field of precision agriculture model. *Korean Soc. Remote Sens.* **2006**, *48*, 17–22.
108. Yuan, L.; Pu, R.; Zhang, J.; Wang, J.; Yang, H. Using high spatial resolution satellite imagery for mapping powdery mildew at a regional scale. *Precis. Agric.* **2016**, *17*, 332–348. [CrossRef]
109. Ferguson, R.; Rundquist, D. Remote sensing for site-specific crop management. *Precis. Agric. Basics* **2018**. [CrossRef]
110. Sidike, P.; Sagan, V.; Maimaitijiang, M.; Maimaitiyiming, M.; Shakoor, N.; Burken, J.; Fritschi, F.B. dPEN: Deep progressively expanded network for mapping heterogeneous agricultural landscape using WorldView-3 satellite imagery. *Remote Sens. Environ.* **2018**, *221*, 756–772. [CrossRef]
111. Martínez-Casasnovas, J.A.; Uribeetxebarría, A.; Escolà, A.; Arnó, J. Sentinel-2 vegetation indices and apparent electrical conductivity to predict barley (*Hordeum vulgare* L.) yield. In *Precision Agriculture*; Wageningen Academic Publishers: Wageningen, The Netherlands, 2019; pp. 415–421.
112. Wolters, S.; Söderström, M.; Piikki, K.; Stenberg, M. Near-real time winter wheat N uptake from a combination of proximal and remote optical measurements: How to refine Sentinel-2 satellite images for use in a precision agriculture decision support system. In Proceedings of the 12th European Conference on Precision Agriculture, Montpellier, France, 8–11 July 2019; Wageningen Academic Publishers: Wageningen, The Netherlands, 2019; pp. 415–421.
113. Bajwa, S.G.; Rupe, J.C.; Mason, J. Soybean disease monitoring with leaf reflectance. *Remote Sens.* **2017**, *9*, 127. [CrossRef]
114. El Sharif, H.; Wang, J.; Georgakakos, A.P. Modeling regional crop yield and irrigation demand using SMAP type of soil moisture data. *J. Hydrometeorol.* **2015**, *16*, 904–916. [CrossRef]
115. Hao, Z.; Zhao, H.; Zhang, C.; Wang, H.; Jiang, Y. Detecting winter wheat irrigation signals using SMAP gridded soil moisture data. *Remote Sens.* **2019**, *11*, 2390. [CrossRef]
116. Chua, R.; Qingbin, X.; Bo, Y. Crop Monitoring Using Multispectral Optical Satellite Imagery. Available online: https://www.21at.sg/publication/publication/cotton-crop-monitoring-using-multispectral-optical-satellite-ima/ (accessed on 23 September 2020).

117. Fisher, J.B.; Lee, B.; Purdy, A.J.; Halverson, G.H.; Dohlen, M.B.; Cawse-Nicholson, K.; Wang, A.; Anderson, R.G.; Aragon, B.; Arain, M.A.; et al. ECOSTRESS: NASA's next generation mission to measure evapotranspiration from the international space station. *Water Resour. Res.* **2020**, *56*, e2019WR026058. [CrossRef]
118. Navrozidisa, I.; Alexandridisa, T.K.; Dimitrakosb, A.; Lagopodic, A.L.; Moshoud, D.; Zalidisa, G. Identification of purple spot disease on asparagus crops across spatial and spectral scales. *Comput. Electron. Agric.* **2018**, *148*, 322–329. [CrossRef]
119. Bannari, A.; Mohamed, A.M.A.; El-Battay, A. Water stress detection as an indicator of red palm weevil attack using worldview-3 data. In Proceedings of the 2017 IEEE International Geoscience and Remote Sensing Symposium (IGARSS), Fort Worth, TX, USA, 23 July 2017; pp. 4000–4003.
120. Salgadoe, A.S.A.; Robson, A.J.; Lamb, D.W.; Dann, E.K.; Searle, C. Quantifying the severity of phytophthora root rot disease in avocado trees using image analysis. *Remote Sens.* **2018**, *10*, 226. [CrossRef]
121. Zhu, X.; Cai, F.; Tian, J.; Williams, T.K.A. Spatiotemporal fusion of multisource remote sensing data: Literature survey, taxonomy, principles, applications, and future directions. *Remote Sens.* **2018**, *10*, 527.
122. Knipper, K.R.; Kustas, W.P.; Anderson, M.C.; Alfieri, J.G.; Prueger, J.H.; Hain, C.R.; Gao, F.; Yang, Y.; McKee, L.G.; Nieto, H.; et al. Evapotranspiration estimates derived using thermal-based satellite remote sensing and data fusion for irrigation management in California vineyards. *Irrig. Sci.* **2019**, *37*, 431–449. [CrossRef]
123. Katsigiannis, P.; Galanis, G.; Dimitrakos, A.; Tsakiridis, N.; Kalopesas, C.; Alexandridis, T.; Chouzouri, A.; Patakas, A.; Zalidis, G. Fusion of spatio-temporal UAV and proximal sensing data for an agricultural decision support system. In Proceedings of the Fourth International Conference on Remote Sensing and Geoinformation of the Environment RSCy 2016, Paphos, Cyprus, 12 August 2016; Volume 9688, p. 96881R.
124. Primicerio, J.; Di Gennaro, S.F.; Fiorillo, E.; Genesio, L.; Lugato, E.; Matese, A.; Vaccar, F.P. A flexible unmanned aerial vehicle for precision agriculture. *Precis. Agric.* **2012**, *13*, 517–523. [CrossRef]
125. Huang, W.; Lu, J.; Ye, H.; Kong, W.A.; Mortimer, H.; Shi, Y. Quantitative identification of crop disease and nitrogen-water stress in winter wheat using continuous wavelet analysis. *Int. J. Agric. Biol. Eng.* **2018**, *11*, 145–151. [CrossRef]
126. Ehsani, R.; Maja, J.M. The rise of small UAVs in precision agriculture. *Resour. Mag.* **2013**, *20*, 18–19.
127. USDA. *Farms and Land in Farms: 2017 Summary. United States Department of Agriculture (USDA)*; National Agricultural Statistics Service: Washington, DC, USA, 2019; p. 19.
128. Honrado, J.L.E.; Solpico, D.B.; Favila, C.M.; Tongson, E.; Tangonan, G.L.; Libatique, N.J.C. UAV Imaging with low-cost multispectral imaging system for precision agriculture applications. In Proceedings of the 2017 IEEE Global Humanitarian Technology Conference (GHTC), San Jose, CA, USA, 19 October 2017.
129. Abdullahi, H.S.; Mahieddine, F.; Sheriff, R.E. Technology impact on agricultural productivity: A review of precision agriculture using unmanned aerial vehicles. In *Proceedings of the International Conference on Wireless and Satellite Systems*; Springer: Cham, Switzerland, 2015; pp. 388–400.
130. Zhang, S.; Zhao, G.; Lang, K.; Su, B.; Chen, X.; Xi, X.; Zhang, H. Integrated satellite, Unmanned Aerial Vehicle (UAV) and ground inversion of the SPAD of winter wheat in the reviving stage. *Sensors* **2019**, *19*, 1485. [CrossRef]
131. Xue, J.; Su, B. Significant remote sensing vegetation indices: A review of developments and application. *J. Sens.* **2017**. [CrossRef]
132. McKinnon, T.; Hoff, P. Comparing RGB-based vegetation indices with NDVI for drone based agricultural sensing. *AGBX* **2017**, *021*, 1–8. Available online: https://agribotix.com/wp-content/uploads/2017/05/Agribotix-VARI-TGI-Study.pdf (accessed on 23 September 2020).
133. Rondeaux, G.; Steven, M.; Baret, F. Optimization of soil-adjusted vegetation indices. *Remote Sens. Environ.* **1996**, *55*, 95–107. [CrossRef]
134. Carlson, T.N.; Ripley, D.A. On the relation between NDVI, fractional vegetation cover, and leaf area index. *Remote Sens. Environ.* **1997**, *62*, 241–252. [CrossRef]
135. Chen, S.; She, D.; Zhang, L.; Guo, M.; Liu, X. Spatial downscaling methods of soil moisture based on multisource remote sensing data and its application. *Water* **2019**, *11*, 1401.
136. Hashimoto, N.; Saito, Y.; Maki, M.; Homma, K. Simulation of reflectance and vegetation indices for Unmanned Aerial Vehicle (UAV) monitoring of paddy fields. *Remote Sens.* **2019**, *11*, 2119.
137. Tan, C.; Zhang, P.; Zhou, X.; Wang, Z.; Xu, Z.; Mao, W.; Li, W.; Huo, Z.; Guo, W.; Yun, F. Quantitative monitoring of leaf area index in wheat of different plant types by integrating nDVi and Beer-Lambert law. *Sci. Rep.* **2020**, *10*, 929.

138. Sun, Y.; Ren, H.; Zhang, T.; Zhang, C.; Qin, Q. Crop leaf area index retrieval based on inverted difference vegetation index and NDVI. *IEEE Geosci. Remote* **2018**, *15*, 1662–1666.

139. LI, F.; Miao, Y.; Feng, G.; Yuan, F.; Yue, S.; Gao, X.; Liu, Y.; Liu, B.; Ustin, S.L.; Chen, X. Improving estimation of summer maize nitrogen status with red edge-based spectral vegetation indices. *Field Crops Res.* **2014**, *157*, 111–123.

140. Shaver, T.M.; Kruger, G.R.; Rudnick, D.R. Crop canopy sensor orientation for late season nitrogen determination in corn. *J. Plant Nutr.* **2017**, *40*, 2217–2223.

141. Xie, Q.; Dash, J.; Huang, W.; Peng, D.; Qin, Q.; Mortimer, H.; Casa, R.; Pignatti, S.; Laneve, G.; Pascucci, S.; et al. Vegetation indices combining the red and red-edge spectral information for leaf area index retrieval. *IEEE J. Sel. Top. Appl. Earth Obs. Remote Sens.* **2018**, *11*, 1482–1493.

142. Lu, J.; Miao, Y.; Huang, Y.; Shi, W.; Hu, X.; Wang, X.; Wan, J. Evaluating an Unmanned Aerial Vehicle-based Remote Sensing System for Estimation of Rice Nitrogen Status. In Proceedings of the Fourth International Conference on Agro-Geoinformatics (Agro-geoinformatics), Istanbul, Turkey, 20 July 2015; pp. 198–203.

143. Govaerts, B.; Verhulst, N. *The Normalized Difference Vegetation Index (NDVI) GreenSeekerTM Handheld Sensor: Toward the Integrated Evaluation of Crop Management*; CIMMYT: Mexico City, Mexico, 2010; p. 13.

144. Schaefer, M.T.; Lamb, D.W. A combination of plant NDVI and LiDAR measurements improve the estimation of pasture biomass in tall fescue (Festuca arundinacea var. Fletcher). *Remote Sens.* **2016**, *8*, 109.

145. Duan, T.; Chapman, S.C.; Guo, Y.; Zheng, B. Dynamic monitoring of NDVI in wheat agronomy and breeding trials using an unmanned aerial vehicle. *Field Crops Res.* **2017**, *210*, 71–80.

146. Hassan, M.A.; Yang, M.; Rasheed, A.; Yang, G.; Reynolds, M.; Xia, X.; Xiao, Y.; He, Z. A rapid monitoring of NDVI across the wheat growth cycle for grain yield prediction using a multi-spectral UAV platform. *Plant Sci.* **2019**, *282*, 95–103.

147. Amaral, L.R.; Molin, J.P.; Portz, G.; Finazzi, F.B.; Cortinov, L. Comparison of crop canopy reflectance sensors used to identify sugarcane biomass and nitrogen status. *Precis. Agric.* **2015**, *16*, 15–28. [CrossRef]

148. Ihuoma, S.O.; Madramootoo, C.A. Sensitivity of spectral vegetation indices for monitoring water stress in tomato plants. *Comput. Electron. Agric.* **2019**, *163*, 104860. [CrossRef]

149. Ballester, C.; Zarco-Tejada, P.J.; Nicolás, E.; Alarcón, J.J.; Fereres, E.; Intrigliolo, D.S.; Gonzalez-Dugo, V.J.P.A. Evaluating the performance of xanthophyll, chlorophyll and structure-sensitive spectral indices to detect water stress in five fruit tree species. *Precis. Agric.* **2018**, *19*, 178–193. [CrossRef]

150. Zhou, J.; Khot, L.R.; Boydston, R.A.; Miklas, P.N.; Porter, L. Low altitude remote sensing technologies for crop stress monitoring: A case study on spatial and temporal monitoring of irrigated pinto bean. *Precis. Agric.* **2018**, *19*, 555–569.

151. Cao, Q.; Miao, Y.; Shen, J.; Yu, W.; Yuan, F.; Cheng, S.; Huang, S.; Wang, H.; Yang, W.; Liu, F. Improving in-season estimation of rice yield potential and responsiveness to topdressing nitrogen application with Crop Circle active crop canopy sensor. *Precis. Agric.* **2016**, *17*, 136–154. [CrossRef]

152. Lukas, V.; Novák, J.; Neudert, L.; Svobodova, I.; Rodriguez-Moreno, F.; Edrees, M.; Kren, J. The combination of UAV survey and landsat imagery for monitoring of crop vigor in precision agriculture. In *The International Archives of the Photogrammetry, Remote Sensing and Spatial Information Sciences, Proceedings of the 2016 XXIII ISPRS Congress, Prague, Czech Republic, 12–19 July 2016*. Available online: https://www.int-arch-photogramm-remote-sens-spatial-inf-sci.net/XLI-B8/953/2016/ (accessed on 23 September 2020). [CrossRef]

153. Khan, M.S.; Semwal, M.; Sharma, A.; Verma, R.K. An artificial neural network model for estimating Mentha crop biomass yield using Landsat 8 OLI. *Precis. Agric.* **2020**, *21*, 18–33.

154. Pourazar, H.; Samadzadegan, F.; Javan, F.D. Aerial multispectral imagery for plant disease detection: Radiometric calibration necessity assessment. *Eur. J. Remote Sens.* **2019**, *52*, 17–31.

155. Kanke, Y.; Tubana, B.; Dalen, M.; Harrell, D. Evaluation of red and red-edge reflectance-based vegetation indices for rice biomass and grain yield prediction models in paddy fields. *Precis. Agric.* **2016**, *17*, 507–530. [CrossRef]

156. DadrasJavan, F.; Samadzadegan, F.; Pourazar, S.H.S.; Fazeli, H. UAV-based multispectral imagery for fast Citrus Greening detection. *J. Plant Dis. Protect.* **2019**, *126*, 307–318. [CrossRef]

157. Phadikar, S.; Goswami, J. Vegetation indices based segmentation for automatic classification of brown spot and blast diseases of rice. In Proceedings of the 3rd International Conference on Recent Advances in Information Technology (RAIT), Dhanbad, India, 3 March 2016; pp. 284–289.

158. Marino, S.; Cocozza, C.; Tognetti, R.; Alvino, A. Use of proximal sensing and vegetation indexes to detect the inefficient spatial allocation of drip irrigation in a spot area of tomato field crop. *Precis. Agric.* **2015**, *16*, 613–629. [CrossRef]

159. Ranjan, R.; Chandel, A.K.; Khot, L.R.; Bahlol, H.Y.; Zhou, J.; Boydston, R.A.; Miklas, P.N. Irrigated pinto bean crop stress and yield assessment using ground based low altitude remote sensing technology. *Inf. Process. Agric.* **2019**, *6*, 502–514. [CrossRef]
160. Tahir, M.N.; Naqvi, S.Z.A.; Lan, Y.; Zhang, Y.; Wang, Y.; Afzal, M.; Cheema, M.J.M.; Amir, S. Real time estimation of chlorophyll content based on vegetation indices derived from multispectral UAV in the kinnow orchard. *IJPAA* **2018**. [CrossRef]
161. Maresma, Á.; Ariza, M.; Martínez, E.; Lloveras, J.; Martínez-Casasnovas, J.A. Analysis of vegetation indices to determine nitrogen application and yield prediction in maize (*Zea mays* L.) from a standard UAV service. *Remote Sens.* **2016**, *8*, 973.
162. Towers, P.C.; Strever, A.; Poblete-Echeverría, C. Comparison of vegetation indices for leaf area index estimation in vertical shoot positioned vine canopies with and without grenbiule hail-protection netting. *Remote Sens.* **2019**, *11*, 1073.
163. Mudereri, B.T.; Dube, T.; Adel-Rahman, E.M.; Niassy, S.; Kimathi, E.; Khan, Z.; Landmann, T. A comparative analysis of PlanetScope and Sentinel-2 space-borne sensors in mapping Striga weed using Guided Regularised Random Forest classification ensemble. *Int. Arch. Photogramm. Remote Sens. Spat. Inf. Sci.* **2019**, *42*, 701–708.
164. Khosravirad, M.; Omid, M.; Sarmadian, F.; Hosseinpour, S. Predicting sugarcane yields in khuzestan using a large time-series of remote sensing imagery region. *Int. Arch. Photogramm. Remote Sens. Spat. Inf. Sci.* **2019**, *42*, 645–648. [CrossRef]
165. Marino, S.; Alvino, A. Hyperspectral vegetation indices for predicting onion (*Allium cepa* L.) yield spatial variability. *Comput. Electron. Agric.* **2015**, *116*, 109–117. [CrossRef]
166. Das, P.K.; Laxman, B.; Rao, S.K.; Seshasai, M.V.R.; Dadhwal, V.K. Monitoring of bacterial leaf blight in rice using ground-based hyperspectral and LISS IV satellite data in Kurnool, Andhra Pradesh, India. *Int. J. Pest Manag.* **2015**, *61*, 359–368. [CrossRef]
167. Zhang, P.; Zhou, X.; Wang, Z.; Mao, W.; Li, W.; Yun, F.; Guo, W.; Tan, C. Using HJ-ccD image and pLS algorithm to estimate the yield of field-grown winter wheat. *Sci. Rep.* **2020**, *10*, 5173. [CrossRef]
168. Klem, K.; Záhora, J.; Zemek, F.; Trunda, P.; Tůma, I.; Novotná, K.; Hodaňová, P.; Rapantová, B.; Hanuš, J.; Vavříková, J.; et al. Interactive effects of water deficit and nitrogen nutrition on winter wheat. Remote sensing methods for their detection. *Agric. Water Manag.* **2018**, *210*, 171–184.
169. Liu, P.; Shi, R.; Gao, W. Estimating leaf chlorophyll contents by combining multiple spectral indices with an artificial neural network. *Earth Sci. Inf.* **2018**, *11*, 147–156.
170. Meng, J.; Xu, J.; You, X. Optimizing soybean harvest date using HJ-1 satellite imagery. *Precis. Agric.* **2015**, *16*, 164–179.
171. Taskos, D.G.; Koundouras, S.; Stamatiadis, S.; Zioziou, E.; Nikolaou, N.; Karakioulakis, K.; Theodorou, N. Using active canopy sensors and chlorophyll meters to estimate grapevine nitrogen status and productivity. *Precis. Agric.* **2015**, *16*, 77–98.
172. Abdulridha, J.; Ampatzidis, Y.; Kakarla, S.C.; Roberts, P. Detection of target spot and bacterial spot diseases in tomato using UAV-based and benchtop-based hyperspectral imaging techniques. *Precis. Agric.* **2019**, *21*, 955–978.
173. Rapaport, T.; Hochberg, U.; Cochavi, A.; Karnieli, A.; Rachmilevitch, S. The potential of the spectral 'water balance index'(WABI) for crop irrigation scheduling. *New Phytol.* **2017**, *216*, 741–757. [PubMed]
174. Gao, Y.; Walker, J.P.; Allahmoradi, M.; Monerris, A.; Ryu, D.; Jackson, T.J. Optical sensing of vegetation water content: A synthesis study. *IEEE J. STARS* **2015**, *8*, 1456–1464.
175. Ma, B.; Pu, R.; Zhang, S.; Wu, L. Spectral identification of stress types for maize seedlings under single and combined stresses. *IEEE Access* **2018**, *6*, 13773–13782.
176. DeJonge, K.C.; Taghvaeian, S.; Trout, T.J.; Comas, L.H. Comparison of canopy temperature-based water stress indices for maize. *Agric. Water Manag.* **2015**, *156*, 51–62.
177. Kullberg, E.G.; DeJonge, K.C.; Chávez, J.L. Evaluation of thermal remote sensing indices to estimate crop evapotranspiration coefficients. *Agric. Water Manag.* **2017**, *179*, 64–73.
178. Mahlein, A.K. Plant disease detection by imaging sensors–parallels and specific demands for precision agriculture and plant phenotyping. *Plant Dis.* **2016**, *100*, 241–251.
179. Prashar, A.; Jones, H.G. Assessing drought responses using thermal infrared imaging. In *Environmental Responses in Plants*; Humana Press: New York, NY, USA, 2016; pp. 209–219.
180. Uphoff, N. *Improving International Irrigation Management with Farmer Participation: Getting the Process Right*; Routledge: London, UK, 2018.

181. Pardossi, A.; Incrocci, L.; Incrocci, G.; Malorgio, F.; Battista, P.; Bacci, L.; Rapi, B.; Marzialetti, P.; Hemming, J.; Balendonck, J. Root zone sensors for irrigation management in intensive agriculture. *Sensors* **2009**, *9*, 2809–2835. [CrossRef] [PubMed]
182. Boland, A.; Bewsell, D.; Kaine, G. Adoption of sustainable irrigation management practices by stone and pome fruit growers in the Goulburn/Murray Valleys. *Aust. Irrig. Sci.* **2006**, *24*, 137–145. [CrossRef]
183. Thompson, R.B.; Gallardo, M.; Valdez, L.C.; Fernadez, M.D. Using plant water status to define threshold values for irrigation management of vegetable crops using moisture sensors. *Agric. Water Manag.* **2007**, *88*, 147–158. [CrossRef]
184. Holt, N.; Sishodia, R.P.; Shukla, S.; Hansen, K.M. Improved water and economic sustainability with low-input compact bed plasticulture and precision irrigation. *J. Irrig. Drain. Eng.* **2019**, *145*, 04019013. [CrossRef]
185. Eching, S. Role of technology in irrigation advisory services: The CIMIS experience. In Proceedings of the 18th Congress and 53rd IEC meeting of the International Commission on Irrigation and Drainage (ICID), FAO/ICID International Workshop on Irrigation Advisory Services and Participatory Extension Management, Montreal, QC, Canada, 24 July 2002; Volume 24. Available online: http://www.ipcinfo.org/fileadmin/user_upload/faowater/docs/ias/paper24.pdf (accessed on 23 September 2020).
186. Smith, M.; Munoz, G. Irrigation advisory services for effective water use: A review of experiences. In Proceedings of the Irrigation Advisory Services and Participatory Extension in Irrigation Management Workshop Organized by FAO-ICID, Montreal, QC, Canada, 24 July 2002.
187. Evans, R.G.; LaRue, J.; Stone, K.C.; King, B.A. Adoption of site-specific variable rate sprinkler irrigation systems. *Irrig. Sci.* **2013**, *31*, 871–887. [CrossRef]
188. McDowell, R.W. Does variable rate irrigation decrease nutrient leaching losses from grazed dairy farming? *Soil Use Manag.* **2017**, *33*, 530–537. [CrossRef]
189. Amani, M.; Parsian, S.; MirMazloumi, S.M.; Aieneh, O. Two new soil moisture indices based on the NIR-red triangle space of Landsat-8 data. *Int. J. Appl. Earth Obs. Geoinf.* **2016**, *50*, 176–186. [CrossRef]
190. Egea, G.; Padilla-Díaz, C.M.; Martinez-Guanter, J.; Fernández, J.E.; Pérez-Ruiz, M. Assessing a crop water stress index derived from aerial thermal imaging and infrared thermometry in super-high density olive orchards. *Agric. Water Manag.* **2017**, *187*, 210–221. [CrossRef]
191. Quebrajo, L.; Perez-ruiz, M.; Perez-Urrestarazu, L.; Martinez, G.; Egea, G. Linking thermal imaging and soil remote sensing to enhance irrigation management of sugar beet. *Biosyst. Eng.* **2018**, *165*, 77–87. [CrossRef]
192. Liou, Y.; Kar, S.K. Evapotranspiration estimation with remote sensing and various surface energy balance algorithms—A review. *Energies* **2014**, *7*, 2821–2849. [CrossRef]
193. Verstraeten, W.W.; Veroustraete, F.; Feyen, J. Assessment of evapotranspiration and soil moisture content across different scales of observation. *Sensors* **2008**, *8*, 70–117. [CrossRef] [PubMed]
194. Mendes, R.W.; Araújo, F.M.U.; Dutta, R.; Heeren, D.M. Fuzzy control system for variable rate irrigation using remote sensing. *Expert Syst. Appl.* **2019**, *124*, 13–24. [CrossRef]
195. Barker, J.B.; Neale, C.M.; Heeren, D.M.; Suyker, A.E. Evaluation of a hybrid reflectance-based crop coefficient and energy balance evapotranspiration model for irrigation management. *Trans. ASABE* **2018**, *61*, 533–548. [CrossRef]
196. Calera, A.; Campos, I.; Osann, A.; D'Urso, G.; Menenti, M. Remote sensing for crop water management: From ET modelling to services for the end users. *Sensors* **2017**, *17*, 1104. [CrossRef] [PubMed]
197. McShane, R.R.; Driscoll, K.P.; Sando, R. A review of surface energy balance models for estimating actual evapotranspiration with remote sensing at high spatiotemporal resolution over large extents. In *U.S. Geological Survey Scientific Investigations Report, 2017–5087*; US Geological Survey: Reston, VA, USA, 2017; Volume 19, pp. 1–30.
198. Zhang, K.; Kimball, J.S.; Running, S.W. A review of remote sensing based actual evapotranspiration estimation. *WIREs Water* **2016**, *3*, 834–853. [CrossRef]
199. Gaur, N.N.; Mohanty, B.P.; Kefauver, S.C. Effect of observation scale on remote sensing based estimates of evapotranspiration in a semi-arid row cropped orchard environment. *Precis. Agric.* **2017**, *18*, 762–778. [CrossRef]
200. Bhattarai, N.; Shaw, S.B.; Quackenbush, L.J.; Im, J.; Niraula, R. Evaluating five remote sensing based single-source surface energy balance models for estimating daily evapotranspiration in a humid subtropical climate. *Int. J. Appl. Earth Obs. Geoinf.* **2016**, *49*, 75–86. [CrossRef]
201. Neale, C.M.U.; Geli, H.M.E.; Kustas, W.P.; Alfieri, J.G.; Gowda, P.H.; Evett, S.R.; Prueger, J.H.; Hipps, L.E.; Dulaney, W.P.; Chávez, J.L.; et al. Soil water content estimation using a remote sensing based hybrid evapotranspiration modeling approach. *Adv. Water Res.* **2012**, *50*, 152–161. [CrossRef]

202. Gobbo, S.; Presti, S.L.; Martello, M.; Panunzi, L.; Berti, A.; Morari, F. Integrating SEBAL with in-field crop water status measurement for precision irrigation applications—A case study. *Remote Sens.* **2019**, *11*, 2069. [CrossRef]
203. Madugundu, R.; Al-Gaadi, K.A.; Tola, E.; Hassaballa, A.A.; Patil, V.C. Performance of the METRIC model in estimating evapotranspiration fluxes over an irrigated field in Saudi Arabia using Landsat-8 images. *Hydrol. Earth Syst. Sci.* **2017**, *21*, 6135–6151. [CrossRef]
204. Campos, I.; Neale, C.M.; Suyker, A.E.; Arkebauer, T.J.; Gonçalves, I.Z. Reflectance-based crop coefficients REDUX: For operational evapotranspiration estimates in the age of high producing hybrid varieties. *Agric. Water Manag.* **2017**, *187*, 140–153. [CrossRef]
205. Bhatti, S.; Heeren, D.M.; Barker, J.B.; Neale, C.M.; Woldt, W.E.; Maguire, M.S.; Rudnick, D.R. Site-specific irrigation management in a sub-humid climate using a spatial evapotranspiration model with satellite and airborne imagery. *Agric. Water Manag.* **2020**, *230*, 105950. [CrossRef]
206. Vanella, D.; Ramírez-Cuesta, J.M.; Intrigliolo, D.S.; Consoli, S. Combining electrical resistivity tomography and satellite images for improving evapotranspiration estimates of Citrus orchards. *Remote Sens.* **2019**, *11*, 373. [CrossRef]
207. Barker, J.B.; Heerenb, D.M.; Nealec, C.M.U.; Rudnick, D.R. Evaluation of variable rate irrigation using a remote-sensing-based model. *Agric. Water Manag.* **2018**, *203*, 63–74. [CrossRef]
208. Vuolo, F.; D'Urso, G.; De Michele, C.; Bianchi, B.; Cutting, M. Satellite based irrigation advisory services: A common tool for different experiences from Europe to Australia. *Agric. Water Manag.* **2015**, *147*, 82–95. [CrossRef]
209. Stone, K.C.; Bauer, P.J.; Sigua, G.C. Irrigation management using an expert system, soil water potentials, and vegetative indices for spatial applications. *Trans. ASABE* **2016**, *59*, 941–948.
210. Bonfante, A.; Monaco, E.; Manna, P.; De Mascellis, R.; Basile, A.; Buonanno, M.; Cantilena, G.; Esposito, A.; Tedeschi, A.; De Michele, C.; et al. LCIS DSS—An irrigation supporting system for water use efficiency improvement in precision agriculture: A maize case study. *Agric. Syst.* **2019**, *176*, 102646. [CrossRef]
211. French, A.N.; Hunsaker, D.J.; Thorp, K.R. Remote sensing of evapotranspiration over cotton using the TSEB and METRIC energy balance models. *Remote Sens. Environ.* **2015**, *158*, 281–294. [CrossRef]
212. Reichstein, M.; Camps-Valls, G.; Stevens, B.; Jung, M.; Denzler, J.; Carvalhais, N.; Prabhat. Deep learning and process understanding for data-driven Earth system science. *Nature* **2019**, *566*, 195–204.
213. Zhang, D.; Zhou, G. Estimation of soil moisture from optical and thermal remote sensing: A review. *Sensors* **2016**, *16*, 1308. [CrossRef]
214. Carlson, T. An overview of the "Triangle Method" for estimating surface evapotranspiration and soil moisture from satellite imagery. *Sensors* **2007**, *7*, 1612–1629. [CrossRef]
215. Zhu, W.; Jia, S.; Lv, A. A universal Ts-VI triangle method for the continuous retrieval of evaporative fraction from MODIS products. *J. Geophys. Res. Atmos.* **2017**, *122*, 206–227. [CrossRef]
216. Babaeian, E.; Sadeghi, M.; Franz, T.E.; Jones, S.; Tuller, M. Mapping soil moisture with the OPtical TRApezoid Model (OPTRAM) based on long-term MODIS observations. *Remote Sens. Environ.* **2018**, *211*, 425–440. [CrossRef]
217. Petropoulos, G.; Carlson, T.N.; Wooster, M.J.; Islam, S. A review of Ts/VI remote sensing based methods for the retrieval of land surface energy fl uxes and soil surface moisture. *Prog. Phys. Geogr.* **2009**, *33*, 224–250. [CrossRef]
218. Carlson, T.N.; Petropoulos, G.P. A new method for estimating of evapotranspiration and surface soil moisture from optical and thermal infrared measurements: The simplified triangle. *Int. J. Remote Sens.* **2019**, *40*, 7716–7729. [CrossRef]
219. Wagner, W.; Bloschl, G.; Pampaloni, P.; Calvet, J.; Bizzarri, B.; Wigneron, J.; Kerr, Y. Operational readiness of microwave remote sensing of soil moisture for hydrologic applications. *Nord. Hydrol.* **2007**, *38*, 1–20. [CrossRef]
220. Mohanty, B.P.; Cosh, M.H.; Lakshmi, V.; Montzka, C. Soil moisture remote sensing: State-of-the-science. *Vadose Zone J.* **2017**, *16*, 1–9. [CrossRef]
221. Peng, J.; Loew, A.; Merlin, O.; Verhoest, N.E.C. A review of spatial downscaling of satellite remotely sensed soil moisture. *Rev. Geophys.* **2017**, *55*, 341–366. [CrossRef]
222. Shin, Y.; Mohanty, B.P. Development of a deterministic downscaling algorithm for remote sensing soil moisture footprint using soil and vegetation classifications. *Water Resour. Res.* **2013**, *49*, 6208–6228. [CrossRef]
223. Ray, R.L.; Jacobs, J.M.; Cosh, M.H. Landslide susceptibility mapping using downscaled AMSR-E soil moisture: A case study from Cleveland Corral, California, US. *Remote Sens. Environ.* **2010**, *114*, 2624–2636. [CrossRef]
224. Molero, B.; Merlin, O.; Malbéteau, Y.; Al Bitar, A.; Cabot, F.; Stefan, V.; Kerr, Y.; Bacon, S.; Cosh, M.H.; Bindlish, R.; et al. SMOS disaggregated soil moisture product at 1 km resolution: Processor overview and first validation results. *Remote Sens. Environ.* **2016**, *180*, 361–376.

225. Montzka, C.; Jagdhuber, T.; Horn, R.; Bogena, H.R.; Hajnsek, I.; Reigber, A.; Vereecken, H. Investigation of SMAP Fusion Algorithms With Airborne Active and Passive L-Band Microwave Remote Sensing. *IEEE Trans. Geosci. Remote Sens.* **2016**, *54*, 3878–3889.
226. Bai, J.; Cui, Q.; Zhang, W.; Meng, L. An approach for downscaling SMAP soil moisture by combining sentinel-1 SAR and MODIS data. *Remote Sens.* **2019**, *11*, 2736.
227. He, L.; Hong, Y.; Wu, X.; Ye, N.; Walker, J.P.; Chen, X. Investigation of SMAP active–passive downscaling algorithms using combined sentinel-1 SAR and SMAP radiometer data. *IEEE Trans. Geosci. Remote Sens.* **2018**, *56*, 4906–4918.
228. Lievens, H.; Tomer, S.K.; Bitar, A.A.; De Lannoy, G.J.M.; Drusch, M.; Dumedah Franssen, H.J.H.; Kerr, Y.H.; Martens, B.; Pan, M.; Roundy, J.K.; et al. SMOS soil moisture assimilation for improved hydrologic simulation in the Murray Darling Basin, Australia. *Remote Sens. Environ.* **2015**, *168*, 146–162.
229. Liu, C.; Chen, Z.; Shao, Y.; Chen, J.; Tuya, H.; Pan, H. Research advances of SAR remote sensing for agriculture applications: A review. *J. Integr. Agric.* **2019**, *18*, 506–525.
230. Baghdadi, N.; Hajj, M.E.; Zribi, M.; Fayad, I. Coupling SAR C-band and optical data for soil moisture and leaf area index retrieval over irrigated grasslands. *IEEE J. Sel. Top. Appl. Earth Obs. Remote Sens.* **2015**, *9*, 1–15.
231. Hassan-Esfahani, L.; Torres-Rua, A.; Ticlavilca, A.M.; Jensen, A.; McKee, M. Topsoil moisture estimation for precision agriculture using unmmaned aerial vehicle multispectral imagery. In Proceedings of the 2014 IEEE Geoscience and Remote Sensing Symposium, Quebec City, QC, Canada, 13 July 2014; pp. 3263–3266.
232. Lyalin, K.S.; Biryuk, A.A.; Sheremet, A.Y.; Tsvetkov, V.K.; Prikhodko, D.V. UAV synthetic aperture radar system for control of vegetation and soil moisture. In Proceedings of the 2018 IEEE Conference of Russian Young Researchers in Electrical and Electronic Engineering, Moscow, Russia, 29 January–1 February 2018; pp. 1673–1675.
233. Wigmore, O.; Mark, B.; McKenzie, J.; Baraerd, M.; Lautz, L. Sub-metre mapping of surface soil moisture in proglacial valleys of the tropical Andes using a multispectral unmanned aerial vehicle. *Remote Sens. Environ.* **2019**, *222*, 104–118.
234. Melkonian, J.J.; ES, H.M.V. Adapt-N: Adaptive nitrogen management for maize using high resolution climate data and model simulations. In Proceedings of the 9th International Conference on Precision Agriculture, Denver, CO, USA, 20–23 July 2008.
235. Ali, M.M.; Al-Ani, A.; Eamus, D.; Tan, D.K.Y. Leaf nitrogen determination using non-destructive techniques—A review. *J. Plant Nut.* **2017**, *40*, 928–953.
236. Raun, W.R.; Solie, J.B.; Stone, M.L.; Martin, K.L.; Freeman, K.W.; Mullen, R.W.; Zhang, H.; Schepers, J.S.; Johnson, G.V. Optical sensor-based algorithm for crop nitrogen fertilization. *Commun. Soil Sci. Plant Anal.* **2005**, *36*, 2759–2781.
237. Bushong, J.T.; Mullock, J.L.; Miller, E.C.; Raun, W.R.; Arnall, D.B. Evaluation of mid-season sensor based nitrogen fertilizer recommendations for winter wheat using different estimates of yield potential. *Precis. Agric.* **2016**, *17*, 470–487. [CrossRef]
238. Franzen, D.; Kitchen, N.; Holland, K.; Schepers, J.; Raun, W. Algorithms for in-season nutrient management in cereals. *Agron. J.* **2016**, *108*, 1775–1781. [CrossRef]
239. Scharf, P.C.; Shannon, D.K.; Palm, H.L.; Sudduth, K.A.; Drummond, S.T.; Kitchen, N.R.; Mueller, L.J.; Hubbard, V.C.; Oliveira, L.F. Sensor-based nitrogen applications out-performed producer-chosen rates for corn in on-farm demonstrations. *Agron. J.* **2011**, *103*, 1684–1691. [CrossRef]
240. Higgins, S.; Schellberg, J.; Bailey, J.S. Improving productivity and increasing the efficiency of soil nutrient management on grassland farms in the UK and Ireland using precision agriculture technology. *Eur. J. Agron.* **2019**, *106*, 67–74. [CrossRef]
241. Colaço, A.F.; Bramley, R.G. Do crop sensors promote improved nitrogen management in grain crops? *Field Crops Res.* **2018**, *218*, 126–140. [CrossRef]
242. Cao, Q.; Miao, Y.; Li, F.; Gao, X.; Liu, B.; Lu, D.; Chen, X. Developing a new crop circle active canopy sensorbased precision nitrogen management strategy for winter wheat in North China Plain. *Precis. Agric.* **2017**, *18*, 2–18.
243. Blasch, G.; Spengler, D.; Hohmann, C.; Neumann, C.; Itzerott, S.; Kaufmann, H. Multitemporal soil pattern analysis with multispectral remote sensing data at the field scale. *Comput. Electron. Agric.* **2015**, *113*, 1–13. [CrossRef]
244. Kalambukattu, J.G.; Kumar, S.; Raj, R.A. Digital soil mapping in a Himalayan watershed using remote sensing and terrain parameters employing artificial neural network model. *Environ. Earth Sci.* **2018**, *77*, 203. [CrossRef]

245. Castaldi, F.; Hueni, A.; Chabrillat, S.; Ward, K.; Buttafuoco, G.; Bomans, B.; Vreys, K.; Brell, M.; Van Wesemael, B. Evaluating the capability of the Sentinel 2 data for soil organic carbon prediction in croplands. *ISPRS J. Photogramm.* **2019**, *147*, 267–282. [CrossRef]
246. Khanal, S.; Fulton, J.; Klopfenstein, A.; Douridas, N.; Shearer, S. Integration of high resolution remotely sensed data and machine learning techniques for spatial prediction of soil properties and corn yield. *Comput. Electron. Agric.* **2018**, *153*, 213–225. [CrossRef]
247. Sladojevic, S.; Arsenovic, M.; Anderla, A.; Culibrk, D.; Stefanovic, D. Deep neural networks based recognition of plant diseases by leaf image classification. *Comput. Intell. Neurosci.* **2016**. [CrossRef]
248. Di Gennaro, S.F.; Battiston, E.; Di Marco, S.; Facini, O.; Matese, A.; Nocentini, M.; Palliotti, A.; Mugnai, L. Unmanned Aerial Vehicle (UAV)-based remote sensing to monitor grapevine leaf stripe disease within a vineyard affected by esca complex. *Phytopathol. Mediterr.* **2016**, *55*, 262–275.
249. Mahlein, A.-K.; Rumpf, T.; Welke, P.; Dehne, H.-W.; Plümer, L.; Steiner, U.; Oerke, E.-C. Development of spectral indices for detecting and identifying plant diseases. *Remote Sens. Environ.* **2013**, *128*, 21–30. [CrossRef]
250. AL-Saddik, H.; Simon, J.; Cointault, F. Development of spectral disease indices for 'Flavescence Dorée' grapevine disease identification. *Sensors* **2017**, *17*, 2772. [CrossRef]
251. Liang, Q.; Xiang, S.; Hu, Y.; Coppola, G.; Zhang, D.; Sun, W. PD2SE-Net: Computer-assisted plant disease diagnosis and severity estimation network. *Comput. Electron. Agric.* **2019**, *157*, 518–529. [CrossRef]
252. Davis, A.M.; Pradolin, J. Precision herbicide application technologies to decrease herbicide losses in furrow irrigation outflows in a Northeastern Australian cropping system. *J. Agric. Food Chem.* **2016**, *64*, 4021–4028. [CrossRef]
253. Lameski, P.; Zdravevski, E.; Kulakov, A. Review of automated weed control approaches: An environmental impact perspective. In *International Conference on Telecommunications*; Springer: Cham, Switzerland, 2018; pp. 132–147.
254. Huang, H.; Lan, Y.; Yang, A.; Zhang, Y.; Wen, S.; Deng, J. Deep learning versus Object-based Image Analysis (OBIA) in weed mapping of UAV imagery. *Int. J. Remote Sens.* **2020**, *41*, 3446–3479. [CrossRef]
255. Partel, V.; Kakarla, S.C.; Ampatzidis, Y. Development and evaluation of a low-cost and smart technology for precision weed management utilizing artificial intelligence. *Comput. Electron. Agric.* **2019**, *157*, 339–350. [CrossRef]
256. De Castro, A.I.; Peña, J.M.; Torres-Sánchez, J.; Jiménez-Brenes, F.; López-Granados, F. Mapping cynodon dactylon in vineyards using UAV images for site-specific weed control. *Adv. Anim. Biosci.* **2017**, *8*, 267–271. [CrossRef]
257. Huang, Y.; Reddy, K.N.; Fletcher, R.S.; Pennington, D. UAV low-altitude remote sensing for precision weed management. *Weed Technol.* **2018**, *32*, 2–6. [CrossRef]
258. Hunter, J.E.; Gannon, T.W.; Richardson, R.J.; Yelverton, F.H.; Leon, R.G. Integration of remote-weed mapping and an autonomous spraying unmanned aerial vehicle for site-specific weed management. *Pest Manag. Sci.* **2019**, *76*, 1386–1392. [CrossRef] [PubMed]
259. Peng, Y.; Li, Y.; Dai, C.; Fang, S.; Gong, Y.; Wu, X.; Zhu, R.; Liu, K. Remote prediction of yield based on LAI estimation in oilseed rape under different planting methods and nitrogen fertilizer applications. *Agric. For. Meteorol.* **2019**, *271*, 116–125. [CrossRef]
260. Kross, A.; McNairn, H.; Lapen, D.; Sunohara, M.; Champagne, C. Assessment of RapidEye vegetation indices for estimation of leaf area index and biomass in corn and soybean crops. *Int. J. Appl. Earth Obs. Geoinf.* **2015**, *34*, 235–248. [CrossRef]
261. Kang, Y.; Özdoğan, M.; Zipper, S.C.; Román, M.O.; Walker, J.; Hong, S.Y.; Marshall, M.; Magliulo, V.; Moreno, J.; Alonso, L.; et al. How universal is the relationship between remotely sensed vegetation indices and crop leaf area index? A global assessment. *Remote Sens.* **2016**, *8*, 597. [CrossRef]
262. Yue, J.; Yang, G.; Li, C.; Li, Z.; Wang, Y.; Feng, H.; Xu, B. Estimation of winter wheat above-ground biomass using unmanned aerial vehicle-based snapshot hyperspectral sensor and crop height improved models. *Remote Sens.* **2017**, *9*, 708. [CrossRef]
263. Campos, I.; González-Gómez, L.; Villodre, J.; Calera, M.; Campoy, J.; Jiménez, N.; Plaza, C.; Sánchez-Prieto, S.; Calera, A. Mapping within-field variability in wheat yield and biomass using remote sensing vegetation indices. *Precis. Agric.* **2019**, *20*, 214–236. [CrossRef]
264. Yeom, J.; Jung, J.; Chang, A.; Ashapure, A.; Maeda, M.; Maeda, A.; Landivar, J. Comparison of vegetation indices derived from UAV data for differentiation of tillage effects in agriculture. *Remote Sens.* **2019**, *11*, 1548. [CrossRef]
265. Salas, E.A.L.; Subburayalu, S.K. Modified shape index for object-based random forest image classification of agricultural systems using airborne hyperspectral datasets. *PLoS ONE* **2019**, *14*, e0213356.

266. Hively, W.D.; Lamb, B.T.; Daughtry, C.S.T.; Shermeyer, J.; McCarty, G.W.; Quemada, M. Mapping crop residue and tillage intensity using worldview-3 satellite shortwave infrared residue indices. *Remote Sens.* **2018**, *10*, 1657. [CrossRef]
267. Jin, X.; Yang, G.; Xu, X.; Yang, H.; Feng, H.; Li, Z.; Shen, J.; Lan, Y.; Zhao, C. Combined multi-temporal optical and radar parameters for estimating LAI and biomass in winter wheat using HJ and RADARSAR-2 data. *Remote Sens.* **2015**, *7*, 13251–13272. [CrossRef]
268. Kalisperakis, I.; Stentoumis, C.; Grammatikopoulos, L.; Karantzalos, K. Leaf area index estimation in vineyards from UAV hyperspectral data, 2D image mosaics and 3D canopy surface models. *Int. Arch. Photogramm. Remote Sens. Spat. Inf. Sci.* **2015**, *40*, 299. [CrossRef]
269. Ali, A.; Imran, M.M. Evaluating the potential of red edge position (REP) of hyperspectral remote sensing data for real time estimation of LAI & chlorophyll content of kinnow mandarin (Citrus reticulata) fruit orchards. *Sci. Hortic. Amst.* **2020**, *267*, 109326.
270. Zhen, Z.; Chen, S.; Qin, W.; Yan, G.; Gastellu-Etchegorry, J.P.; Cao, L.; Murefu, M.; Li, J.; Han, B. Potentials and limits of vegetation indices with brdf signatures for soil-noise resistance and estimation of leaf area index. *IEEE Trans. Geosci. Remote Sens.* **2020**, *58*, 5092–5108. [CrossRef]
271. Dong, T.; Liu, J.; Shang, J.; Qian, B.; Ma, B.; Kovacs, J.M.; Walters, D.; Jiao, X.; Geng, X.; Shi, Y. Assessment of red-edge vegetation indices for crop leaf area index estimation. *Remote Sens. Environ.* **2019**, *222*, 133–143. [CrossRef]
272. Candiago, S.; Remondino, F.; De Giglio, M.; Dubbini, M.; Gattelli, M. Evaluating multispectral images and vegetation indices for precision farming applications from UAV images. *Remote Sens.* **2015**, *7*, 4026–4047. [CrossRef]

2

Soil Moisture Analysis by Means of Multispectral Images According to Land use and Spatial Resolution on Andosols in the Colombian Andes

Maria Casamitjana [1,2,*], Maria C. Torres-Madroñero [3], Jaime Bernal-Riobo [1] and Diego Varga [2,*]

1 Corporación Colombiana de Investigación Agropecuaria—AGROSAVIA, Bogotá 250047, Colombia; jhbernal@agrosavia.co
2 Geography Department, University of Girona, Plaça Ferrater Mora, 1, 17004 Girona, Spain
3 Research Group on Smart Machine and Pattern Recognition, Instituto Tecnológico Metropolitano, Calle 54A No. 30 - 01, Barrio Boston, Medellin 050012, Colombia; mariatorres@itm.edu.co
* Correspondence: mcasamitjana@agrosavia.co (M.C.); diego.varga@udg.edu (D.V.)

Abstract: Surface soil moisture is an important hydrological parameter in agricultural areas. Periodic measurements in tropical mountain environments are poorly representative of larger areas, while satellite resolution is too coarse to be effective in these topographically varied landscapes, making spatial resolution an important parameter to consider. The Las Palmas catchment area near Medellin in Colombia is a vital water reservoir that stores considerable amounts of water in its andosol. In this tropical Andean setting, we use an unmanned aerial vehicle (UAV) with multispectral (visible, near infrared) sensors to determine the correlation of three agricultural land uses (potatoes, bare soil, and pasture) with surface soil moisture. Four vegetation indices (the perpendicular drought index, PDI; the normalized difference vegetation index, NDVI; the normalized difference water index, NDWI, and the soil-adjusted vegetation index, SAVI) were applied to UAV imagery and a 3 m resolution to estimate surface soil moisture through calibration with in situ field measurements. The results showed that on bare soil, the indices that best fit the soil moisture results are NDVI, NDWI and PDI on a detailed scale, whereas on potatoes crops, the NDWI is the index that correlates significantly with soil moisture, irrespective of the scale. Multispectral images and vegetation indices provide good soil moisture understanding in tropical mountain environments, with 3 m remote sensing images which are shown to be a good alternative to soil moisture analysis on pastures using the NDVI and UAV images for bare soil and potatoes.

Keywords: soil moisture; andosols; remote sensing

---

1. Introduction

In the area of agriculture, surface water content is known as soil moisture and is an important variable to consider and study to improve crops and yield. Depending on the soil moisture percentages, plant growth will be optimized, increasing nutrient absorption and the presence of microorganisms, regulating soil temperature, and affecting the speed of matter degradation and weathering processes. From a chemical point of view, soil moisture is essential for plants to undergo photosynthesis [1].

The Andes mountain range is a contrasting region with microclimates associated with its relief, where soil moisture is an important hydrological parameter that plays a vital role in the complex and vulnerable ecohydrology [2]. In agriculture, soil moisture is a complex parameter that can support soil sustainability [3]. In tropical countries such as Colombia, understanding soil moisture behavior is important to control plant growth, particularly in drought periods. The existent inter-annual climate variability and consequent soil moisture changes can affect agricultural production and by extension planting dates, varieties, and other agricultural management practices [4]. The effect of land use on andosol water storage is poorly understood and implies a high variability of soil moisture surfaces. Wigmore et al. [5] recently stated that high-resolution remote sensing images are a good alternative to study large areas of land in the tropical Andes, providing unique insights into the surface and subsurface hydrologic processes that move and store water within these heterogeneous mountain environments.

One of the main challenges in agriculture and hydrology is estimating soil moisture content by means of remote sensing [6–8]. Remote sensing techniques can be categorized by the sensor—optical (visible and infrared), thermal, or microwave (active or passive)—and depending on the range of the electromagnetic spectrum monitor [9]. These sensors are placed in private and public satellites in space. For instance, SAR (synthetic aperture radar) sensors work by means of microwave pulses that are transmitted towards the Earth's surface by an antenna, measuring the microwave energy scattered back to the sensor, in addition to the time delay between the emission and the backscattered reception signal [10]. Remote optically sensed images are obtained by measuring the solar radiation reflected by targets on the ground. Radiation reflects, transmits, and absorbs differently at different wavelengths depending on the features of the materials on the Earth's surface. When the optical sensors have several channel detectors (3 to 15) sensitive to radiation within a narrow wavelength band, the result is a multispectral image based on multiple layers containing the brightness and spectral information of the observed Earth surface at each specific wavelength band. There are satellite initiatives that periodically capture multispectral images of the Earth's surface, including Landsat, Sentinel, Spot, and Ikonos [11]. While there have been several relevant initiatives to analyze soil moisture using satellite optical and radar sensors, which provide soil moisture products on several scales, the number of in situ soil moisture networks that are accessible and suited for satellite soil moisture evaluation is especially low for tropical regions [7].

Spatial soil moisture approximations are based on the indirect climate approach, with specific programs launched by ESA (European Space Agency) and NASA (National Aeronautics and Space Administration) in the USA. An important remote sensing project specific to the study of global soil moisture is the European Space Agency's Climate Change Initiative for Soil Moisture (ESA CCI SM), with a resolution of 25 km [12]. NASA launched its SMAP (Soil Moisture Active Passive) mission on January 2015, consisting of a radar and radiometer to monitor the amount of water in the top 5 cm of soils worldwide [6]. However, the radar failed in September of the same year, with the mission continuing to the present only with the radiometer data, which involve a resolution of 40 km.

There is an especially high variability of soil moisture in zones where land use, topography, and soil type are also highly variable [13]. Knowledge of soil moisture and its spatial distribution is of considerable importance to economic, social, hydrological, and agronomical planning. The scale required for each purpose varies, with initiatives at a resolution on a medium or global scale unsuitable for working on precision agriculture at a plot scale.

Over the last decade, the data obtained by unmanned aerial vehicles (UAV) have been intensively studied for agricultural applications given their flexibility of image acquisition and high spatial resolution, with customized cameras installed depending on the band (spectral resolution) and resolution requirements [14], especially including infrared bands used in several vegetation indices to monitor their states. Advances in UAV technology and sensor size, lower costs, global positioning systems (GPS), and pre-programmed flights have led to this knowledge gap being filled and a reduction in the spatial resolution of the most common current remote sensing systems. However, UAVs currently have several limitations related to weather conditions and re-visit times, in addition to being costly [15].

Soil mapping and image analysis are recent tools to simulate and monitor soil moisture [16]. The topographic wetness index (TWI) enables potential sites where moisture or water is accumulated to be identified by means of the geomorphologic analysis of the land using the DEM (digital elevation model), considering that topography is a first-order control of the spatial variation of hydrological conditions [17]. The TWI is effective for studying soil moisture on a coarse scale with slope variability and is dependent on geology and the possible divergence between surface and subsurface conditions [17].

Multispectral satellite imagery is another approach to estimate soil moisture content [18] by means of the reflectance of the Earth's surface, although pixel spatial resolution is too coarse to be used on agriculture on a plot scale. Satellite measurements are also limited by their return period and are often impacted by cloud cover, particularly in tropical mountainous regions [5], reducing the available images to study the landscape. Soil reflectance is influenced by soil moisture and other intrinsic parameters, such as soil texture, mineral composition, and organic matter [19], affecting the absorption of different wavelengths. Recent laboratory studies have demonstrated the effect of soil moisture on reflectance for different orders of soils [20]. Organic matter and mineral composition affect short visible wavelengths and soil moisture in the NIR (near-infrared) and SWIR (shortwave infrared) spectral bands [19].

Regarding the spectral variations of water absorption, several multispectral indices using NIR and SWIR to analyze water content and soil moisture by means of optical sensors from space have been studied over the last decade [21–23]. For instance, the soil moisture of land covered by vegetation has been studied using indices such as the vegetation dryness index (VDI), the temperature vegetation dryness index (TVDI) [5], the enhanced vegetation index (EVI), the green coverage index (GCI) and, most commonly, the normalized difference vegetation index (NDVI), an enhanced vegetation index to determine vegetation status using drought as an indicator of soil moisture, and the normalized difference water index (NDWI), used to determine water bodies and areas where soil is saturated and additionally used to determine the vegetation hydric index, maximizing water reflectance. There are several methodologies to determine the NDWI. The Mc Feeters [24] equation uses the green band and the NIR band, optimizing vegetation moisture reflection and minimizing water bodies, whereas Dr Gao [25] determines the NDWI by means of the relationship between NIR and SWIR. Xu [26] later proposed the modified normalized difference water index (MNDWI), considering the green and SWIR bands. However, Chen et al. [27] state that soil moisture can cause side effects when using the SWIR band because its absorption is constrained to a reasonable extent. A soil-adjusted vegetation index such as the SAVI (soil-adjusted vegetation index) is used to reduce the soil effect, minimizing the related brightness by considering first-order soil vegetation interaction with soil-adjustment parameters [28]. Jeihouny et al. [29] use this index to map soil moisture by means of data mining, finding that SAVI is an important covariate in predicting soil moisture retention properties.

Another common methodology to estimate soil moisture by means of remote sensing is the trapezoid method, based on thermic and optical data regarding the Earth's surface [30]. This methodology has the problem that land surface temperature varies significantly with the ambient atmospheric parameters, while optical reflectance does not [31]. Starting from this assumption, some indices using optical observations have been proposed for soil moisture and drought monitoring based on triangular spaces from pixel distributions of optical observations in different electromagnetic frequency bands [31]. One of these triangular indices is the PDI (perpendicular drought index), designed by Ghulam et al. [32], which determines soil moisture for bare soils and low covers by means of the near infrared correlation of pixels. Amani and Parsian [22] evaluated the PDI, finding that it has some limitations that challenge its performance in areas with dense vegetation, but that it is highly effective for bare soils.

In this study, four indices (NDVI, NDWI, SAVI, and PDI) are evaluated to estimate soil moisture (SM) from high resolution images obtained by means of remote optical sensors and UAV flights in the highest part of the Las Palmas catchment area in Envigado, Colombia (See Figure 1). Soil moisture was evaluated according its land use on Andosol to determine an algorithm to correlate the studied indices with the soil moisture field data at different spatial resolutions. The four indices were evaluated to estimate soil moisture for three land uses (potatoes, bare soil and pasture). In addition, we analyze these indices in several spatial resolution using re-sampled imagery from UAV. We demonstrate that the performance of these indices is conditioned to both land uses and spatial imagery resolution.

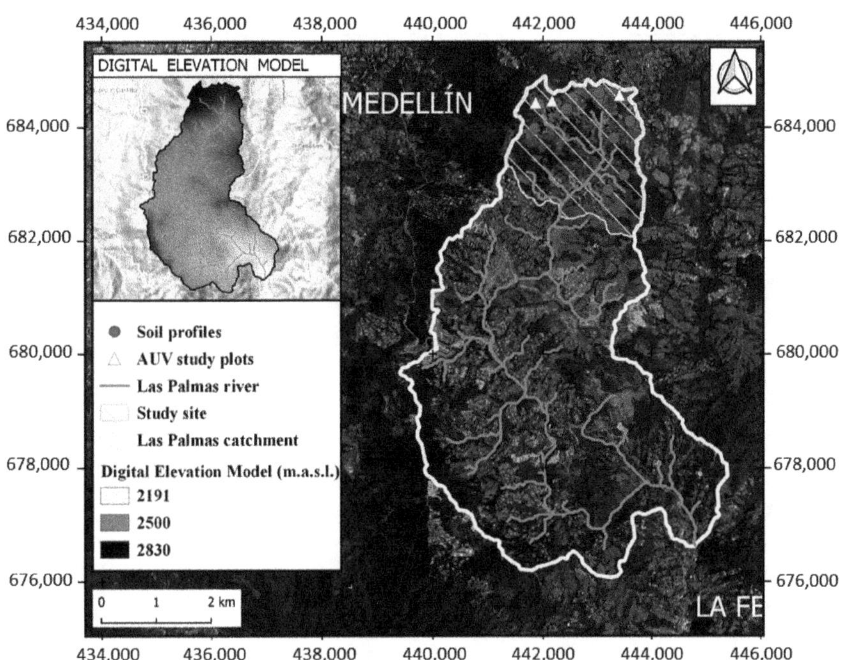

Figure 1. Study site location and unmanned aerial vehicle (UAV) flight sites by the digital elevation model (DEM 12.5 m) of the Las Palmas catchment area, Antioquia, Colombia, South America. Source: Current research.

2. Materials and Methods

2.1. Study Area

The study site is located in the Las Palmas catchment area in the central Andes mountain range. This catchment area supplies the water for La Fe reservoir, which guarantees the drinking water supply for the three million inhabitants of the Aburrá Valley metropolitan region [32]. This study site was selected to characterize soil moisture according to land use in an agricultural microcatchment area located in the upper section of Las Palmas catchment area in Envigado, Colombia (Figure 1).

There is an automated climatic EPM (Empresas Publicas Medellin) station in the upper part of the basin (44,3831, 68,4977 elevation: 2820 m.a.s.l.). The total annual precipitation average is 2500 mm/year (1980–2020), with a minimum annual precipitation in 1980 (1379.4 mm) and a maximum annual precipitation in 1999 (2837.2 mm). There are usually two dry seasons, from December to March and from June to August. The mean temperature for the same period was 18 °C (min 10.3 °C, max 22.3 °C).

The soil type in the study site is Andosol with its associated physical properties, making good water reservoirs with fluctuant hydrological properties [33]. Andosol is an unfertile soil due to its high degree of meteorization and the fact that it is derived from volcanic ashes that physically condition its porous system and structure, resulting in a high variation of soil moisture. Furthermore, the soil moisture regime in the study site is udic [34], meaning fewer than 90 cumulative days each year when water is not available in the rooting zone in normal years. Perennial plants are adequately supplied with water most years. In most similar areas, two crops can be grown each year, but the available water is less reliable for some of the year and farmers often plant more drought-tolerant crops [35].

2.2. Procedure

The workflow used in this study is shown in Figure 2. It consisted of four steps: (a) preprocessing of datasets; (b) determination of vegetation indices; (c) analysis of the optimal resolutions; and (d) comparison of remote sensing variables for SM retrieval according to land use.

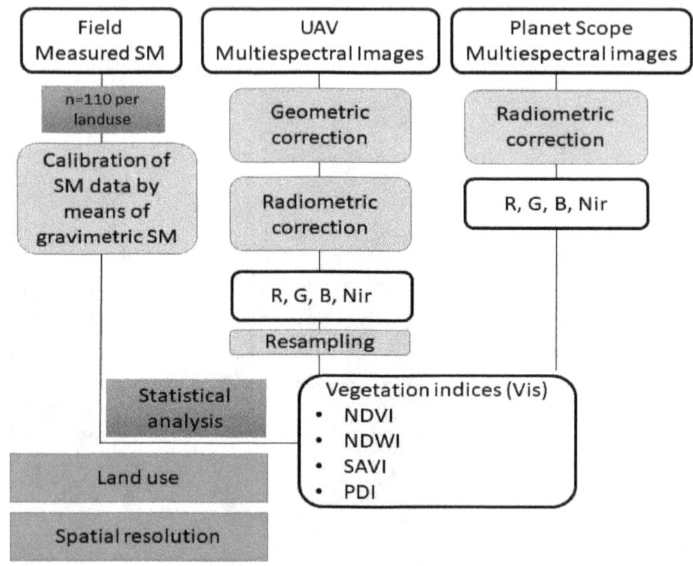

Figure 2. Workflow of the pursued methodology. R (Red), G (Green), B (Blue), NIR (Near Infra-Red). Source: Current research.

The field campaigns were carried out during the dry season on 5, 6, and 7 February 2019 to evaluate the soil moisture of three study plots measuring 1 ha per land use evaluated (i.e., pasture (*Pennisetum clandestinum*), potatoes (*Solanum tuberosum*), and bare soil) located in the highest part of the Las Palmas catchment area, Envigado, Colombia (Figure 1). Soil characterization of the study site was determined by means of 7 soil profile descriptions and pedologic and hydrological measurements (Figure 1, Table 1), analyzing NaF (sodium fluoride) reaction and pH, profile depth, volcanic ashes depth, infiltration, and field-saturated soil hydraulic conductivity (Kfs) in the upper soil layer. The reaction of sodium fluoride solution with soils and soil minerals is used as a parameter to determine the presence of amorphous minerals and hydromorphic soil conditions.

Table 1. Soil profile descriptions and associated hydrological and pedological variables. Kfs: field-saturated soil hydraulic conductivity.

Soil Profile	x	y	pH (0–10 cm)	Soil Texture (0–10 cm)	NaF Reaction	Depth A Profile	Depth Volcanic Ashes	K_{fs}	Infiltration T_{10}
1	841,661.7	1,174,433.7	5.1	Loam	Moderate	47	125	0.01574	39.3
2	841,812.6	1,174,370.8	5.2	Silty loam	Strong	34	106	0.01168	15
3	840,311.1	1,176,556.6	5.9	Loam	Strong	44	72	0.01815	7.33
4	840,298.9	1,176,382.1	5.5	Loam	Strong	44	150	0.00406	16.39
5	841,312.4	1,176,329.7	5.6	Loam	Strong	33	100	0.00899	134.33
6	841,552.3	1,176,249.2	5.3	Loam	Strong	26	120	0.015583	44.67
7	841,476.8	1,176,338.2	5	Loam	Strong	26	92	0.032536	123.33

The ground data used for the calibration and validation of the regression models were collected from 110 sampling points on each study plot, previously marked using 25 cm diameter polystyrene dishes and forming a regular grid with a distance of 10 m × 10 m between them (Figure 3). To verify the exact location of the sampling points, 5 sub-metric high-precision GPS spots were georeferred by means of a Topcon© Hiper V RTK, Livermore, CA, USA (Figure 3). On each studied plot, 110 sampling points were considered, and the soil moisture and temperature data were collected using a TDR sensor.

(a) (b) (c)

Figure 3. UAV flight over the study plots of pasture, potatoes, and bare soil. (**a**) UAV hexacopter on the pasture study plot, (**b**) potatoes study site, (**c**) Topcon submetric device on bare soil study site.

Simultaneously with the ground measurement, aerial images were acquired using a hexacopter UAV and a multispectral RedEdge camera, Micasense©, Seattle, WA, USA obtaining multiple sets of images in five spectral bands, blue (475 nm), green (560 nm), red (668 nm), red edge (717 nm) from the visible rank, and NIR (840 nm), to determine soil moisture reflectance (Figure 4). UgCS software, Riga, Latvia, Europe, was used for the automated drone mission planning. The images were later merged and postprocessed in the laboratory for geometric correction and calibration using the Pix4D© software, Prilly, Switzerland, Europe. Radiometric correction of the images PlanetScope©, San Francisco, CA, USA was carried out by means of the Qgis software, Gossau, Switzerland, Europe and the required parameters were obtained from the image metadata.

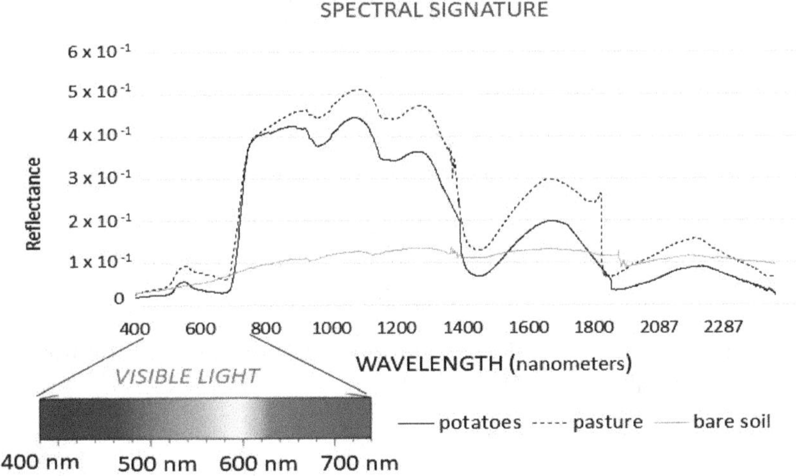

Figure 4. Reflectance according to the spectral region and wavelengths calculated on the study plots by means of a ASD FieldSpec 4 Hi-Res NG Spectroradiometer, Malvern Panalytical Ltd, Cambridge, UK (n = 50 per land use: potatoes, bare soil and pasture). Source: Current research.

The climatological information for the month prior to the sampling for the field experiment was collected at the EPM meteorological station located 450 m from the study plots, considering rainfall, temperature, and wind as influent parameters.

Optical Planet Scope 3m resolution images in four bands (R, G, B and NIR) were obtained for the same week as the ground measurements were taken. The images used were divided by bands and subsequently multiplied by the reflectance coefficient to convert the Digital number radiance, reescaled into an 8-bit digital number (DN) with a range between 0 and 255, into Top of Atmosphere (TOA) Reflectance.

The vegetation indices were computed using both the UAV and the planet scope images. According to the literature, the NDVI is defined as Equation (1), NDWI (Equation (2)), SAVI (Equation (3)) and PDI (perpendicular drought index) (Equation (4))

$$\text{NDVI} = (\text{NIR} - \text{R})/(\text{NIR} + \text{R}) \tag{1}$$

$$\text{NDWI} = (\text{GREEN} - \text{NIR})/(\text{GREEN} + \text{NIR}) \tag{2}$$

$$\text{SAVI} = [(\text{NIR} - \text{R})/(\text{NIR} + \text{R} + \text{L})]\,(1 + \text{L}) \tag{3}$$

$$\text{PDI} = 1/\sqrt{}(\text{M}^2 + 1)(\text{R} - (\text{M} \times \text{NIR})) \tag{4}$$

To determine the PDI, a soil line was built by means of red and NIR reflectivity correlation of pixels on bare soil, where red was the independent variable and NIR the dependent variable [30]. This drought index was compiled using spatial characteristics of the soil moisture in red and NIR feature spaces to assess soil moisture stress. M is the slope of the soil line in the red–NIR spectral feature space, forming one edge of the triangle in the NIR–red spectral space represented by the soil line (Figure 5).

After extracting the pixel information from the spectral vegetation indices calculated from the UAV and satellite images, a regression analysis was carried out using the obtained field data.

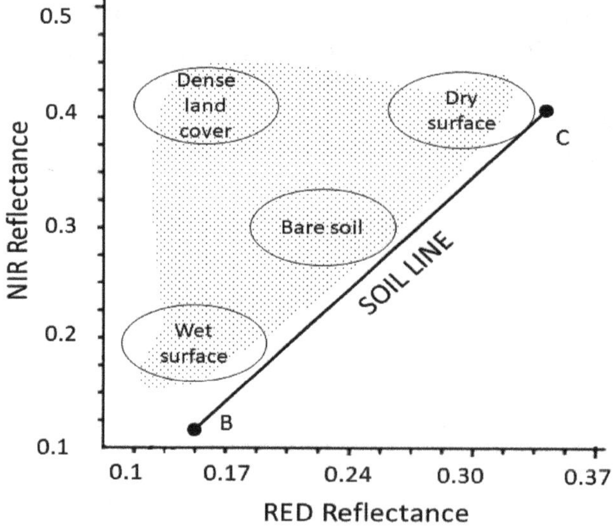

Figure 5. Near infrared versus red (NIR/R) correlation to obtain the soil line for the perpendicular drought index (PDI) calculation. Adapted from [32].

3. Results

3.1. Pedo-Hydrological Characterization of the Study Area

The study plots were located in areas with a udic soil moisture regime, a deep soil profile, 0–5% flat topography, and an isothermal temperature regime with well drained soils. The soils in the study site have loam textures in the upper layers and a mean depth of volcanic ashes of 109.29 cm before saprolite presence. Table 1 shows the pedologic and hydrologic variables analyzed to determine the homogeneity of the study plots (located near to soil profiles 3 and 4 in the case of pasture and potatoes, and near to soil profiles 5, 6, and 7 in the case of the bare soil study plot).

3.2. The PDI According the Spatial Resolution

The NIR–red linear regression was obtained to calculate the soil line (Figure 5), and the M value was determined (Table 2) by means of Equation (4) to determine the PDI (perpendicular drought index).

Table 2. NIR–red linear equations to obtain the M value to calculate the PDI at 4 cm.

Land Use	NIR – Red Equation	M	PDI – SM (R^2)
Pastures	y = −0.4276x + 0.5824	−0.4276	0.4392
Potatoes	y = −3.9215x + 0.65	−3.9215	0.002
Bare soil	y = 0.5921x + 0.165	0.5921	0.5062

To validate the PDI, the in situ SM (soil moisture) data measurements every 10 m were compared with the PDI, obtaining the results shown in Figure 6 according to land use. Among these results, correlation is strongest between PDI and soil moisture under bare soil (R2 = 0.5062), followed by pasture and then potatoes.

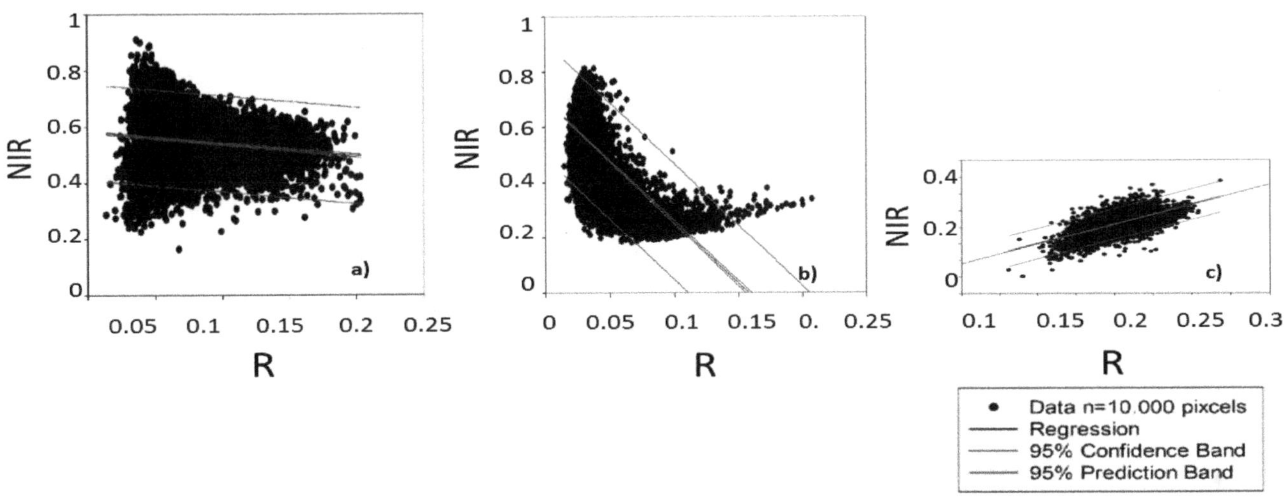

Figure 6. Polynomial linear correlations of NIR–R reflectance at 10,000 random points on the study site according to land use. (**a**) pasture study site, (**b**) potatoes study site, (**c**) bare soil study site. Source: Results of the current research.

Ghulam et al. [32] state that visible and near infrared spectral data are closely related to soil moisture at a soil depth of 10 cm.

The results obtained from repeating the same process at 3 m spatial resolution using the Planet Scope images are shown in Table 3. They show that there is a high correlation between the red and the NIR bands on satellite images with a spatial resolution of 300 cm, whereas the correlation between the PDI and soil moisture is lower than the UAV (unmanned aerial vehicle) 4 cm spatial resolution correlation.

A comparison of the PDI and soil moisture can be influenced by plant albedo and shade. The results shown in Figure 7 clearly demonstrate that potatoes at 4 cm resolution correlate less than bare soil and pastures at the same resolution.

Table 3. NIR– Red linear equations to obtain the M value to calculate the PDI at 300 cm spatial resolution.

Land Use	NIR – Red Equation (R^2)	M	PDI – SM (R^2)
Pastures	y = 0.9083x + 4.2942(R2 = 0.214)	0.9083	R2 = 0.141
Potatoes	y = −1.3897x + 6.562(R2 = 0.322)	−1.3897	R2 = 0.0191
Bare Soil	y = 1.6884x − 1.8164 (R2 = 0.465)	1.6884	R2 = 0.1137

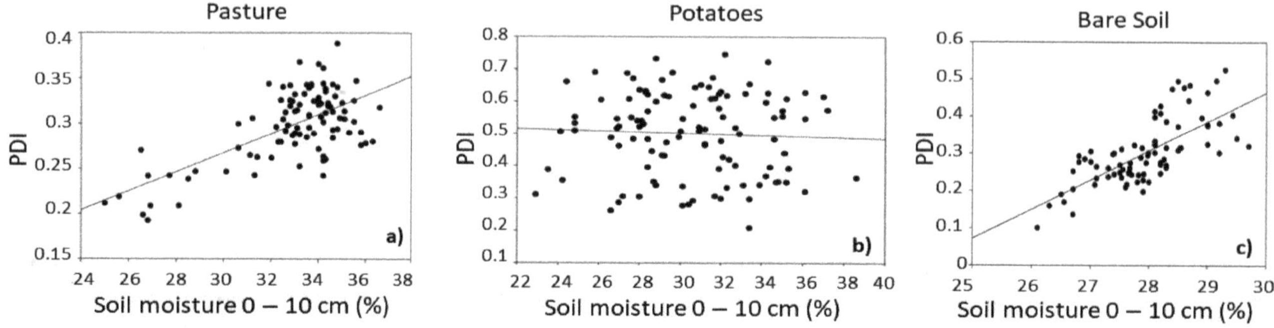

Figure 7. The PDI vs. soil moisture at 0–10 cm depth under different soil uses: (**a**) pasture R2 = 0.4392; (**b**) potatoes R2 = 0.002; (**c**) bare soil R2 = 0.5062) at UAV resolution (4 cm pixel). n = 110. Source: Current research.

3.3. Soil Moisture vs. Vegetation Indices

For each land use, the SAVI, NDVI, and NDWI were determined from the UAV images (4 cm pixel), as can be seen in Figure 8.

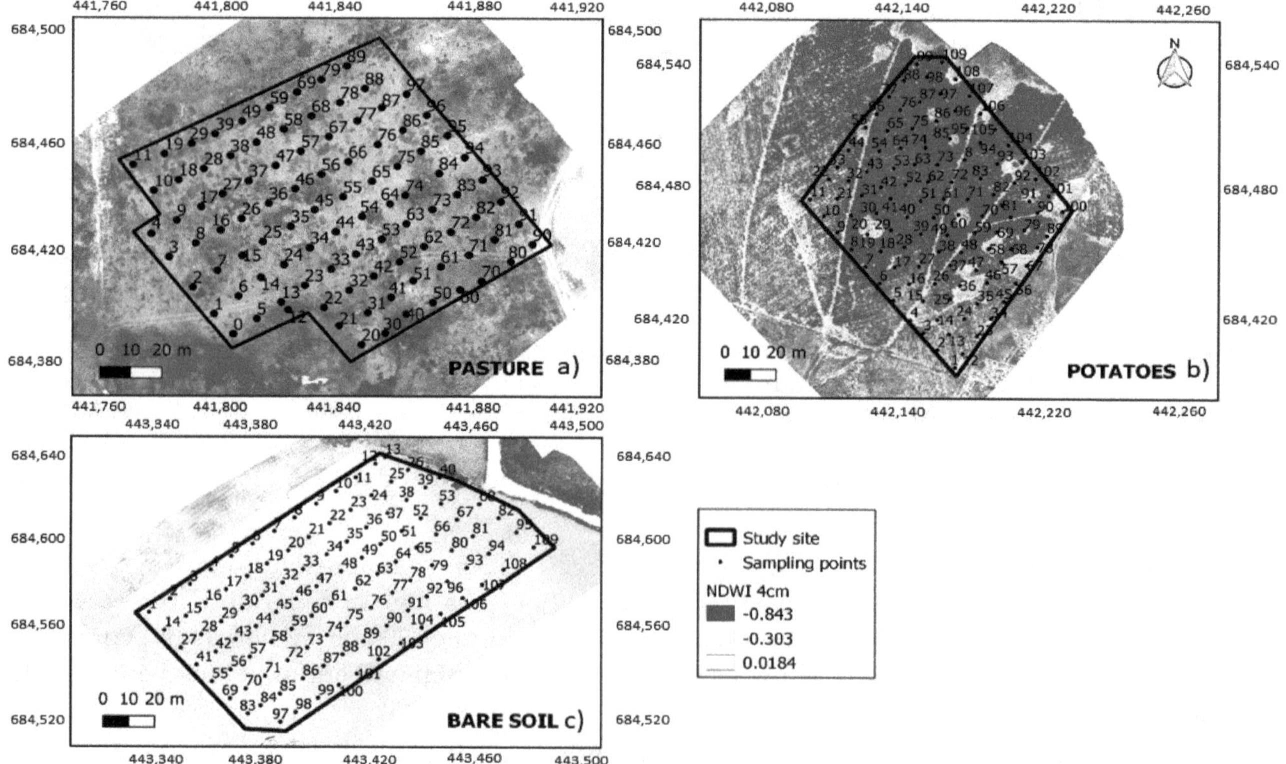

Figure 8. Normalized difference wetness index (NDWI) data obtained by means UAV composite images at 4 cm pixel resolution with the ground measurement location per land use. (**a**) pasture study site, (**b**) potatoes study site, (**c**) bare soil study site. Source: Current research.

Satellite Planet Analyst Scope images (300 cm pixel) were processed and then the same indices were determined per studied land use. Data for 12, 40, 100, 300 cm were obtained by means of an oversampling of the pixels of the UAV images on several scales, and from the means of the index values for each buffer zone of the sampling points. Posteriorly the georeferred data of each index were correlated with the ground soil moisture measurements.

The soil moisture data obtained from the sampling plots did not have a normal statistical distribution. The correlation between the measured soil moisture data and the indices obtained by means of the obtained images was analyzed by applying Spearman's rank correlation rho test according to the spatial resolution and land use. The following table shows these correlations (Table 4), where the triangle symbols denote the significant correlations.

It can be seen that there is an index that fits better, or presents a better correlation, with soil moisture for each of the land uses and resolutions studied.

Regarding pasture land use, soil moisture analysis by means of satellite images at 3 m resolution only had significant correlations with the NDVI. Pasture land use at a detailed 4 cm resolution scale showed a significant correlation between the PDI and soil moisture.

Under land use for potatoes, all the indices showed a positive correlation with soil moisture (Figure 9). Satellite images at 3 m resolution can be used to determine the soil moisture of potatoes land use using the NDWI and NDVI, that is, the indices that showed the best correlations (Table 4). At detailed resolution, only the NDWI showed a significant positive correlation with soil moisture (Figure 9).

The best representation to analyze soil moisture under bare soil is by means of the PDI with UAV images at high resolution, whereas the same index with a coarser satellite resolution (3 m) cannot be directly correlated with surface soil moisture (0 to 10 cm). At 3 m resolution, the NDVI and NDWI show the best significant correlations with soil moisture under bare soils, showing negative correlations (Figure 9).

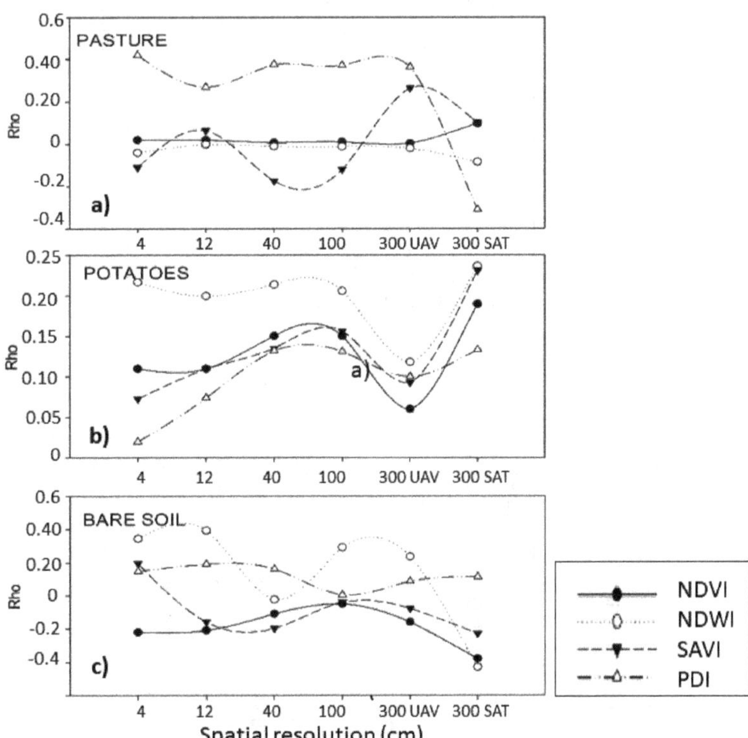

Figure 9. Rho correlation between the index and soil moisture according to land use and spatial resolution. (**a**) pasture study site, (**b**) potatoes study site, (**c**) bare soil study site. Source: Current research.

Table 4. Spearman's correlation coefficients (rho, *p*-value) for the normalized difference vegetation index (NDVI), the normalized difference water index (NDWI), the soil-adjusted vegetation index (SAVI), and the perpendicular drought index (PDI) regarding soil moisture at 0–10 cm depth under pasture, bare soil, and potatoes (n = 110).

Spatial Resolution	Pastures				Potatoes				Bare Soil			
	NDVI	NDWI	SAVI	PDI	NDVI	NDWI	SAVI	PDI	NDVI	NDWI	SAVI	PDI
4	0.02 (0.77)	−0.04 (0.67)	−0.11 (0.28)	0.42 (1.649×10^{-5})▲▲	0.11 (0.24)	0.21 (0.02)▲	0.07 (0.44)	0.002 (0.34)	−0.22 (0.03)▲	0.34 (0.0002)▲▲	0.19 (0.04)▲	0.14 (0.02)▲
12	0.02 (0.85)	−0.002 (0.97)	0.06 (0.55)	0.26 (0.007)▲▲	0.11 (0.28)	0.20 (0.03)▲	0.11 (0.23)	0.074 (0.47)	−0.21 (0.04)▲	0.39 (2.21×10^{-5})▲▲	−0.16 (0.09)	0.19 (0.04)▲
40	0.007 (0.94)	−0.01 (0.89)	−0.17 (0.10)	0.37 (0.0001)▲▲	0.15 (0.11)	0.21 (0.02)▲	0.13 (0.16)	0.13 (0.16)	−0.11 (0.25)	−0.02(0.79)	−0.19 (0.03)▲	0.16 (0.09)
100	0.01 (0.91)	−0.01 (0.90)	−0.12 (0.25)	0.37 (0.0001)▲▲	0.15 (0.11)	0.20 (0.03)▲	0.15 (0.10)	0.13 (0.17)	−0.05 (0.55)	0.29 (0.002)▲▲	−0.04 (0.64)	0.005 (0.09)
300	0.004 (0.96)	−0.02 (0.83)	0.26 (0.01)	0.36 (0.0002)▲▲	0.06 (0.51)	0.11 (0.02)▲	0.09 (0.33)	0.10 (0.29)	−0.16 (0.36)	0.23 (0.01)▲	−0.07 (0.41)	0.08 (0.36)
300sat	0.597 (0.0001)▲▲	−0.08 (0.40)	0.10 (0.39)	−0.31 (0.19)	0.19 (0.04)▲	0.23 (0.01)▲	0.23 (0.54)	0.13 (0.19)	−0.38 (5.6×10^{-5})▲▲	0.43 (9.16×10^{-6})▲▲	−0.23 (0.01)▲	0.11 (0.54)

Statistical significance: *p*-value is significant at 5% (▲) when it is lower than 0.05, and significant at 1% (▲▲) when it is lower than 0.01.

On bare soil land use, any resolution can be used to estimate soil moisture by means of optical images with the NDWI.

The index that performs the best on bare soil is the NDWI at any spatial resolution, the NDVI at 4 or 12 cm resolution or from satellite images, and the PDI at 4 cm resolution. On bare soil, the reflectance effect of the existent furrows every 2 m must be considered on coarser scales (Figure 3c), because the land roughness could cause differences on the averaged land reflectance.

4. Discussion

Farhan and Al Bakri [36] report that the NDVI mainly reflects seasonal vegetation conditions, showing higher correlations with seasonal soil moisture stress, whereas the PDI does not show this relationship.

The NDVI in this research was the vegetation index that performed better on coarser resolution than thinner spatial resolution, regardless of land use.

Both sensing drought indices, the NDVI and the PDI can explain soil moisture variability in all the studied land uses. One study [37] showed significant negative correlations in spring, summer, and autumn between the NDVI and soil moisture, whereas farmland showed a significant positive correlation between NDVI and soil moisture in winter. In the current research, NDVI positively correlates with potato and pasture land uses and negatively correlates with bare soil, possibly due to the higher evaporation on bare soils (Figure 9).

Bare soils are not affected by vegetation cover, so their reflectivity in red and NIR bands is only affected by the soil moisture content. If there is a decrease in soil moisture, the reflective+ity of the red and NIR bands increases [38]. When vegetation cover increases, reflectance in the NIR band is higher than in the red band. Where land use in the current study plot includes both soil and vegetation, the points scatter inside a triangular region in the NIR–R, as shown in Figure 5. These results on bare soils concur with laboratory reflectance studies [20].

Spatial resolution clearly determines the ability of a sensor to generate the indices that can successfully approximate soil moisture.

The NDVI produced no significant correlations with soil moisture on UAV images, whereas Planet Scope NDVI variants with their higher spectral and spatial resolution positively correlated with bare soils, concurring with [39]. The ease of calculating the NDVI and the high temporal resolution of the data may mean that Sentinel-2 Planet Scope may play a future role in early warning systems of drought, as it enables high-resolution vegetation condition monitoring, which may be useful in detecting the onset of agricultural drought.

In regard to the SAVI, this can only be used with a significant correlation to estimate the soil moisture of bare soils.

The NDWI was the index that performed best on detailed resolutions, especially to study the soil moisture of land use for vegetables such as potatoes, which is useful when considering precision agriculture.

Observing Table 4, it can be seen that on pasture, the most significant correlations are found on coarser scales, whereas bare soil and potatoes have better results on detailed resolutions. These results show that UAV with multispectral cameras are useful to evaluate bare soil and potato soil moisture at detailed scales, and, above all, with the NDWI, SAVI and PDI.

5. Conclusions

On bare soil, the indices that best fit with the soil moisture results were the NDVI, the NDWI, and the PDI on a detailed scale. In contrast, Amani et al. [21] found that bare soils have good significance on a coarse scale with Landsat8 images in arid environments. These results are in line with those of a recent sub-metric soil moisture study using UAV and multispectral images in tropical conditions in Peru [5].

Under potato crops, the NDWI correlates significantly with soil moisture irrespective of the scale of the analyzed image under potato land use.

The PDI is the index that correlates the highest with detailed scales, showing better results on pasture than on potatoes or bare soil. In regard to pastures at a coarser resolution, the NDVI showed the best correlation with soil moisture. These results are relevant due to the fact that the pasture is an extensive crop in Colombia and soil moisture monitoring can be useful to realize environmental studies of multitemporal changes of this important hydrological parameter.

A UAV soil moisture study [40] on Karst heterogeneous landscapes determined that the optimal resolution to analyze soil moisture by means of DEMs is 7 m, and that soil moisture variability is mainly explained by the vegetation type (35.7%), which concurs with the results of the current research.

The study of soil moisture with UAVs study presents several advantages over conventional platforms such as satellites, including the fact that they fly at lower altitudes, increasing the spatial resolution of the images, and cost less than private remote sensing images, allowing for more frequent monitoring. For average-size farms in Colombia, high-resolution remote sensing at 3 m such as Planet Scope combined with UAV data can be used to estimate soil moisture for the evaluated land uses. Remote sensing indices are currently being tested and improved to propose proxies that reflect the physiological status of crops under changing environmental conditions, and they can be used to determine plant water status for several crop species.

The best scale to study soil surface moisture with optical images is at 3m resolution, which can determine soil moisture at a depth of 0 to 10 cm using either the NDWI or NDVI according to its land use. None of the indices can be used for all crops or land uses with the same resolution. A prior classification of land use is needed to study soil moisture effectively due to the effect of vegetation on soil moisture at depths of 0 to 10cm, as supported by Ghulam et al. [32], who state that visible and near infrared spectral data have a close relationship with soil moisture at a soil depth of 10 cm.

According to land use as a means to determine soil moisture, a different index and resolution were found to provide the most accurate results; that is, resolutions of 3 m appropriate to study soil moisture under pasture, potatoes and bare soil using NDVI correlations with soil moisture in Andosols.

Author Contributions: Conceptualization, M.C. and D.V.; methodology, M.C. and D.V.; software, J.B.-R.; validation, M.C.T.-M., M.C.; formal analysis, M.C.; investigation, M.C.; data curation, M.C.T.-M.; writing—original draft preparation, M.C.; writing—review and editing, D.V.; visualization, J.B.-R.; supervision, D.V. All authors have read and agreed to the published version of the manuscript.

Acknowledgments: The authors would like to thank the continuous support to this research to the Secretaría de Medioambiente from Envigado municipality, Colombia. This research was achieved thanks to the valuable support of AGROSAVIA. We would like to thank the engineers Javier Salcedo and Andrés Berbesí for their technical and professional assistance at field. Finally, the main author would like to specially thank Juan C. Loaiza for his contribution in soil sciences description and his invaluable support, guidance and motivation to complete this research.

References

1. Li, S.; Pezeshki, S.R.; Goodwin, S. Effects of soil moisture regimes on photosynthesis and growth in cattail (Typha latifolia). *Acta Oecologica* **2004**, *25*, 17–22. [CrossRef]
2. Wright, C.; Kagawa-Viviani, A.; Gerlein-Safdi, C.; Mosquera, G.M.; Poca, M.; Tseng, H.; Chun, K.P. Advancing ecohydrology in the changing tropics: Perspectives from early career scientists. *Ecohydrology* **2018**, *11*, 112–125. [CrossRef]
3. Firman Ghazali, M.; Wikantika, K.; Budi Harto, A.; Kondoh, A. Generating soil salinity, soil moisture, soil pH from satellite imagery and its analysis. *Inf. Process. Agric.* **2019**. [CrossRef]
4. Esquivel, A.; Llanos-Herrera, L.; Agudelo, D.; Prager, S.D.; Fernandes, K.; Rojas, A.; Valencia, J.J.; Ramirez-Villegrasa, J. Predictability of seasonal precipitation across major crop growing areas in Colombia. *Clim. Serv.* **2018**, *12*, 36–47. [CrossRef]

5. Wigmore, O.; Marka, B.; McKenzie, J.; Baraerd, M.; Lautze, L. Sub-metre mapping of surface soil moisture in proglacial valleys of the tropical Andes using a multispectral unmanned aerial vehicle. *Remote. Sens. Environ.* **2019**, *222*, 104–118. [CrossRef]
6. Entekhabi, D.; Nioku, E.G.; O'Neill, P.E.; Kellogg, K.H.; Crow, W.T.; Edelstein, W.N.; Entin, J.K.; Goodman, S.D.; Jackson, T.J.; Johnson, J.; et al. The Soil Moisture Active Passive (SMAP) Mission. *Proc. IEEE* **2010**, *98*, 704–716. [CrossRef]
7. Gruber, A.; De Lannoy, G.; Albergel, C.; Al-Yaari, A.; Brocca, L.; Calvet, J.C.; Colliander, A.; Cosh, M.; Crow, W.; Dorigo, W.; et al. Validation practices for satellite soil moisture retrievals: What are (the) errors? *Remote Sens. Environ.* **2020**, *244*, 111806. [CrossRef]
8. Albergel, C.; de Rosnay, P.; Gruhier, C.; Muñoz-Sabater, J.; Hasenauer, S.; Isaksen, L.; Kerr, Y.; Wagner, W. Evaluation of remotely sensed and modelled soil moisture products using global ground-based in situ observations. *Remote Sens. Environ.* **2020**, *118*, 215–226. [CrossRef]
9. Wang, L.; Qu, J.J. Satellite remote sensing applications for surface soil moisture monitoring: A review. *Front. Earth Sci. China* **2009**, *3*, 237–247. [CrossRef]
10. Lievens, H.; Martens, B.; Verhoest, N.E.C.; Hahn, S.; Reichle, R.H.; Miralles, D.G. Assimilation of global radar backscatter and radiometer brightness temperature observations to improve soil moisture and land evaporation estimates. *Remote Sens. Environ.* **2017**, *18*, 194–210. [CrossRef]
11. He, Y.; Weng, Q. *High Spatial Resolution Remote Sensing. Data, Analysis and Applications 2018. Taylor & Francis Series in Imaging Science*; CRC Press: New York, NY, USA, 2018; ISBN 9781498761682.
12. Gruber, A.; Scanlon, T.; van der Schalie, R.; Wagner, W.; Dorigo, W. Evolution of the ESA CCI Soil Moisture climate data records and their underlying merging methodology. *Earth Syst. Sci. Data* **2019**, *11*, 717–739. [CrossRef]
13. Vicente-Serrano, S.; Begueria, S.; Lorenzo, J.; Camarero, J.; Lopez.Moreno, J.; Azorin-Molina, C.; Revuelto, J.; Moran-Tejeda, E.; Sanchez-Lorenzo, A. Performance of Drought Indices for Ecological, Agricultural, and Hydrological Applications. *Earth Interact.* **2012**, *16*, 1–27. [CrossRef]
14. Jelev, G.; Dimitrov, P.; Kamenova, I.; Ilieva, I.; Roumenina, E.; Filchev, L.; Gikov, A.; Banov, M.; Krasteva, V.; Kolchakov, V.; et al. Using UAV Spectral Vegetation Indices for Estimation and Mapping of Biophysical Variables in Winter Wheat. In *Digital Earth Observation, Proceedings of the 39th European Association of Remote Sensing Laboratories Symposium & 43rd General Assembly, Salzburg, Austria, 1–4 July 2019*; Riedler, B., Lettner, S., Lang, S., Tiede, D., Eds.; University of Salzburg: Salzburg, Austria, 2019.
15. Gago, J.; Douthe, C.; Coopman, R.E.; Gallego, P.P.; Ribas-Carbo, M.; Flexas, J.; Escalona, J.; Medrano, H. UAVs challenge to assess water stress for sustainable agriculture. *Agric. Water Manag.* **2015**, *153*, 9–19. [CrossRef]
16. Montoani Silva, B.; Godinho Silva, S.H.; de Oliveira, G.H.; Caspar Rosa Peters, P.H.; Reis dos Santos, W.J.; CuriI, N. Soil Moisture Assessed by Digital Mapping Techniques and Its Field Validation. *Ciênc. Agrotec.* **2014**, *38*, 140–148. [CrossRef]
17. Sørensen, R.; Zinko, U.; Seibert, J. On the calculation of the topographic wetness index: Evaluation of different methods based on field observations. *Hydrol. Earth Syst. Sci.* **2006**, *10*, 101–112. [CrossRef]
18. Shafian, S.; Maas, S. Index of Soil Moisture Using Raw Landsat Image Digital Count Data in Texas High Plains. *Remote Sens.* **2015**, *7*, 2352–2372. [CrossRef]
19. Wang, L.; Qu, J.J.; Hao, X.; Zhu, Q. Sensitivity studies of the moisture effects on MODIS SWIR reflectance and vegetation water indices. *Int. J. Remote Sens.* **2008**, *29*, 7065–7075. [CrossRef]
20. Fabre, S.; Briottet, X.; Lesaignoux, A. Estimation of Soil Moisture Content from the Spectral Reflectance of Bare Soils in the 0.4–2.5 μm Domain. *Sensors (Basel)* **2015**, *15*, 3262–3281. [CrossRef]
21. Lobell, D.B.; Asner, G.P. Moisture Effects On Soil Reflectance. *Soil Sci. Soc. Am. J.* **2002**, *66*, 722–727. [CrossRef]
22. Amani, M.; Parsian, S.; MirMazloumi, S.M.; Aieneh, O. Two new soil moisture indices based on the NIR-red triangle space of Landsat-8 data. *Int. J. Appl. Earth Obs. Geoinf.* **2016**, *50*, 176–186. [CrossRef]
23. Mohseni, F.; Mokhtarzade, M. A new soil moisture index driven from an adapted long-term, temperature-vegetation scatter plot using MODIS data. *J. Hydrol.* **2020**, 581. [CrossRef]
24. McFeeter, S.K. The use of the Normalized Difference Water Index (NDWI) in the delineation of open water features. *Int. J. Remote Sens.* **1996**, *17*, 7. [CrossRef]
25. Gao, A. NDWI—A normalized difference water index for remote sensing of vegetation liquid water from space. *Remote Sens. Environ.* **1996**, *58*, 257–266. [CrossRef]

26. Xu, H. Modification of normalised difference water index (NDWI) to enhance open water features in remotely sensed imagery. *Int. J. Remote Sens.* **2006**, *27*, 3025–3033. [CrossRef]
27. Chen, X.; Guo, Z.; Chen, J.; Yang, W.; Yao, Y.; Zhang, C.; Cui, X.; Cao, X. Replacing the Red Band with the Red-SWIR Band (0.74ρred+0.26ρswir) Can Reduce the Sensitivity of Vegetation Indices to Soil Background. *Remote Sens.* **2019**, *11*, 851. [CrossRef]
28. Qi, J.; Chehbouni, A.; Huete, A.R.; Kerr, Y.H.; Sorooshian, S.A. Modified soil adjusted vegetation index. *Remote Sens. Environ.* **1994**, *48*, 119–126. [CrossRef]
29. Jeihounia, M.; Alavipanah, S.K.; Toomaniana, A.; Jafarzadeh, A.A. Digital mapping of soil moisture retention properties using solely satellite based data and data mining techniques. *J. Hydrol.* **2019**, *585*, 124786. [CrossRef]
30. Nemani, R.; Pierce, L.; Running, S.; Goward, S. Developing satellite-derived estimates of surface moisture status. *J. Appl. Meteorol.* **1993**, *32*, 548–557. [CrossRef]
31. Sadeghi, M.; Babaeian, E.; Tuller, M.; Jones, S.B. The optical trapezoid model: A novel approach to remote sensing of soil moisture applied to Sentinel-2 and Landsat-8 observations. *Remote Sens. Environ.* **2017**, *198*, 52–68. [CrossRef]
32. Ghulam, A.; Qin, Q.; Zhan, Z. Designing of the perpendicular drought index. *Environ. Geol.* **2007**, *52*, 1045–1052. [CrossRef]
33. Salazar, M.P. Water Distribution and Drainage Systems of Aburrá Valley, Colombia—Empresas Públicas de Medellín, E.S.P. *Procedia Eng.* **2017**, *186*, 1877–7058.
34. Casamitjana, M.; Loaiza, J.C. Propiedades físicas e hidrológicas en suelos derivados de cenizas volcánicas. In *Movimientos en Masa*; Casamitjana, Sidle, Eds.; Fondo Editorial EIA: Medellín, Colombia, 2019; ISBN 978-958-52367-0-7.
35. Abera, G.; Wolde-Meskel, A. Soil properties *[35*and soil organic carbon stocks of tropical andosol under different land uses. *Open J. Soil Sci.* **2013**, *3*, 153–162. [CrossRef]
36. Farhan, I.; Al-Bakri, J. Detection of a Real Time Remote Sensing Indices and Soil Moisture for Drought Monitoring and Assessment in Jordan. *Open J. Geol.* **2019**, *9*, 1048–1068. [CrossRef]
37. West, H.; Quinn, N.; Horswell, M.; White, P. Assessing Vegetation Response to Soil Moisture Fluctuation under Extreme Drought Using Sentinel-2. *Water* **2018**, *10*, 838. [CrossRef]
38. Zhang, J.; Zhang, Q.; Bao, A.; Wang, Y. A New Remote Sensing Dryness Index Based on the Near-Infrared and Red Spectral Space. *Remote Sens.* **2019**, *11*, 456. [CrossRef]
39. Chi, Y.; Sun, Y.; Sun, Y.; Liu, S.; Fu, Z. Multi-temporal characterization of land surface temperature and its relationships with normalized difference vegetation index and soil moisture content in the Yellow River Delta, China. *Glob. Ecol. Conserv.* **2020**, *23*, e01092. [CrossRef]
40. Luo, W.; Xianli, X.; Wen, X.; Wen, L.; Meixian, L.; Zhenwei, L.; Tao, P.; Chaohao, X.; Yaohuam, Z.; Ronfei, Z. UAV based soil moisture remote sensing in a karst mountainous catchment. *CATENA* **2019**, *174*, 478–489. [CrossRef]

3

A Machine Learning Method to Estimate Reference Evapotranspiration using Soil Moisture Sensors

Antonio Fernández-López [1], Daniel Marín-Sánchez [2,3], Ginés García-Mateos [2,*], Antonio Ruiz-Canales [1], Manuel Ferrández-Villena-García [1] and José Miguel Molina-Martínez [4]

[1] Engineering Department, Miguel Hernandez University of Elche, 03312 Orihuela, Spain; antonio.fernandez83@gmail.com (A.F.-L.); acanales@umh.es (A.R.-C.); m.ferrandez@umh.es (M.F.-V.-G.)
[2] Computer Science and Systems Department, University of Murcia, 30100 Murcia, Spain; daniel.marin.94@gmail.com
[3] Google, Brandschenkestrasse 110, 8002 Zürich, Switzerland (current affiliation)
[4] Food Engineering and Agricultural Equipment Department, Technical University of Cartagena, 30203 Cartagena, Spain; josem.molina@upct.es
[*] Correspondence: ginesgm@um.es

Featured Application: The proposed approach for estimating reference evapotranspiration allows obtaining accurate approximations of this important crop parameter in an inexpensive way by using moisture sensors, which can be translated to an optimization of water resources.

Abstract: One of the most important applications of remote imaging systems in agriculture, with the greatest impact on global sustainability, is the determination of optimal crop irrigation. The methodology proposed by the Food and Agriculture Organization (FAO) is based on estimating crop evapotranspiration (ETc), which is done by computing the reference crop evapotranspiration (ETo) multiplied by a crop coefficient (Kc). Some previous works proposed methods to compute Kc using remote crop images. The present research aims at complementing these systems, estimating ETo with the use of soil moisture sensors. A crop of kikuyu grass (*Pennisetum clandestinum*) was used as the reference crop. Four frequency-domain reflectometry sensors were installed, gathering moisture information during the study period from May 2015 to September 2016. Different machine learning regression algorithms were analyzed for the estimation of ETo using moisture and climatic data. The values were compared with respect to the ETo computed in an agroclimatic station using the Penman–Monteith method. The best method was the randomizable filtered classifier technique, based on the K* algorithm. This model achieved a correlation coefficient, *R*, of 0.9936, with a root-mean-squared error of 0.183 mm/day and 6.52% mean relative error; the second-best model used artificial neural networks, with an R of 0.9470 and 11% relative error. Thus, this new methodology allows obtaining accurate and cost-efficient prediction models for ETo, as well as for the water balance of the crops.

Keywords: reference evapotranspiration; moisture sensors; machine learning regression; frequency-domain reflectometry; randomizable filtered classifier

---

1. Introduction

In agricultural sciences, the optimal determination of crop water needs over time is based on measuring the soil water balance and the evaporative demand of the plants. The use of soil moisture measurements was adopted as a suitable strategy for soil water balance estimation. Several methods for computing this balance were developed by different authors [1–3]. These techniques were applied in agriculture to obtain the water needs in conjunction with other methods based on remote image sensing. The ultimate objective is to provide farmers with information on the appropriate irrigation volumes to apply in every phenological period of the crop, depending on the desired levels of yield and other parameters. Different physical principles are applied to determine soil moisture, such as gamma-ray spectroscopy [4], synthetic aperture radar [5], and others [6]. Furthermore, there is a wide range of techniques for measuring soil moisture based on electricity, which are applied in geophysical prospecting [7,8] and agronomy [9,10], among other areas. In these measuring techniques, capacitive methods such as frequency-domain reflectometry (FDR) are included [11–13]. The accuracy of such sensors varies due to the employed techniques and working conditions.

The key of these techniques is to model the relationships among soil water balance, crop yield, and water use efficiency (WUE) in order to develop better semiarid crops and water management practices [14]. In Mediterranean agriculture, particularly in the southeast of Spain, soil water availability is one of the main limitations for the practice of an economically sustainable agriculture. For this reason, using a suitable irrigation management is critical in the quantity and quality of the obtained harvests. This involves the determination of crop water needs and an optimal irrigation scheduling [15–19]. Although yield reduction is generally expected when crops are subject to limited irrigation, a well-designed limited irrigation system can minimize the impact on yield and still lead to grower profitability.

On the other hand, according to the FAO (Food and Agriculture Organization)-56 methodology [20], crop evapotranspiration (ETc) can be obtained as the product of the reference crop evapotranspiration (ETo) and a crop coefficient (Kc). This Kc coefficient takes into account the development season of the cultivated species, the type of irrigation (by sprinkler, trickle, etc.), and the cultivation techniques (plantation density, pruning, etc.) [21]. Allen et al. [20] proposed Kc values for a great number of species, in standard crop conditions; an adjustment of Kc is necessary when the actual conditions are different from this standard scenario [22,23].

The infrastructure required for a direct measurement of ETc involves high-cost equipment such as lysimeters and Bowen-ratio stations [24]. Therefore, this methodology is unacceptable for small farms. However, both ETo and ETc can also be estimated indirectly, through their relationship with the values obtained using other inexpensive sensors. For example, in the Penman–Monteith method [20], ETo is estimated based on solar radiation, air temperature, humidity, wind speed, atmospheric pressure, site elevation above sea level, Julian day, and latitude degree of the study site. Thus, the main parameters of the soil water state (water content, water potential, and water balance, among others) can be estimated in a cost-efficient way. This methodology is widely extended and used by farmers because of its simplicity. However, it has the disadvantage of giving isolated measurements; in some cases, the obtained parameters are not representative of the entire plot.

To overcome this drawback, remote image capture systems offer a promising alternative to traditional water status measurements [25]. They can provide a snapshot of the whole crop over a reduced period. The advent of unmanned aerial vehicles (UAVs) offers an opportunity to develop remote sensing-based methodologies for precision irrigation [26,27]; they are more affordable than the costly systems based on manned aircrafts, and they provide higher spatial and temporal resolutions than those normally offered by satellites. Various sources of remotely sensed imagery, with differences in spectral, spatial, radioactive, and temporal characteristics, are applied to different purposes of vegetation mapping [28].

Soil moisture sensors are also used to measure the content of water in the soil and provide an estimation of ETc. For example, Sharma et al. [29] applied two different types of sensors based on time-domain reflectometry and capacitance, to estimate the actual evapotranspiration of a greenhouse crop of chili peppers. The final purpose was to reduce the amount of water needed, achieving a reduction of 30%. Other works used soil moisture sensors as a direct way to define irrigation decisions, for example, in crops of tomato [30], turfgrass [31], citrus orchards [32], and other kinds of vegetables [33].

All these facts suggest that it is recommended to combine the FAO-56 methodology with other techniques based on remote images and measurements of the soil water state [34]. Efforts should focus on developing new methods that are more robust, reliable, and sustainable [35,36]. In this paper, the feasibility of estimating ETo using remote sensing techniques based on soil moisture sensors is analyzed. The experiment was performed in a plot in the southeast of Spain, which is an arid zone with a great shortage of water. The proposed methodology integrates different types of inexpensive sensors: meteorological data (daily temperature and rainfall), soil moisture sensors, and irrigation volumes. In order to estimate ETo from the sensor data with high accuracy, a comparison of some advanced pattern recognition techniques is presented. The results of these models are analyzed in detail, selecting the most accurate algorithm. Finally, this estimation of ETo is integrated with an existing methodology for the computation of Kc using remote image sensing [37–39], thereby obtaining the daily water balance to adopt the most adequate irrigation decision.

2. Materials and Methods

2.1. Data Acquisition

Data collection for this study covered a long period of more than one year, from May 2015 to September 2016. The moisture sensing devices were tested in an experimental plot of 34 m^2 located in the Higher Polytechnic School of Orihuela (EPSO) of the Miguel Hernández University of Elche (UMH), Spain. Crop rows had a north–south (N–S) orientation. A crop of kikuyu grass (*Pennisetum clandestinum*) with total coverage of the soil was cultivated on this plot. This species of grass was previously used by many authors in different studies on evapotranspiration (ET) [40], and it was employed in our case as a reference model for directly determining the reference ET (ETo). According to Allen et al. [20], the reference crop is "a hypothetical crop with an assumed height of 0.12 m, a surface resistance of 70 $s{\cdot}m^{-1}$, and an albedo of 0.23, closely resembling the evaporation from an extensive surface of green grass of uniform height, actively growing and adequately watered". All these characteristics are approximately met by the selected species.

The geographical location of the plot is shown in Figure 1. It has a latitude of 38°4′10.17″ N, longitude 0°59′6.81″ west (W), and an altitude of 19 m above sea level. The climate in this region is semiarid Mediterranean type, with mild winters and scarce rainfall.

Figure 1. Location of the experimental plot in the Higher Polytechnic School of Orihuela (EPSO) of the Miguel Hernández University of Elche (UMH), Spain, and the agroclimatic station of La Murada, Orihuela. Aerial images were extracted from Google Maps.

The soil of the experimental plot was loamy in the most superficial layer, with a water field capacity of 0.27 m^3/m^3 and a permanent wilting point of 0.15 m^3/m^3. More detailed information of the main soil characteristics is shown in Table 1. The employed irrigation water had an average quality, with a slightly moderate electrical conductivity and a moderate content of total salts. The plot included a pressure irrigation system with a programmer that provided optimal irrigation to the crop during all the experiment, to meet the requirement of "actively growing and adequately watered".

Table 1. Main characteristics of the soil in the experimental plot at different depths.

Property	0–10 cm	10–20 cm	20–30 cm
Sand (%)	45	61	75
Silt (%)	30	22	16
Clay (%)	25	17	9
Water field capacity (m^3/m^3)	0.27	0.22	0.18
Permanent wilting point (m^3/m^3)	0.15	0.12	0.08

The agroclimatic data for the study of the water balance were obtained from the website of the Valencian Institute of Agricultural Research (http://www.ivia.gva.es/en), which uses information taken from the climatic station of La Murada, Orihuela. This station has a latitude of 38°10′51.8′′ N, longitude 0°57′30.8′′ W, and it is installed at 96 m above sea level, as shown in Figure 1; the distance from this station to the experimental crop is approximately 9.5 km. It is a Model 3 station, which is the predominant scheme of the SIAR Network (Agroclimatic Information System for Irrigation) adopted by the Spanish Ministry of Agriculture.

This model contains a temperature–humidity sensor, a radiation sensor, a wind speed and direction sensor, a pluviometer, and a datalogger. The main variables used in the current research during the study period are shown in Figure 2, including the daily ETo, mean daily temperature, daily rainfall, and daily irrigation by drip and sprinkler. The last two values were directly obtained from flow meters in the experimental plot. The ETo values from the station were estimated using the Penman–Monteith method [20], and the results were considered as the ground truth.

It can be observed in Figure 2 that daily temperature and ETo are highly correlated, although the ETo cannot be simply deduced from the temperature. The highest values of temperature were obtained in the summer months, from June to September, and the lowest values corresponded to the winter months, from December to March, while the cycle of ETo was slightly displaced, with the highest values from May to July, and the lowest from November to February. Regarding the water supply, it was mostly uniform in the study period, although it decreased around the spring season, with irrigations of less than 7 mm and only one every two days. The total rainfall in this period was 269 mm.

To measure the soil moisture in the study plot, commercial frequency-domain reflectometry (FDR) sensors were used. These FDR sensors consist of several cylindrical rings located at four depths (10, 20, 30, and 40 cm) in an isolated gauge, with a polyvinyl chloride (PVC) access tube of type EnviroSCAN® (Sentek Sensor Technologies, Stepney, Australia). Four of these gauges were installed spaced in the plot during the experiment. Soil moisture data were collected every 5 min from 30 May 2015 to 30 September 2016. The sensors were firstly installed in the laboratory, where they were tested and calibrated. Afterward, they were installed in the experimental plot. The data obtained from the sensors were stored in data loggers, and then they were collected weekly.

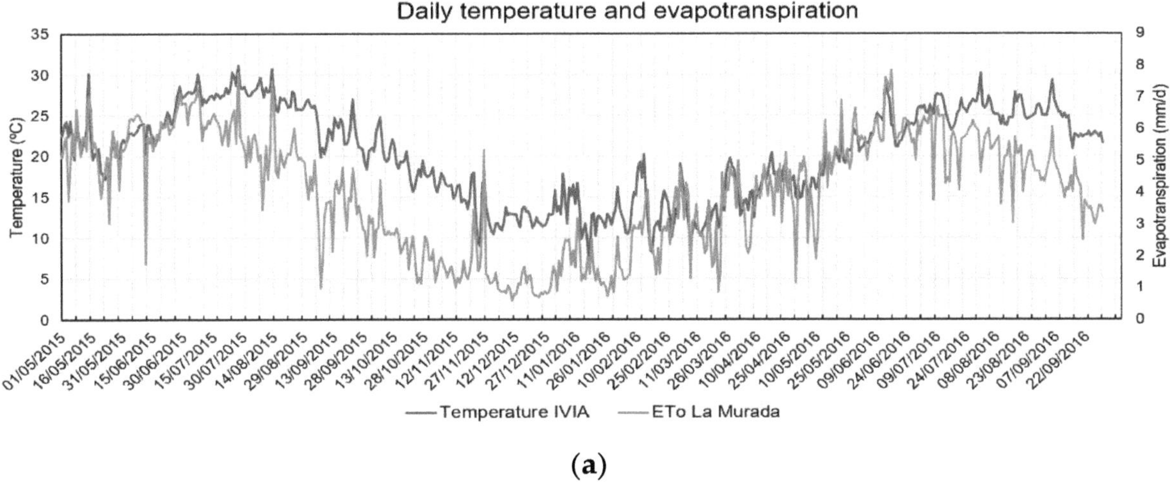

(a)

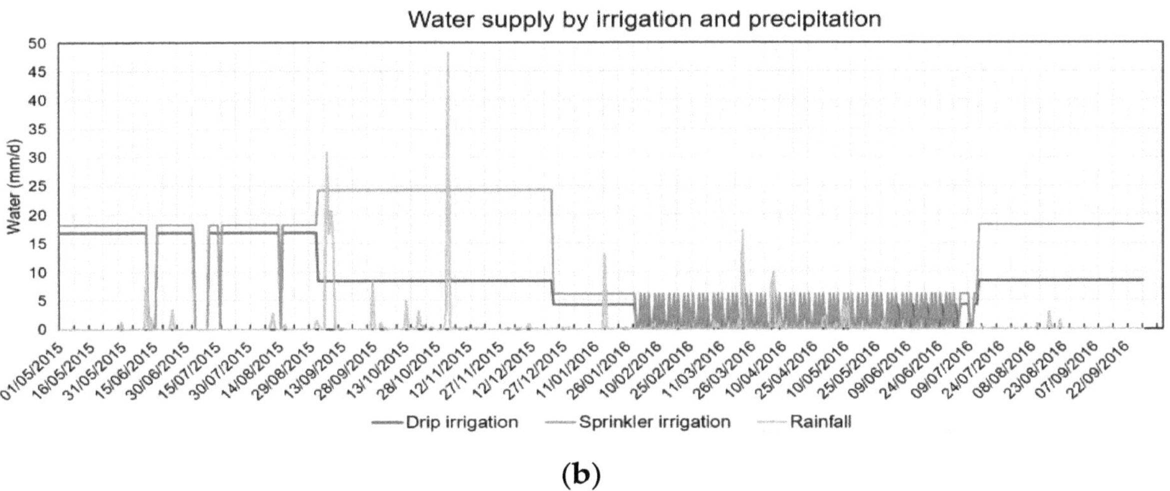

(b)

Figure 2. Main agroclimatic and irrigation variables during the study period. The agroclimatic variables were obtained in the station of La Murada; irrigation was obtained in the crop. (**a**) Daily values of the temperature and reference evapotranspiration (ETo). (**b**) Daily values of the water supply by irrigation and rainfall.

2.2. Data Preparation and Preprocessing

The information considered in the present research consists of the following variables: the measures obtained from the four soil moisture sensors at four different depths; the daily rain, temperature, and ETo measured in the agroclimatic station of La Murada; the amount of irrigation (by sprinkler and drip) applied every day to the reference crop. Recall that the purpose is to estimate ETo using the soil moisture sensors and agroclimatic data. In this way, the additional information required in the proposed approach can be easily obtained using a thermometer and a pluviometer. The total amount of information collected during the period of study contained more than 100 MB. However, some sensors had failures on certain days, providing no information. Therefore, data preprocessing was a very important preliminary step.

On the one hand, it has to be observed that moisture measurements were taken very frequently (288 times per day), while ETo, rainfall, irrigation, and temperature were collected only daily. This can be seen in the sample depicted in Figure 3. Therefore, for each of the four different depths of the sensors, the daily median, average, and standard deviation were computed and used as input features. This way, all the features had daily frequency. In total, there were full data available for 187 days.

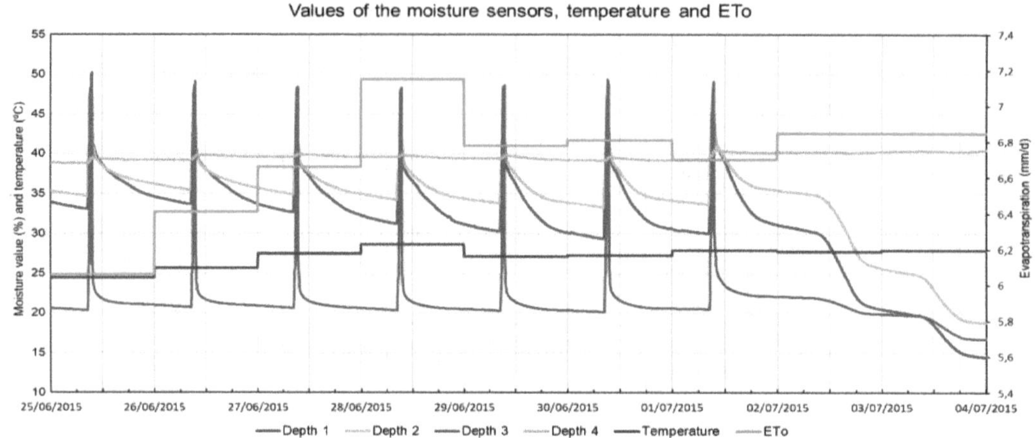

Figure 3. Sample values of the moisture sensors and agroclimatic data for a period of eight days from 26 June 2015 to 4 July 2015. The values of sensor 1 are shown in this plot. Depth 1: 10 cm; depth 2: 20 cm; depth 3: 30 cm; depth 4: 40 cm.

In the sample days shown in Figure 3, irrigation was done at approximately 8:30 a.m. every day. At these moments, moisture sensors reached their highest values before quickly decreasing. The rate of decrease was greater with a higher temperature and ETo, as seen in the days from 28 to 30 June 2015. It can also be observed in Figure 3 that no irrigation was applied in the last two days; thus, the soil moisture values continued decreasing, mainly from 12:00 to 6:00 p.m. on both days. To avoid the effect of peaks in the regression algorithms, before computing the statistics, the logarithm of the measures was firstly taken, and then the resulting values were normalized between 0 and 1.

Moreover, since many days were removed because of the missing data, another dataset was generated to test the machine learning algorithms. In this case, we took into account that the evapotranspiration, temperature, moisture, etc. of the crops were not the same throughout the day, but they changed cyclically. Thus, this new dataset was obtained computing the same statistics (median, average, and standard deviation) from the moisture sensors in periods of 6 h, instead of 24 h. This way, for each day of the study, four different samples were generated: 0 to 6 h; 6 to 12 h; 12 to 18 h; 18 to 24 h. Some days presented missing data in these periods; thus, the total number of samples for the 187 days was 682 (i.e., 66 samples were discarded). Since the data of temperature, ETo, and irrigation were available as the mean/total of each day, these values were the same for the four tuples created for each day.

Finally, it was observed that moisture sensor number 3 was responsible for the most missing data. In order to increase the number of available samples, a third dataset was generated. The procedure was the same as in the second dataset, with four periods per day, but without using the data from sensor 3. Consequently, in this dataset, the total number of samples increased to 930.

2.3. Regression Algorithms Used

In general, a regression method is a process to estimate the value of a numerical dependent variable (the output) given a set of independent variables (the input). In machine learning, this regression is not necessarily calculated by a mathematical equation, but by an algorithmic process. In our case, more than 20 machine learning algorithms and variants were applied to train the regression models for the estimation of the reference evapotranspiration, ETo, using the other attributes available: soil moisture sensors at depths 1, 2, 3, and 4; temperature; rainfall; irrigation; day of the year. The last parameter allows the algorithms to model the relationships between the season of the year and the other variables.

Since the purpose of the present research is to select the most accurate regression technique, a detailed description of all the algorithms tested would be outside the scope of this paper, and only the most relevant methods are briefly described. Three scientific tools (one commercial tool and two free tools) were used to test the models before selecting the best one.

- MATLAB 9 (MathWorks Inc., Natick, MA, USA) was used to validate the regression algorithms based on artificial neural networks (ANNs), since it has a powerful ANN toolkit. Specifically, the models used were classical multilayer perceptron ANNs [41]. These networks consist of several input neurons (one for each input variable), one output neuron (the estimation of ETo), and several hidden layers with some neurons per layer. Different configurations were tested in the experiments.
- R 3.4 (R Foundation, Vienna, Austria) is a free programming language and software environment that is very common in scientific computation and statistical analysis. This tool was used to test the algorithms based on support vector regression (SVR) and regression trees (RT). SVR is an adaptation of support vector machines (SVM) to regression problems [42], where a number of relevant samples are selected from the training set (the support vectors) to minimize the error. Similarly, RTs are an adaptation of decision trees [43], where the intermediate nodes perform decisions based on the input variables, and the terminal nodes contain the predicted output values.
- Weka 3.8.1 (The University of Waikato, Hamilton, New Zealand) is a very popular, complete, and free machine learning tool. It contains many classification and regression algorithms that were included in this research. Some of the main techniques are as follows: linear regression; k-nearest neighbors; Bayes networks; logistic regression; K* algorithm; locally weighted learning; rule-based methods; different types of decision trees. Moreover, some meta-algorithms are included; these are algorithms that use other algorithms as parameters. For example, it is worth mentioning the randomizable filtered classifier (RFC) [44], a variant of the filtered classifier that applies an arbitrary transformation to the input data, and then executes another base algorithm on this transformed input. In this way, the regression accuracy could be greater in the filtered input than in the original one.

2.4. Model Validation Measures

As already described, many different combinations of regression methods, variants, and configurations were applied in the experiments to the available datasets. Their results were analyzed using the root-mean-squared error (RMSE) of the obtained predictions for the test set. This error is defined as

$$RMSE(m) = \sqrt{\frac{1}{n} \cdot \sum_{i=1}^{n} (y(i) - y_m(i))^2}, \quad (1)$$

where m is a given regression model, n is the number of test samples, $y(i)$ is the ETo calculated with the Penman–Monteith method for the i-th sample, and $y_m(i)$ is the estimated value for that sample using the proposed model m. This parameter is a common way to compare the accuracy of different methods, where a model is better when its RMSE is lower. As an alternative, the mean absolute error (MAE) can also be used to assess the accuracy of a method, which is given by

$$MAE(m) = \frac{1}{n} \cdot \sum_{i=1}^{n} abs(y(i) - y_m(i)). \quad (2)$$

However, both RMSE and MAE by themselves are difficult to interpret unless they are compared with a range of values for predicted variables. For this reason, the mean relative error (MRE) is another good accuracy measure, since it considers the errors with respect to the ground-truth values. It is defined as

$$MRE(m) = \frac{1}{n} \cdot \sum_{i=1}^{n} \frac{abs(y(i) - y_m(i))}{y(i)}. \quad (3)$$

Finally, another frequently used measure is the correlation coefficient, R, that expresses the linearity in a scatter plot of measured and predicted values of ETo. It can be computed as

$$R(m) = \frac{\sum_{i=1}^{n}(y(i) - \overline{y})\cdot(y_m(i) - \overline{y_m})}{\sqrt{\sum_{i=1}^{n}(y(i) - \overline{y})}\cdot\sqrt{\sum_{i=1}^{n}(y_m(i) - \overline{y_m})}}, \quad (4)$$

where $\overline{y}$ is the average of $y(i)$ for i in $(1 \ldots n)$, and $\overline{y_m}$ is the average of $y_m(i)$ for i in $(1 \ldots n)$.

2.5. Remote Image System for the Estimation of the Water Balance

The ultimate goal of the present research is to predict the actual values of the crop evapotranspiration, ETc, which is a part of the water balance equation [37].

$$\Delta W = P + I - ET_c - D - R, \quad (5)$$

where ΔW is the water balance of the crop of interest in a given period, P is the rainfall, I is the irrigation, D is the drainage, and R is the surface runoff. According to the FAO-56 methodology [20], ETc can be calculated as the product of the reference evapotranspiration, ETo, and a crop coefficient, Kc, which is specific to the type of crop and its growth state. Therefore, the proposed method was designed to be integrated with the remote image capture system presented in Reference [39] to create a complete infrastructure for the computation of the water balance. Some sample views of this system are shown in Figure 4 for a crop of lettuce (*Lactuca sativa* L) which was used as the crop of interest, while kikuyu grass was used as the reference crop for the estimation of ETo.

(a) (b) (c)

Figure 4. Sample images of the remote image system where the proposed method was integrated. (**a**) Global view of the lettuce crop and a remote image capture module. (**b**) Sample image captured by the remote module. (**c**) Segmentation of the previous image in plant and soil.

The integration was done as follows: firstly, the meteorological data (daily temperature and rainfall), the soil moisture values and the irrigation of the reference crop, and the images of the crop of interest were captured with the corresponding remote modules and transmitted to the local coordinator node via XBee wireless connection [39]. A segmentation algorithm was applied to the images in the coordinator node to separate plants and background [38], as shown in Figure 4c, obtaining the percentage of green cover (PGC). Then, a regression model was applied to estimate Kc from the PGC, as defined in Reference [37]. Additionally, using the present method, ETo was calculated for the reference crop using the obtained data. Additionally, the estimation of ETc for the crop of interest was given by Kc × ETo. The resulting value was applied in Equation (5) to compute the daily water balance in the crop of interest and adopt the most adequate irrigation decisions.

3. Results and Discussion

The tests performed to obtain an accurate estimation technique of ETo using the moisture sensors consisted of the application of all the machine learning methods described in Section 2.3. These methods were applied to the three datasets defined: (i) daily data; (ii) data at 6-h intervals; (iii) data at 6-h intervals but without taking sensor 3 into account. In all the experiments, the data separation was 80% samples for training and the remaining 20% samples for validation. Therefore, in set (i), there were 150 training samples and 37 test samples; in set (ii), there were 546 training and 136 test samples; in set (iii), there were 744 training and 186 test samples.

The description of the experimental results is presented in two parts: first the comparison of the different models, and then the detailed analysis of the best model selected.

3.1. Comparison of the Regression Models

All the machine learning algorithms studied have several configurable hyperparameters, which were adjusted by trial and error to select their optimal configurations. The optimal configuration of the regression models was as follows:

- In the case of the RTs, the algorithm used for construction of the decision trees was Classification and Regression Trees (CART) [43]. Figure 5a shows a graphical representation of the tree obtained for dataset (iii). It can be observed that the temperature is a very important variable in the regression, combined with the average information of some sensors.
- For the SVR algorithm, the kernel function was a radial basis function, while the cost parameter was 4, with a value of epsilon 0.03 and gamma 0.1. These values were obtained using the *tune* function of R, which performs an optimization of the hyperparameters of the algorithm.
- The best architecture found using the ANN for dataset (iii) is shown in Figure 5b. In the three datasets, the network had an input layer, an output layer with one neuron, and a hidden layer. The ANN for dataset (i) had 10 hidden neurons, and the backpropagation algorithm was a scale conjugate gradient; for datasets (ii) and (iii), there were 15 hidden neurons and the backpropagation algorithm was Levenberg–Marquardt. In the training process, the datasets were divided into training (60%), validation (20%), and testing (20%). In this way, consistency was maintained in the comparison with the other regression methods.
- Regarding the regression models tested in Weka, in dataset (i), the best algorithm was M5 Rules, while, for the other two datasets, the best method was the randomizable filtered classifier (RFC). M5 Rules is an algorithm that uses a separate-and-conquer strategy to construct a list of decisions or rules [45]. In the case of the RFC, the filter applied to the input data is a random projection to a sub-space of less dimensionality, and the base method of RFC is the K* algorithm [46]. This K* (or K-star) algorithm is an instance-based regressing method, where the estimation for a given input is calculated from samples more similar to it, according to a certain similarity function (normally using entropy-based distance functions). A global blend value of 15 was used in this algorithm.

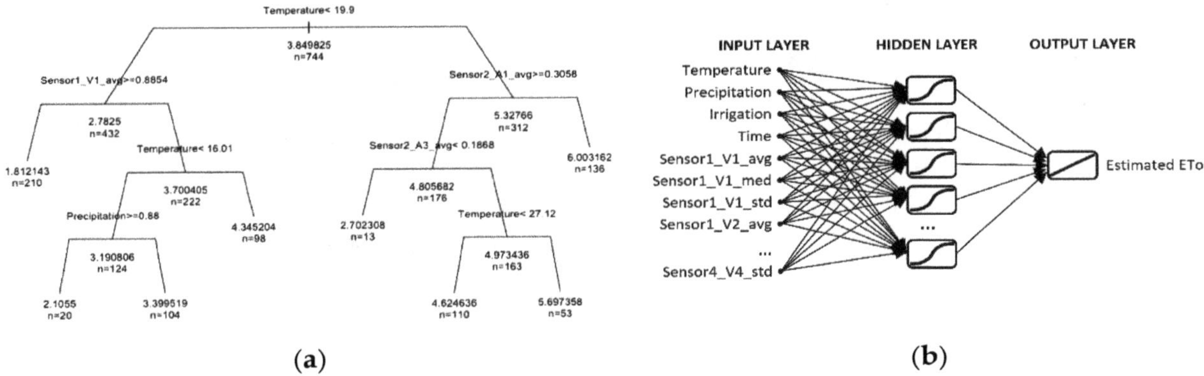

Figure 5. Two of the regression methods studied. (**a**) Trained structure of the regression tree for dataset (iii); in each leaf node, the regression value and the number n of samples is shown. (**b**) Scheme of the artificial neural network designed for dataset (iii).

Table 2 shows the RMSE values obtained for all these regression models. In the case of Weka, only the error of the best model is shown for concision; the rest of models produced worse results than those presented in Table 2.

Table 2. Root-mean-squared error (RMSE) in mm/day, for the prediction of the daily ETo using the different regression algorithms on the three datasets defined. RT: regression tree; SVR: support vector regression; ANN: artificial neural network.

Algorithm	(i) Daily Data	(ii) Data at 6-h Intervals	(iii) Data at 6-h Intervals without Sensor 3
RT	0.7000	0.6700	0.6600
Best Weka model	0.4973	0.3567	0.1829
SVR	0.5354	0.4994	0.4043
ANN	0.5481	0.3037	0.2972

Globally, the smallest error was obtained using Weka's RFC algorithm with dataset (iii), combined with the K* algorithm, achieving a very low RMSE of 0.1829 mm/day. Concerning the datasets, it can be observed that the data with daily information led to very poor results in all the methods. The optimal RMSE was almost 0.5 mm/day, using M5 Rules in Weka, which was equivalent to a relative error, MRE, of 33.96%. Thus, this arrangement of the data was not adequate to produce good estimations of the reference evapotranspiration. There can be several reasons for this fact. Firstly, the number of samples of this dataset could be insufficient for most of the machine learning methods used. A much larger number of samples could be necessary to improve the accuracy of the methods. However, this would be difficult for practical use of this methodology, requiring too many days of experimentation (in our case, after more than one year of data collection, the training dataset (i) contained only 150 samples). Moreover, since the daily data accumulated information of the peaks and valleys of each day, the loss of information did not allow producing good estimations.

Both datasets at 6-h intervals achieved better results for all the regression methods studied. This indicates that splitting up the data had a positive effect in the accuracy. The number of samples was more than 3 times larger in dataset (ii), and five times larger in the case of dataset (iii). The difference between the results of sets (ii) and (iii) was negligible for the regression trees, SVR, and ANN. That is, these methods were more robust to missing data due to sensor failures. However, for the optimal method, RFC + K*, the RMSE was reduced to half by discarding sensor 3. This fact shows that, since this method is an instance-based regression, the existence of erroneous or incomplete data had a bigger effect on its accuracy. For this reason, it worked much better in dataset (iii).

The second-best method was the ANN, which achieved the lowest RMSE for dataset (ii), with a value of 0.3037 mm/day, improving the accuracy of RFC + K* in this set. This was equivalent to an MRE above 11.3% and an R of 0.924, which is also a very accurate result. However, the improvement in the ANN upon removing sensor 3 was insignificant, reducing RMSE by only 0.0065. The results of the remaining methods were always worse than those of RFC + K* and ANN. This proves the complexity of the problem, indicating that the relationship between ETo and the input parameters cannot be captured with a simple model. Furthermore, the other techniques applied in Weka, such as linear regression, k-nearest neighbors, Bayes networks, and logistic regression, produced even worse results. Figure 4a shows that temperature is a very important parameters, but there are other factors not related to the temperature.

3.2. Accuracy Analysis of the Selected Model

As a result of the previous experiments, the optimal method selected for the estimation of ETo using the moisture sensors and meteorological data was the combination of RFC and the K* algorithm. Recall that this method consists of two steps: firstly, the input tuples are projected into a random subspace of lower dimensionality; then, the K nearest training samples to the input tuple are used to estimate the value of ETo. In this subsection, the results of this method are analyzed in more detail and compared with those of the ANN. The main accuracy parameters are presented in Table 3. These measures used the dataset at 6-h intervals removing sensor 3.

Table 3. Root-mean-squared error (RMSE) in mm/day, mean absolute error (MAE) in mm/day, mean relative error (MRE), and correlation coefficient (*R*) for the prediction of the daily ETo using the two best regression algorithms, randomizable filtered classifier with K* algorithm (RFC + K*) and artificial neural network (ANN).

Algorithm	RMSE	MAE	MRE	*R*
RFC + K*	0.1829	0.0899	6.52%	0.9936
ANN	0.2972	0.1521	11.03%	0.9470

Figure 6 shows a comparison of the ETo values estimated using the two best regression models found, compared to the actual values obtained with the Penman–Monteith method in the agroclimatic station of La Murada, for a random subset of evapotranspiration measurements in dataset (iii).

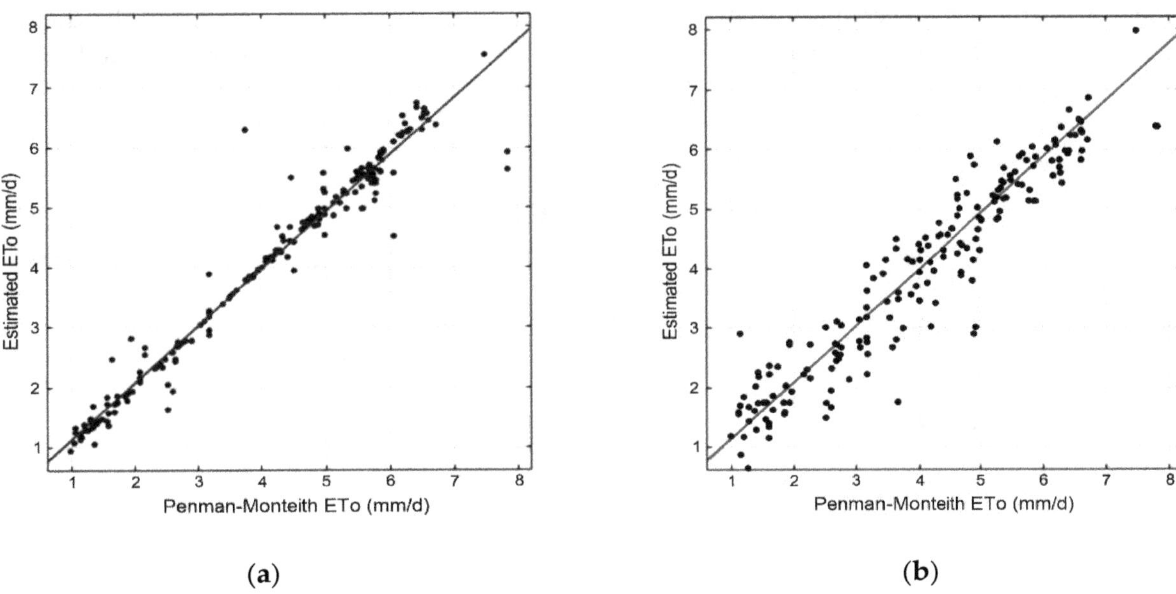

Figure 6. Scatter plots of the reference evapotranspiration (ETo) obtained with the Penman–Monteith method, and the generated predictions using the dataset at 6-h intervals removing sensor 3, applying the two best models. (**a**) randomizable filtered classifier and K* algorithm; (**b**) artificial neural network.

The predictions in Figure 6a are clearly closer to the expected values than those in Figure 6b, in coherence with the error measures. The regression line of the plot in Figure 6a corresponds to the following equation:

$$y = 0.9518x + 0.1699. \tag{6}$$

As indicated in Table 3, the correlation of the regression line is 0.9936. Finally, Figure 7 shows a histogram showing the error distribution of the RFC + K* model for the same dataset. Most of the errors were within −1 and +1, with almost no samples having a greater error.

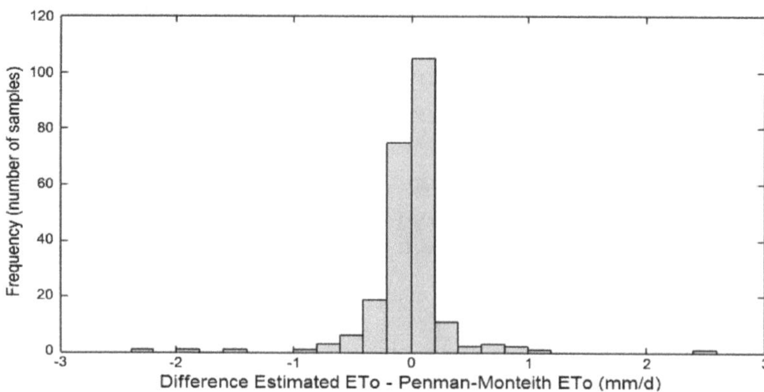

Figure 7. Error histogram of the RFC + K* regression model in the estimation of ETo.

The accuracy measures prove that the proposed method is able to achieve a very precise estimation of ETo, with an average error below 6.5% in relative terms. Thus, a good estimation of ETo can be obtained with a cost-effective system that would only require moisture sensors and agroclimatic data. Since the selected RFC method relies on subspace projections, we performed a principal component analysis (PCA) of the independent variables. It was observed that, from the 43 input variables, the first principal component (PC) accounted for only a 32% of the variance, and the second PC accounted for 13%. This indicates a high complexity in the distribution of the input data. A plot of the dataset (iii) projection in these two components is depicted in Figure 8. The first 10 PCs were required to capture 89% of the variance.

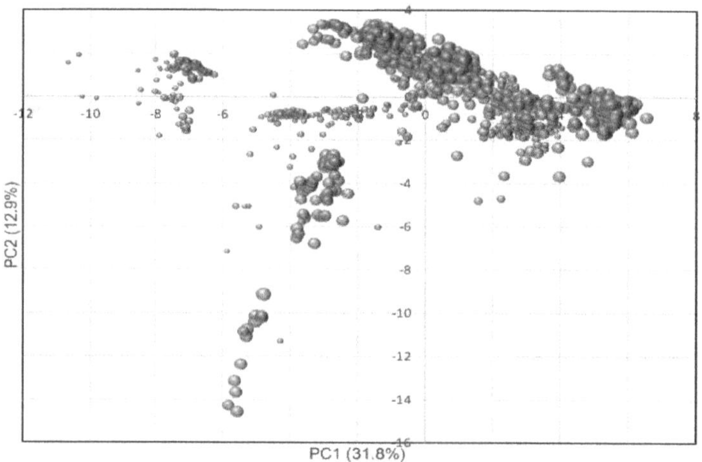

Figure 8. Projection of the samples in dataset (iii) into the subspace of the two first principal components, PC1 and PC2. The size of the points is proportional to the corresponding ETo values.

The predicted ETo values are very accurate compared to other estimation methods based on meteorological information of remote imagery. For example, in Reference [47], ETo was estimated using ANN and climatic data (solar radiation, maximum and minimum temperature, maximum and minimum relative humidity, and wind speed) obtaining a weighted standard error of 0.3 mm/day. The Penman–Monteith method was considered as the ground truth, finding that the error of a lysimeter-based method was 0.74 mm/day. Although they used a different accuracy metric, the obtained value was far from the 0.0899 mm/day MAE error achieved with our proposal. Another technique using minimal climatic data (maximum and minimum air temperatures, extraterrestrial radiation, and daylight hours) and ANN was presented in Reference [48], reporting a mean squared error in the daily estimation of 0.356 $(mm/day)^2$, which corresponds to an RMSE of 0.597 mm/day.

Moreover, Glenn et al. [49] explored several methods of computing ETo using satellite imagery, ground measurements, and meteorological data, observing values of relative root-mean-squared error in the range 10%–30%; again, this is worse than our obtained MRE of 6.5%.

Nevertheless, we consider that more experiments are necessary to verify the excellent results achieved by the proposed method. For example, since the data acquisition was only done in a single configuration, it would be advisable to perform more experiments in different environmental settings, with different climates, crops, ground types, etc. These factors could translate into significant evapotranspiration changes, and this could affect the effectiveness of the generated models. However, even in this case, the same methodology proposed in this article could be followed to obtain and prepare the data, as well as to generate and compare the models.

Finally, it is interesting to observe that the presented model does not consider any prior expert knowledge about the evapotranspiration computation, and it only uses generic machine learning algorithms. Therefore, the application of the proposed methodology to a different environment would be straightforward. It is likely that the results could be further improved in the future upon introducing changes in the model to take into account this expert knowledge.

4. Conclusions

One of the main applications of remote image sensing systems in agriculture is the computation of the optimal irrigation requirements of the crops. For this purpose, estimating the crop evapotranspiration (ETc) is an essential and preliminary step, since it models the water consumption of the crops. In this paper, we presented a new methodology to create precise models that are able to estimate the reference evapotranspiration (ETo) of crops using moisture sensors located in the ground. This estimation of ETo is then integrated with other existing techniques to calculate ETc using remote images and ETo, such as that in Reference [38]. Through the analysis of the aerial images of the crops, the percentage of ground cover is firstly computed, then the crop coefficient is deduced from it, and finally ETc is calculated using ETo and the crop coefficient.

The obtained results indicate that it is possible to obtain a very accurate approximation of ETo using only daily temperature, rainfall, watering and moisture data, and generic machine learning methods. This can be compared with the standard Penman–Monteith method for ETo estimation [20], which requires more climatic data, such as solar radiation, air temperature, humidity, wind speed, and atmospheric pressure, in addition to elevation above sea level, Julian day, and latitude degree of the study site. The high correlation coefficient found between both ETo estimations, over 0.993, and the small mean relative error of 6.5% indicate that this method could be used as an effective substitution of other more expensive solutions, such as lysimeters and Bowen stations, offering a cost-efficient alternative.

As discussed in the paper, this proposal opens new opportunities to experiment with these excellent results under different conditions, for example, in diverse environmental settings, with different climates, varieties of crops, ground types, etc. The proposed methodology consists of obtaining the datasets aggregated at 24- or 6-h intervals, applying different machine learning regression algorithms, and selecting the optimal model and time interval. This procedure would be the same for different conditions, although the best model and the optimal time interval could be dissimilar under those conditions. Another interesting future line is to perform a direct estimation of the ETc of the crop of interest using the moisture sensors, agroclimatic data, and remote image sensing. This would eliminate the need to use a reference crop as a previous step in the process.

Author Contributions: Conceptualization, D.M.-S., A.F.-L., G.G.-M., and A.R.-C.; methodology, J.M.M.-M., M.F.-V.-G., and A.R.-C.; software, D.M.-S.; validation, D.M.-S., A.F.-L., G.G.-M., and A.R.-C.; formal analysis, D.M.-S. and A.F.-L.; investigation, D.M.-S., A.F.-L., G.G.-M., J.M.M.-M., M.F.-V.-G., and A.R.-C.; resources, A.F.-L., M.F.-V.-G., and A.R.-C.; data curation, D.M.-S. and A.F.-L.; writing—original draft preparation, D.M.-S., A.F.-L., and A.R.-C.; writing—review and editing, G.G.-M.; visualization, D.M.-S. and G.G.-M.; supervision, J.M.M.-M., M.F.-V.-G., and A.R.-C.; project administration, G.G.-M., J.M.M.-M., and A.R.-C.; funding acquisition, G.G.-M., J.M.M.-M., M.F.-V.-G., and A.R.-C. All authors read and agreed to the published version of the manuscript.

References

1. Zhao, J.; Huang, S.; Huang, Q.; Leng, G.; Wang, H.; Li, P. Watershed water-energy balance dynamics and their association with diverse influencing factors at multiple time scales. *Sci. Total Environ.* **2020**, *711*, 135189. [CrossRef]
2. Ramos, T.B.; Simionesei, L.; Jauch, E.; Almeida, C.; Neves, R. Modelling soil water and maize growth dynamics influenced by shallow groundwater conditions in the Sorraia Valley region, Portugal. *Agric. Water Manag.* **2017**, *185*, 27–42. [CrossRef]
3. Albaugh, J.M.; Domec, J.C.; Maier, C.A.; Sucre, E.B.; Leggett, Z.H.; King, J.S. Gas exchange and stand-level estimates of water use and gross primary productivity in an experimental pine and switchgrass intercrop forestry system on the Lower Coastal Plain of North Carolina, U.S.A. *Agric. For. Meteorol.* **2014**, *192*, 27–40. [CrossRef]
4. Filippucci, P.; Tarpanelli, A.; Massari, C.; Serafini, A.; Strati, V.; Alberi, M.; Raptis, K.G.C.; Mantovani, F.; Brocca, L. Soil moisture as a potential variable for tracking and quantifying irrigation: A case study with proximal gamma-ray spectroscopy data. *Adv. Water Resour.* **2020**, *136*, 103502. [CrossRef]
5. Hosseini, M.; McNairn, H. Using multi-polarization C- and L-band synthetic aperture radar to estimate biomass and soil moisture of wheat fields. *Int. J. Appl. Earth Obs. Geoinf.* **2017**, *58*, 50–64. [CrossRef]
6. Vienken, T.; Reboulet, E.; Leven, C.; Kreck, M.; Zschornack, L.; Dietrich, P. Field comparison of selected methods for vertical soil water content profiling. *J. Hydrol.* **2013**, *501*, 205–212. [CrossRef]
7. Sena-Lozoya, E.B.; González-Escobar, M.; Gómez-Arias, E.; González-Fernández, A.; Gómez-Ávila, M. Seismic exploration survey northeast of the Tres Virgenes Geothermal Field, Baja California Sur, Mexico: A new Geothermal prospect. *Geothermics* **2020**, *84*, 101743. [CrossRef]
8. Linck, R.; Fassbinder, J.W.E. Determination of the influence of soil parameters and sample density on ground-penetrating radar: A case study of a Roman picket in Lower Bavaria. *Archaeol. Anthropol. Sci.* **2014**, *6*, 93–106. [CrossRef]
9. Banerjee, K.; Krishnan, P. Normalized Sunlit Shaded Index (NSSI) for characterizing the moisture stress in wheat crop using classified thermal and visible images. *Ecol. Indic.* **2020**, *110*, 105947. [CrossRef]
10. Fatás, E.; Vicente, J.; Latorre, B.; Lera, F.; Viñals, V.; López, M.V.; Blanco, N.; Peña, C.; González-Cebollada, C.; Moret-Fernández, D. TDR-LAB 2.0 Improved TDR Software for Soil Water Content and Electrical Conductivity Measurements. *Procedia Environ. Sci.* **2013**, *19*, 474–483. [CrossRef]
11. Chen, L.; Zhangzhong, L.; Zheng, W.; Yu, J.; Wang, Z.; Wang, L.; Huang, C. Data-driven calibration of soil moisture sensor considering impacts of temperature: A case study on FDR sensors. *Sensors* **2019**, *19*, 4381. [CrossRef] [PubMed]
12. Oates, M.J.; Ramadan, K.; Molina-Martínez, J.M.; Ruiz-Canales, A. Automatic fault detection in a low cost frequency domain (capacitance based) soil moisture sensor. *Agric. Water Manag.* **2017**, *183*, 41–48. [CrossRef]
13. Al-Asadi, R.A.; Mouazen, A.M. Combining frequency domain reflectometry and visible and near infrared spectroscopy for assessment of soil bulk density. *Soil Tillage Res.* **2014**, *135*, 60–70. [CrossRef]
14. Wiedenfeld, R.P. Water stress during different sugarcane growth periods on yield and response to N fertilization. *Agric. Water Manag.* **2000**, *43*, 173–182. [CrossRef]
15. Ratnakumar, P.; Khan, M.I.R.; Minhas, P.S.; Farooq, M.A.; Sultana, R.; Per, T.S.; Deokate, P.P.; Khan, N.A.; Singh, Y.; Rane, J. Can plant bio-regulators minimize crop productivity losses caused by drought, salinity and heat stress? An integrated review. *J. Appl. Bot. Food Qual.* **2016**, *89*, 113–125.
16. Osman, H.; Ammar, M.; El-Said, M. Optimal scheduling of water network repair crews considering multiple objectives. *J. Civ. Eng. Manag.* **2017**, *23*, 28–36. [CrossRef]
17. Al-Karadsheh, E. Precision Irrigation: New strategy irrigation water management. In Proceedings of the Conference on International Agricultural Research for Development, Deutscher Tropentag, Wiltzenhausen, Germany, 9–11 October 2002; pp. 1–7.
18. Evans, R.G.; Buchleiter, G.W.; Sadler, E.J.; King, B.A.; Harting, G.B. Controls for precision irrigation with self propelled systems. In Proceedings of the 4th Decennial Symposium, Phoenix, AZ, USA, 14–16 November 2000.
19. Sadler, E.J.; Evans, R.G.; Stone, K.C.; Camp, C.R. Opportunities for conservation with precision irrigation. *J. Soil Water Conserv.* **2005**, *60*, 371–379.
20. Allen, R.G.; Pereira, L.S.; Raes, D.; Smith, M. *Crop evapotranspiration—Guidelines for computing crop water requirements—FAO Irrigation and drainage paper 56*; FAO—Food and Agriculture Organization of the United Nations: Rome, Italy, 1998; ISBN 92-5-104219-5.

21. Doorenboos, J.; Pruitt, W.O. *Guidelines for Predicting Crop Water Requirements, Irrigation and Drainage Paper 24*; Food and Agriculture Organization of the United Nations: Rome, Italy, 1977; ISBN 92-5-100279-7.
22. Aydin, Y. Determination of reference ETo by using different Kp equations based on class a pan evaporation in southeastern anatolia project (GAP) region. *Appl. Ecol. Environ. Res.* **2019**, *17*, 15117–15129.
23. Kato, T.; Kamichika, M. Determination of a crop coefficient for evapotranspiration in a sparse sorghum field. *Int. Comm. Irrig. Drain.* **2006**, *55*, 165–175. [CrossRef]
24. Crago, R.; Brutsaert, W. Daytime evaporation and the self-preservation of the evaporative fraction and the Bowen ratio. *J. Hydrol.* **1996**, *178*, 241–255. [CrossRef]
25. Huang, Y.; Chen, Z.X.; Yu, T.; Huang, X.Z.; Gu, X.F. Agricultural remote sensing big data: Management and applications. *J. Integr. Agric.* **2018**, *17*, 1915–1931. [CrossRef]
26. Mogili, U.R.; Deepak, B.B.V.L. Review on Application of Drone Systems in Precision Agriculture. *Procedia Comput. Sci.* **2018**, *133*, 502–509. [CrossRef]
27. Maes, W.H.; Steppe, K. Perspectives for Remote Sensing with Unmanned Aerial Vehicles in Precision Agriculture. *Trends Plant Sci.* **2019**, *24*, 152–164. [CrossRef] [PubMed]
28. Xie, Y.; Sha, Z.; Yu, M. Remote sensing imagery in vegetation mapping: A review. *J. Plant Ecol.* **2008**, *1*, 9–23. [CrossRef]
29. Sharma, H.; Shukla, M.K.; Bosland, P.W.; Steiner, R. Soil moisture sensor calibration, actual evapotranspiration, and crop coefficients for drip irrigated greenhouse chile peppers. *Agric. Water Manag.* **2017**, *179*, 81–91. [CrossRef]
30. Vázquez, N.; Huete, J.; Pardo, A.; Suso, M.L.; Tobar, V. Use of soil moisture sensors for automatic high frequency drip irrigation in processing tomato. In Proceedings of the XXVIII International Horticultural Congress on Science and Horticulture for People (IHC2010): International Symposium on 922, Lisbon, Portugal, 22–27 August 2010; pp. 229–235.
31. Shedd, M.; Dukes, M.D.; Miller, G.L. Evaluation of evapotranspiration and soil moisture-based irrigation control on turfgrass. In Proceedings of the World Environmental and Water Resources Congress 2007: Restoring Our Natural Habitat, Tampa, FL, USA, 15–19 May 2007; pp. 1–21.
32. O'Connell, N.V.; Snyder, R.L. Monitoring soil moisture with inexpensive dialectric sensors (Echoprobe) in a citrus orchard under low volume irrigation. In Proceedings of the IV International Symposium on Irrigation of Horticultural Crops 664, Davis, CA, USA, 1–6 September 2003; pp. 445–451.
33. Thompson, R.B.; Gallardo, M.; Valdez, L.C.; Fernández, M.D. Using plant water status to define threshold values for irrigation management of vegetable crops using soil moisture sensors. *Agric. Water Manag.* **2007**, *88*, 147–158. [CrossRef]
34. Carlson, T.N.; Petropoulos, G.P. A new method for estimating of evapotranspiration and surface soil moisture from optical and thermal infrared measurements: The simplified triangle. *Int. J. Remote Sens.* **2019**, *40*, 7716–7729. [CrossRef]
35. Krishna, P.R. Evapotranspiration and agriculture—A review. *Agric. Rev.* **2019**, *40*, 1–11.
36. Niu, H.; Zhao, T.; Wang, D.; Chen, Y. Estimating Evapotranspiration with UAVs in Agriculture: A Review. In Proceedings of the 2019 ASABE Annual International Meeting; American Society of Agricultural and Biological Engineers, Boston, MA, USA, 7–10 July 2019; p. 1.
37. Escarabajal-Henarejos, D.; Molina-Martínez, J.M.; Fernández-Pacheco, D.G.; Cavas-Martínez, F.; García-Mateos, G. Digital photography applied to irrigation management of Little Gem lettuce. *Agric. Water Manag.* **2015**, *151*, 148–157. [CrossRef]
38. García-Mateos, G.; Hernández-Hernández, J.L.; Escarabajal-Henarejos, D.; Jaén-Terrones, S.; Molina-Martínez, J.M. Study and comparison of color models for automatic image analysis in irrigation management applications. *Agric. Water Manag.* **2015**, *151*, 158–166. [CrossRef]
39. Mateo-Aroca, A.; García-Mateos, G.; Ruiz-Canales, A.; Molina-García-Pardo, J.M.; Molina-Martínez, J.M. Remote Image Capture System to Improve Aerial Supervision for Precision Irrigation in Agriculture. *Water* **2019**, *11*, 255. [CrossRef]
40. Barton, L.; Wan, G.G.Y.; Buck, R.P.; Colmer, T.D. Nitrogen increases evapotranspiration and growth of a warm-season turfgrass. *Agron. J.* **2009**, *101*, 17–24. [CrossRef]
41. Baum, E.B. On the capabilities of multilayer perceptrons. *J. Complex.* **1988**, *4*, 193–215. [CrossRef]

42. Drucker, H.; Burges, C.J.C.; Kaufman, L.; Smola, A.J.; Vapnik, V. Support vector regression machines. In Proceedings of the Advances in Neural Information Processing Systems, Denver, CO, USA, 5 May 1997; pp. 155–161.
43. Breiman, L. *Classification and Regression Trees*; Routledge: Abingdon, UK, 2017; ISBN 1351460498.
44. Pak, I.; Teh, P.L. Machine learning classifiers: Evaluation of the performance in online reviews. *Indian J. Sci. Technol.* **2016**, *9*, 1–9. [CrossRef]
45. Holmes, G.; Hall, M.; Prank, E. Generating rule sets from model trees. In Proceedings of the Australasian Joint Conference on Artificial Intelligence, Sydney, NSW, Australia, 6–10 December 1999; pp. 1–12.
46. Painuli, S.; Elangovan, M.; Sugumaran, V. Tool condition monitoring using K-star algorithm. *Expert Syst. Appl.* **2014**, *41*, 2638–2643. [CrossRef]
47. Kumar, M.; Raghuwanshi, N.S.; Singh, R.; Wallender, W.W.; Pruitt, W.O. Estimating evapotranspiration using artificial neural network. *J. Irrig. Drain. Eng.* **2002**, *128*, 224–233. [CrossRef]
48. Zanetti, S.S.; Sousa, E.F.; Oliveira, V.P.; Almeida, F.T.; Bernardo, S. Estimating evapotranspiration using artificial neural network and minimum climatological data. *J. Irrig. Drain. Eng.* **2007**, *133*, 83–89. [CrossRef]
49. Glenn, E.P.; Nagler, P.L.; Huete, A.R. Vegetation index methods for estimating evapotranspiration by remote sensing. *Surv. Geophys.* **2010**, *31*, 531–555. [CrossRef]

4

Designing a Fruit Identification Algorithm in Orchard Conditions to Develop Robots using Video Processing and Majority Voting based on Hybrid Artificial Neural Network

Sajad Sabzi [1], Razieh Pourdarbani [1,*], Davood Kalantari [2] and Thomas Panagopoulos [3,*]

1 Department of Biosystems Engineering, College of Agriculture, University of Mohaghegh Ardabili, Ardabil 56199-11367, Iran; s.sabzi@uma.ac.ir
2 Department of Mechanics of Biosystems Engineering, Faculty of Agricultural Engineering, Sari Agricultural Sciences and Natural Resources University, Sari 48181-68984, Iran; d.kalantari@sanru.ac.ir
3 Research Center for Spatial and Organizational Dynamics, University of Algarve, Campus de Gambelas, 8005 Faro, Portugal
* Correspondence: r_pourdarbani@uma.ac.ir (R.P.); tpanago@ualg.pt (T.P.)

Abstract: The first step in identifying fruits on trees is to develop garden robots for different purposes such as fruit harvesting and spatial specific spraying. Due to the natural conditions of the fruit orchards and the unevenness of the various objects throughout it, usage of the controlled conditions is very difficult. As a result, these operations should be performed in natural conditions, both in light and in the background. Due to the dependency of other garden robot operations on the fruit identification stage, this step must be performed precisely. Therefore, the purpose of this paper was to design an identification algorithm in orchard conditions using a combination of video processing and majority voting based on different hybrid artificial neural networks. The different steps of designing this algorithm were: (1) Recording video of different plum orchards at different light intensities; (2) converting the videos produced into its frames; (3) extracting different color properties from pixels; (4) selecting effective properties from color extraction properties using hybrid artificial neural network-harmony search (ANN-HS); and (5) classification using majority voting based on three classifiers of artificial neural network-bees algorithm (ANN-BA), artificial neural network-biogeography-based optimization (ANN-BBO), and artificial neural network-firefly algorithm (ANN-FA). Most effective features selected by the hybrid ANN-HS consisted of the third channel in hue saturation lightness (HSL) color space, the second channel in lightness chroma hue (LCH) color space, the first channel in L*a*b* color space, and the first channel in hue saturation intensity (HSI). The results showed that the accuracy of the majority voting method in the best execution and in 500 executions was 98.01% and 97.20%, respectively. Based on different performance evaluation criteria of the classifiers, it was found that the majority voting method had a higher performance.

Keywords: artificial intelligence; precision agriculture; agricultural robot; optimization algorithm; online operation; segmentation

1. Introduction

Nowadays, agricultural automation is inevitable to reduce costs, minimize labor difficulty, decrease environmental impact, increase timely farming and crop quality, and brink transparency in the supply chain [1,2]. One of the challenges of robots is to identify and harvest the ripe fruits. In this regard, cameras and sensors in the robot's arms have also been used to evaluate the information [3]. Such robots that capture real-time images and touch-based information enable the data to be collected in a variety of ways, helping to expand and improve this approach.

One of the most well-known techniques is image processing, which provides useful, simple, fast, and accurate information over manual techniques. Previous studies have used this technique to extract color, morphological, and texture feature of foods such as fish, fruits, and vegetables. The different pattern recognitions used to classify, analyze, sort, and evaluate foods depend on these features. Thus, image processing [4–8] and pattern recognition [9,10] are the best way to determine the quality of crops.

Since harvesting robots have to work in unstructured environments with natural light conditions, the image segmentation algorithm is extremely important because direct sunlight causes shadows in the image that lead to incorrect diagnosis of the system [11,12]. Cui et al. [13] studied the fruit detection and extraction of color and shape features of kiwifruit under natural light. Until recently, two methods have been used to reduce the effects of natural light on image processing: (a) improvement of imaging conditions before image acquisition and (b) enhancing image contrast such as the use of optical filters on camera lenses or artificial auxiliary sources to improve imaging [14]. Such methods can be effective but require the installation of large structures and high energy consumption that lead to the poor performance of agricultural robots in the orchard [15]. Wang et al. [16] developed a new method for fruit segmentation under different lighting conditions that involved the use of advanced wavelet transforms to normalize object surface lighting.

The Retinex algorithm was used to highlight the fruit object. The image was then analyzed using K-means clustering. Imaging was performed on sunny and cloudy days, and the results showed that the proposed algorithm was able to operate successfully under different lighting conditions. Sabzi et al. [17] proposed a method for segmenting apples under orchard conditions. The goal was to find the color space with the least number of colors to apply the threshold. They examined 17 color spaces, and according to their reports, the best results were obtained by applying segmentation using color, texture, and intensity conversion, and the overall correct classification rate was 98.92%.

Miao et al. [18] categorized farm images under natural light conditions. Based on the color space, eight color features of the images were extracted that included three components of HSL, the second and third components of HSV, and three components of HIS. Then, five specific vectors selected by principal component analysis (PCA) were selected to reduce the dimensionality of the images. The performance of using multiple color spaces in combination with PCA and RBF showed that the correct classification rates of sunny and cloudy days in the same scenes and different scenes were 100%, 87.36%, and 84.58%, 68.11%, respectively. However, this classification accuracy is not enough for practical use in machine vision systems as it results in many errors. Hernández et al. [19] proposed an approach to color processing that could provide optimal color space for plant/soil segmentation. It evaluated all possible options and developed color models in optimal spaces and channels. This prevented dependence on plant type, camera, and illumination conditions. The basis of their proposal was to use non-parametric models for the probability density functions of the plant/soil. They implemented and validated their algorithm with a new software tool called ACPS (Automatic Plant and Soil Classification) and claimed that the algorithm could be used in mobile applications and portable controllers that are currently being developed.

Aquino et al. [20] proposed an algorithm based on mathematical morphology and pixel classification for grape berry counting. Features were extracted using a supervised approach. Eighteen images of seven individual varieties were obtained using a low-cost cell phone camera. The results implied that the ANN method performed better than the support vector machine (SVM) with recall and precision values of 0.9572 and 0.8705, respectively. Kang and Chen [21] presented a deep-learning based fruit detector for harvesting. The developed framework includes an auto label generation module and a deep learning-based fruit detector named "LedNet". LedNet used multi-scale pyramid and clustering classifier to improve model recognition performance. LedNet achieved recall and precision of 0.821 and 0.853 in apple detection and the inference time was 28 ms.

Lin et al. [22] identified guava fruit in outdoor conditions. It is very important to pick the fruit without colliding with its mother branch. Thus, they evaluated the estimation of fruit using the low-cost red–green–blue-depth method. Based on the fruit binary image and RGB-D depth, Euclidean clustering was applied to group the point cloud into a set of individuals. Next, a 3D linear multi-segment detection method was developed to reconstruct the branches. Finally, the 3D position of the fruit was estimated using its center position and nearest branch information. Results showed that the accuracy of guava detection was 0.983. The run time of algorithm was 0.565 s.

Estimating nitrogen content in wheat plant is a very crucial task in the application of precision farming. For this reason, Refs. [23–25] presented a method for predicting nitrogen content in wheat plant. Their methods were based on a segmentation algorithm that was trained in three light intensities for separating wheat plants from the background. Since light intensity changes during the day, training the segmentation algorithm based on only three light intensities can be a weakness for the segmentation algorithm.

As observed, different researchers have focused on segmentation as an important step in designing a different machine vision. In fact, the performance of each machine vision system has a direct relation with the segmentation accuracy. Most segmentation algorithms work based on high quality images. This is the first problem for machine vision systems that work in natural garden conditions, since these methods are not applicable for operations where the camera must be moved in the field such as spraying in proportion to the density of products because when the camera is moving, the quality of the received frames is much less than when images are captured in static mode. The second problem is that most studies focus on images that do not have complicated backgrounds. However, in fruit gardens, there are different backgrounds including the trunks of trees, leaves, different branches with different colors, plants on the surface of the ground, sky in cloudy conditions, sky in clear state, and other objects. The third problem is imaging at one time of day. In fact, little research has been done on images taken throughout the day, so the segmentation algorithm cannot work correctly during the day.

Therefore, a new segmentation algorithm that resolves these problems is necessary. For this reason, the aim of this study was to offer a pixel-by-pixel segmentation algorithm based on an ensemble with a majority voting rule for segmentation of plum fruits in orchards at different ripeness stages under natural conditions. The segmentation methods that are used for voting were the artificial neural network-bees algorithm (ANN-BA), artificial neural network-biogeography-based optimization (ANN-BBO), and artificial neural network-firefly algorithm (ANN-FA).

2. Materials and Methods

The different steps to designing a plum identification algorithm in field conditions are as follows: (1) Video recording in different natural light conditions of plum orchards; (2) Converting the videos to their constituent frames; (3) Extracting different color features from each pixel; (4) Selecting the most effective color features among the total extracted features; (5) Classification using different hybrid artificial neural networks; and (6) Final classification of pixels in two classes of background and plum fruits using the majority voting method.

2.1. Video Recording to Train the Algorithm of Plum Fruit Identification

Since light intensity changes during day and the main aim of the proposed segmentation algorithm is working in natural conditions related to light and background, segmentation algorithms should offer high accuracy in all conditions, so they should be trained under all possible light intensities. For this reason, the videos of the Kermanshah orchards were recorded at 16 different light intensities including 287, 342, 467, 593, 639, 738, 826, 1052, 1296, 1420, 1563, 1689, 1769, 1848, 1963, and 2078 lux and during different stages of plum fruit growth. The camera used in the present study was a color GigE camera DFK 23GM021 (Imaging Source Europe GmbH, Bremen, Germany), with a 1/3-inch Aptina CMOS MT9M021 sensor (ON Semiconductor, Aurora, CO, USA) and a spatial resolution of 1280 × 960 pixels. The mounted lens was a model H0514-MP2 (Computer CBC Group, Tokyo, Japan), with f = 5 mm and

F1.4. From each video, 10 frames were selected randomly and then different objects were detected by a human operator. Finally, 48,000 pixels from different frames were manually extracted by a human operator and classified in the plum or background class (an average of 300 pixels were selected from each frame). Of these, 70% of the data, namely 33,600 pixels, were used to train; 15% of the data, namely 7200 pixels, were used for validation, and the remaining 7200 pixels, namely 15% of the data, were used to test the proposed algorithm. After extracting all pixels from the frames, these divisions were done randomly. Figure 1 shows several frames of the videos produced.

Figure 1. Four sample frames at different light intensities: (**a**) 467 lux, (**b**) 639 lux, (**c**) 826 lux, (**d**) 1848 lux.

2.2. Extraction of Different Color Features from Each Frame

After recording the videos and converting them to frames, different color features were extracted from different pixels of the background and fruits. These features include the first, second, and third channel in the L*C*h*, YCbCr, HSI, CMY, XYZ, HSV, YUV, HLS, L* u*v*, L*a *b*, and YIQ color spaces. Since there are three features and 11 color spaces, thus the total features of each pixel were $33 = 11 \times 3$.

2.3. Selection of the Most Effective Features Using Hybrid Artificial Neural Network-Harmony Search

Since the purpose of this algorithm is to detect fruit on trees by garden robots in a real-time state, the process time should be as short as possible. Therefore, it is not possible to use all of the features in the algorithm because it is time consuming. Thus, the most effective color features must be selected among the extracted features. In this paper, a hybrid artificial neural network-harmony search (ANN-HS) was used for this purpose. The harmony search (HS) algorithm is a meta-heuristic algorithm that imitates the natural process of music optimization. In making a song, the beauty of the song determines the gamut of each musical instrument, in other words, each instrument must be optimized. Therefore, the value of the objective function is determined by the values of the variables [26]. The task of the harmonic search algorithm is to first consider all the extracted features as a vector and then send the vectors of different sizes to the artificial neural network. For example, it sends a vector with five extracted features to the ANN as the input, and the output of the ANN is the two classes of background and fruit. The mean squared error (MSE) of ANN is recorded for each vector of the features. Finally, the vector whose MSE is less will be selected as the optimal vector and the intra-vector features as the most effective features. Table 1 gives the neural network structure used to select the most effective features.

Table 1. The structure used in hybrid artificial neural network-harmony search for selecting effective features.

Number of Layers	Number of Neurons	Transfer Function	Back-Propagation Network Training Function	Back-Propagation Weight/Bias Learning Function
2	First layer: 16 Second layer: 8	First layer: tansig Second layer: tribas	traincgf	learnk

2.4. *Classification of the Pixels Using different Classifiers*

In this paper, the classification was performed using four different classifiers. The main classification was based on the majority voting method based on different hybrid artificial neural networks.

2.4.1. Hybrid Artificial Neural Network-Bees Algorithm (ANN-BA) Classifier

The multilayer perceptron artificial neural network has various adjustable parameters and the performance of an ANN depends on the optimal adjustment of these parameters. These adjustable parameters include the number of layers, the number of neurons per layer, transfer function, the back-propagation network training function, and the back-propagation weight/bias learning function. The bees algorithm has the task of optimally adjusting these parameters. The bees algorithm is a bee swarm optimization algorithm proposed by Pham et al. [27]. This algorithm is inspired by the behavior of bees in search of food resources [28]. The different stages of the bee algorithm are as follows: (1) Generating initial responses and evaluating them; (2) Selecting the best sites (replies) and sending the worker bees to those sites; (3) Returning the bees to the hive by bee dancing (neighboring response); (4) Comparing all the bees in a site and choosing the best one; (5) Non-selected bees are replaced with random answers; (6) Save best position; and (7) Return to step 2, if termination is not fulfilled. The minimum and maximum number of layers that the bees algorithm could select was 1 and 3, respectively. The acceptable number of neurons per layer was between 1 and 25 for the first layer and between 0 and 25 for the other layers. The transfer function was selected from 13 transfer functions such as tansig and logsig. The back-propagation network training function was also selected from 19 functions such as trainrp and traincgb. Finally, back-propagation weight/bias learning function were selected from 15 different functions such as learnis and learncon. The bees algorithm sends these parameters to an MLP neural network in the form of a vector. The input of the ANN is the most effective features and its outputs are the background and fruit classes. Whenever a vector is sent to the ANN by the BA, the MSE corresponding to that vector is recorded. Finally, the vector corresponding to the lowest MSE is considered as the optimal vector and the values of the parameters within the vector are considered as the optimal parameters.

2.4.2. Hybrid Artificial Neural Network-Biogeography Based Optimization (ANN-BBO) Classifier

This classifier is similar to the ANN-BA, except that the values of the adjustable ANN parameters will be selected by the biogeography-based optimization algorithm. The BBO algorithm is inspired by how different animal and plant species are distributed in different parts of the universe [29]. The different steps of the bio-based algorithm are as follows: (1) Generating the initial population or so-called initial random habitat and sorting them; (2) Determining migration and immigration rates; (3) Repeating step 4–8 for each habitat such as j; (4) Steps 5 to 8 are repeated for each variable such as k at location j; (5) Changes are made according to steps 6 to 8 with the probability of migrating to a habitation; (6) Determine the origin of the migration using the migration values on random; (7) Migrating from one habitation to another; (8) Random changes (mutations) are applied to the variable; (9) The set of new responses is evaluated; (10) Combining the original population and the migration-related population and creating a new stage population; and (11) Return to step 3 if termination is not fulfilled.

2.4.3. Hybrid Artificial Neural Network-Firefly Algorithm (ANN-FA) Classifier

The method of this classifier is similar to the two classifiers above-mentioned, except that here, the firefly algorithm has the task of determining the adjustable parameters of the ANN. This algorithm is inspired by the optical communication between the worms. This algorithm can be seen as a manifestation of swarm intelligence, where the cooperation (and possibly competition) of simple and low-intelligence members results in a higher degree of intelligence that is certainly not obtainable by any component [30].

2.4.4. Configuration of the Best Training Mode Based on Artificial Neural Network (ANN)

After the investigation of 3000 different structures by optimization algorithms in hybrid ANN classifiers, namely ANN-BA, ANN-BBO, and ANN-FA, the best ANN structure for each classifier was configured (Table 2). In order to evaluate the reliability of the classifiers, 500 repetitions were performed for each method, that is, 500 independent executions of the training/testing process.

Table 2. The best structure of hidden layers of the artificial neural network (ANN) adjusted by bees algorithm (BA), biogeography-based optimization (BBO) and firefly algorithm (FA).

Classifier	Num. of Layers	Number of Neurons	Transfer Function	Back-Propagation Network Training Function	Back-Propagation Weight/Bias Learning Function
ANN-BA	3	First layer: 9 Second layer: 17 Third layer: 13	First layer: radbas Second layer: radbas Third layer: radbas	learnlv1	traingda
ANN-BBO	3	First layer: 5 Second layer: 14 Third layer: 18	First layer: tansig Second layer: radbas Third layer: satlin	learnk	trainoss
ANN-FA	3	First layer: 7 Second layer: 12 Third layer: 21	First layer: logsig Second layer: satlin Third layer: satlins	learnhd	trains

2.4.5. The Method of Majority Voting (MV)

After the ANN-BA, ANN-BBO, and ANN-FA perform the classification, the final operation will be carried out using the majority voting method. In fact, using the majority voting method, the opinion of the majority of classifiers determines the class of samples.

2.5. Evaluating the Performance of the Different Classifiers

To evaluate the performance of the classifiers, the receiver operating characteristic (ROC) and the area under the curve [31] as well as the criteria for the confusion matrix were used (Table 3).

Table 3. Description of the classifiers used in performance evaluation.

Description	Formula
Percent of the correct samples that have been correctly identified	$\text{Recall} = \frac{TP}{TP+FN} \times 100$
Total percentage of the correct system responses	$\text{Accuracy} = \frac{TP+TN}{TP+FN+FP+TN} \times 100$
Total percentage of the correct system responses	$\text{Specificity} = \frac{TP}{TP+FP} \times 100$
	$Precision = \frac{TP}{TP+FP} \times 100$
	$\text{F\_measure} = \frac{2\times\text{Recall}\times Precision}{\text{Recall}+Precision}$

Here, the positive class is the fruit (the object of interest) and the negative class is the background. Therefore, TP is equal to the number of samples of plum fruit that are correctly classified; TN is the number of samples of the background class that are correctly classified; FN is the number of fruit pixels misclassified as background; and FP is the number of background pixels misclassified as fruit. It has to be noted that some measures should not be analyzed by themselves. For example, a naïve system that always says true would have a recall of 100%, while a system that always says false would have a specificity of 100%.

3. Results

3.1. The Selected Effective Features Using Hybrid ANN-HS

Effective properties selected by the ANN-HS include the third channel of HSL (hue saturation lightness), the second channel of LCH (lightness chroma hue), the first channel of L*a*b*, and the first channel of HIS (hue saturation intensity) color space.

3.2. Performance of ANN-BA Classifier in the Best State of Training

Table 4 gives the confusion matrix, incorrect classification rate, and the correct classification rate (CCR) of the ANN-BA classifier for the best state of training among 500 executions. From the 7200 pixels examined in the fruit and background classes, only 154 samples were incorrectly classified into a class other than their original class, resulting in a classification error of 2.14%. Table 5 evaluates the performance of the ANN-BA classifier in the best state of training, according to different criteria. It shows that in all the criteria, with the exception of the area under the curve (AUC) with a value of 0.9962, the values obtained were above 97%, which is a high value. Therefore, based on Tables 4 and 5, it can be concluded that this classifier has a high performance.

Table 4. Confusion matrices, classification error per class, and correct classification rate (CCR) in the best state of training of the hybrid ANN-BA classifier for the testing data.

Classification Method	Real/ Obtained Class	Fruit	Background	Total Data	Classification Error per Class (%)	Correct Classification Rate (%)
ANN-BA	Fruit	3520	80	3600	2.22	97.86
	Background.	74	3526	3600	2.05	

Table 5. The performance of the hybrid ANN-BA classifier base on different criteria in the best state of training for the testing data.

Class	Recall (%)	Specificity (%)	Precision (%)	F_measure (%)	AUC	Accuracy
Fruit	97.94	97.78	97.78	97.85	0.9962	97.86
Background	97.78	97.94	97.94	97.86		

3.3. Performance of ANN-BBO Classifier in the Best State of Training

Table 6 gives the confusion matrix for the ANN-BBO classifier for the best state of training on the test data. This table shows that 3% of the samples in the fruit class were incorrectly classified in the background class and 2.8% of the samples in the background class were incorrectly classified in the fruit class. The correct classification rate (CCR) of this classification was 97.59% and this value reflects its high performance. Table 7 indicates the performance of the ANN-BBO classifier in the best state of training. As can be seen, all the criteria had values close to 100 and this proves that the classifier performed well.

Table 6. Confusion matrices, classification error per class, and correct classification rate (CCR) in the best state of training of the hybrid ANN-BBO classifier for the testing data.

Classification Method	Real/ Obtained Class	Fruit	Background	Total Data	Classification Error per Class (%)	Correct Classification Rate (%)
ANN-BBO	Fruit	3492	108	3600	3.00	97.59
	Background	65	3535	3600	2.80	

Table 7. The performance of the hybrid ANN-BBO classifier based on different criteria in the best state of training for the testing data.

Class	Recall (%)	Specificity (%)	Precision (%)	F_measure (%)	AUC	Accuracy
Fruit	98.17	97.03	97.00	97.58	0.9965	97.59
Background	97.03	98.17	98.19	97.61		

3.4. Performance of ANN-FA Classifier in the Best State of Training

Tables 8 and 9 give the confusion matrix and criteria evaluating the performance of ANN-FA, respectively. The results for this classifier in the two tables were similar to those for the ANN-BA and ANN-BBO. Correct classification rate (CCR) and the area under curve (AUC) were 97.77% and 0.9778%, respectively, indicating the high performance of the classifier.

Table 8. Confusion matrices, classification error per class, and correct classification rate (CCR) in the best state of training of the hybrid ANN-FA classifier for the testing data.

Classification Method	Real/ Obtained Class	Fruit	Background	Total Data	Classification Error per Class (%)	Correct Classification Rate (%)
ANN-FA	Fruit	3499	101	3600	2.80	97.77
	Background	59	3541	3600	1.64	

Table 9. The performance of the hybrid ANN-FA classifier based on different criteria in the best state of training for the testing data.

Class	Recall (%)	Specificity (%)	Precision (%)	F_Measure (%)	AUC	Accuracy
Fruit	98.34	97.23	97.19	97.76	0.9778	97.77
Background	97.23	98.34	98.36	97.79		

3.5. Performance of the ANN-FA Classifier in the Best State of Training

Table 10 gives the confusion matrices, the classification error per class, and the correct classification rates for the majority voting method in the best state of training. As obvious, only 29 of the 3600 samples of background were incorrectly classified in the fruit class, resulting in an error of 0.81%. This method has been able to perform classification with a high accuracy of 98%. This is much more accurate than the three classifications used for voting. Table 11 evaluates the performance of the majority voting method based on different criteria. As can be seen, the values corresponding to the performance of different criteria were close to 100, which prove that the majority voting method performs better than the other methods.

Table 10. Confusion matrices, classification error per class, and correct classification rate (CCR) in the best state of training of the majority voting (MV) method for the testing data.

Classification Method	Real/ Obtained Class	Fruit	Background	Total Data	Classification Error per Class (%)	Correct Classification Rate (%)
MV	Fruit	3486	114	3600	3.17	98.01
	Background	29	3571	3600	0.81	

Table 11. The performance of the MV method based on different criteria in the best state of training for the testing data.

Class	Recall (%)	Specificity (%)	Precision (%)	F_Measure (%)	AUC	Accuracy
Fruit	99.17	96.91	96.83	97.99	0.9970	98.01
Background	96.91	99.17	99.19	98.04		

3.6. *Comparison of the Performance of Classifiers Used in 500 Iterations*

After evaluating the performance of the various classifiers in the best state of training, the validity of the classifiers should now be evaluated, especially the majority voting method. The validity of the classifiers is determined by the results of them in different iterations, thus, if the classifier results are close together in different iterations, then it can be claimed that the classifier is valid. Table 12 shows the confusion matrix, the classification error per class, and the correct classification rate of the different classifiers at 500 iterations. The highest rate of correct classification was related to the majority voting method, which was 97.20% (Table 13).

Table 12. Confusion matrices, classification error per class, and correct classification rate (CCR) of different classifiers for 500 iterations.

Classification Method	Real/ Obtained Class	Fruit	Background	Total Data	Classification Error per Class (%)	Correct Classification Rate (%)
ANN-BA	Fruit	1,742,085	57,915	1,800,000	3.22	96.47
	Backgr.	69,102	1,730,898	1,800,000	3.84	
ANN-BBO	Fruit	1,741,427	58,573	1,800,000	3.25	96.46
	Backgr.	68,847	1,731,153	1,800,000	3.82	
ANN-FA	Fruit	1,746,422	53,578	1,800,000	2.98	96.91
	Backgr.	57,786	1,742,214	1,800,000	3.21	
Voting	Fruit	1,741,920	58,080	1,800,000	3.23	97.20
	Backgr.	42,643	1,757,357	1,800,000	2.37	

Table 13. The performance of different classifiers based on different criteria for 500 iterations.

Classifier	Class	Recall (%)	Specificity (%)	Precision (%)	F_Measure (%)	AUC (Mean ± Std. dev.)	Accuracy (Mean % ± Std. dev.)
ANN-BA	Fruit	96.18	96.76	96.78	96.48	0.9956 ± 0.0007	96.47 ± 0.5657
	Backgr.	96.76	96.18	96.16	96.46		
ANN-BBO	Fruit	96.19	96.73	96.74	96.47	0.9956 ± 0.0007	96.46 ± 0.5167
	Backgr.	96.73	96.19	96.17	96.45		
ANN-FA	Fruit	96.79	97.02	97.02	96.91	0.9691 ± 0.0046	96.91 ± 0.4572
	Backgr.	97.02	96.79	96.78	96.90		
Voting	Fruit	97.61	96.80	96.77	97.19	0.9958 ± 0.0008	97.20 ± 0.4917
	Backgr.	96.80	97.61	97.63	97.21		

Figure 2 illustrates a box diagram of area under the ROC curve obtained by different classifiers at 500 iterations. As can be seen, with the exception of the hybrid ANN-FA classifier, the other classifiers have fully compressed box diagrams, indicating close proximity to the results in different iterations.

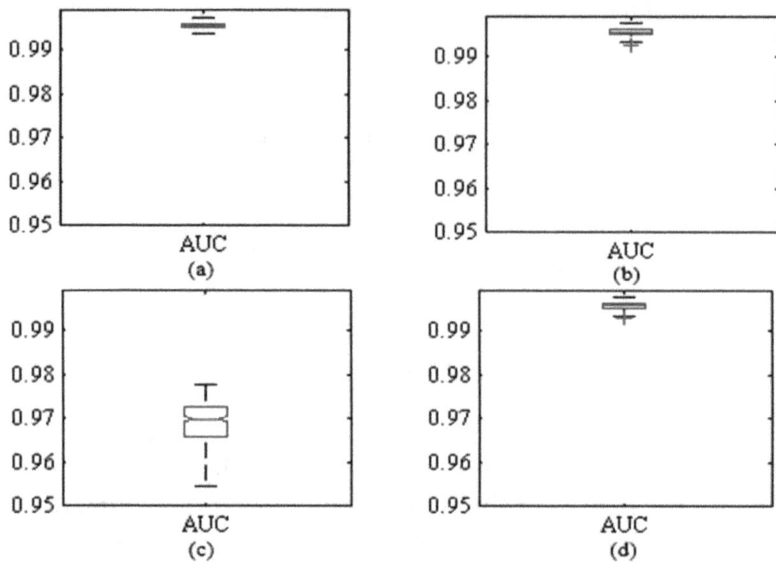

Figure 2. Boxplots of area under the curve (AUC) obtained by the classifiers for the 500 iterations on test data. (**a**) Hybrid ANN-BA; (**b**) Hybrid ANN-BBO; (**c**) Hybrid ANN-FA; (**d**) Voting method.

Figure 3 shows a box diagram of the correct classification rate obtained by different classifiers in 500 iterations. The graph shows that the majority voting method, with the exception of a few repeats, was above 97%. This method also had a more compact diagram than others, thus it can be concluded that the MV method is valid. Figure 4 shows the ROC curves obtained by different classifiers for 500 iterations. The closer the ROC curve is to the vertical, the higher the performance of the classifier.

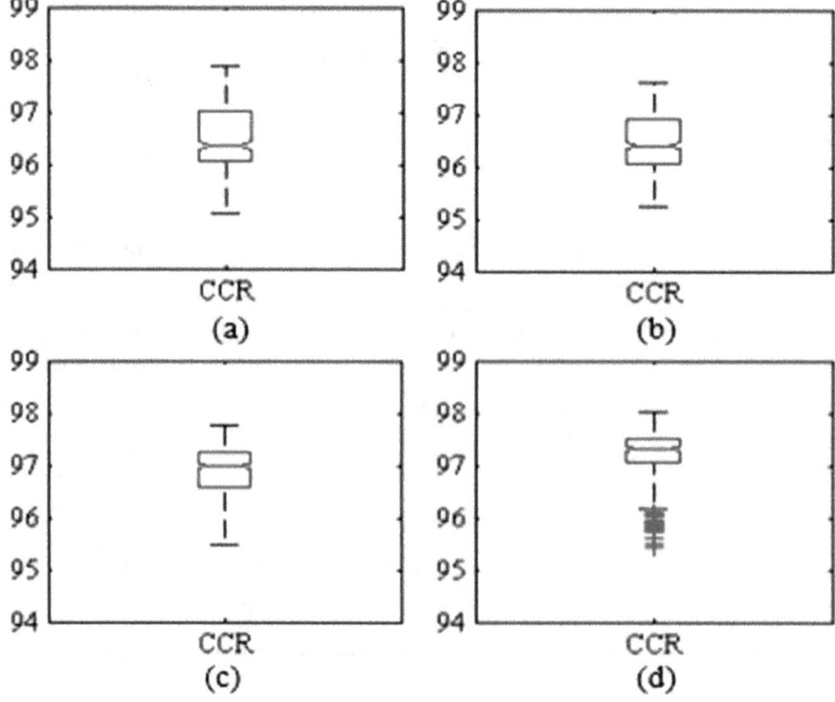

Figure 3. Boxplots of the correct classification rate (CCR) obtained by the classifiers for the 500 iterations on the test data. (**a**) Hybrid ANN-BA; (**b**) Hybrid ANN-BBO; (**c**) Hybrid ANN-FA; (**d**) Voting method.

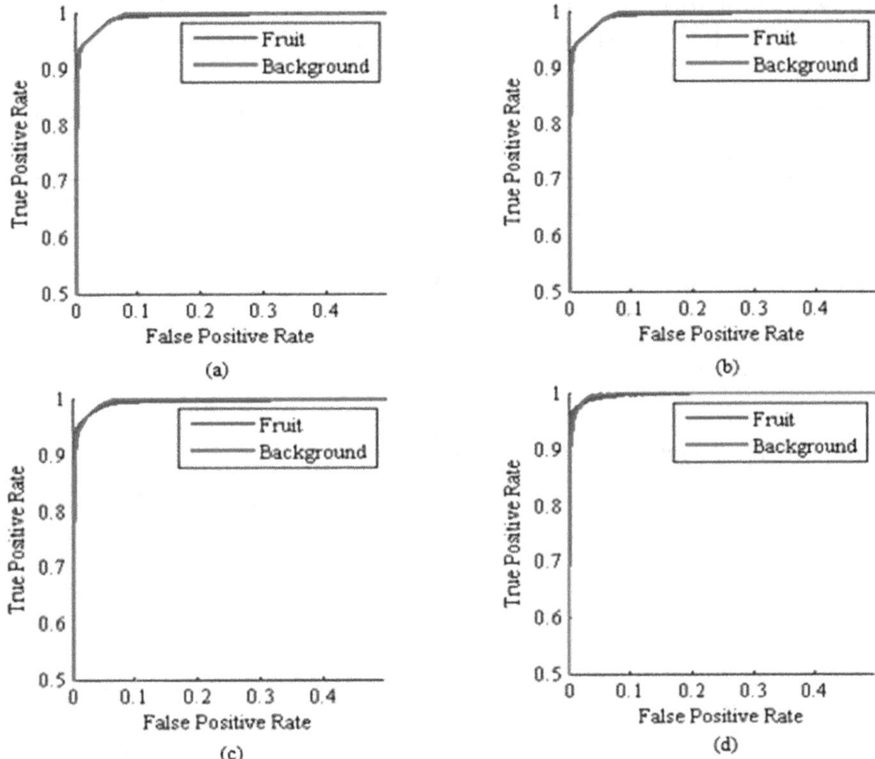

Figure 4. Receiver operating characteristic (ROC) curves obtained by the classifiers for the 500 iterations on the test data. (**a**) Hybrid ANN-BA; (**b**) Hybrid ANN-BBO; (**c**) Hybrid ANN-FA; (**d**) Voting method.

According to Figure 4, three classifiers, namely hybrid ANN-BA, ANN-BBO, and ANN-FA, had a similar performance. The ROC curve of the MV method was closer to orthogonal than the other classifiers, indicating the superiority of this method over others. Figure 5 illustrates the various steps of the computer vision system proposed in this paper.

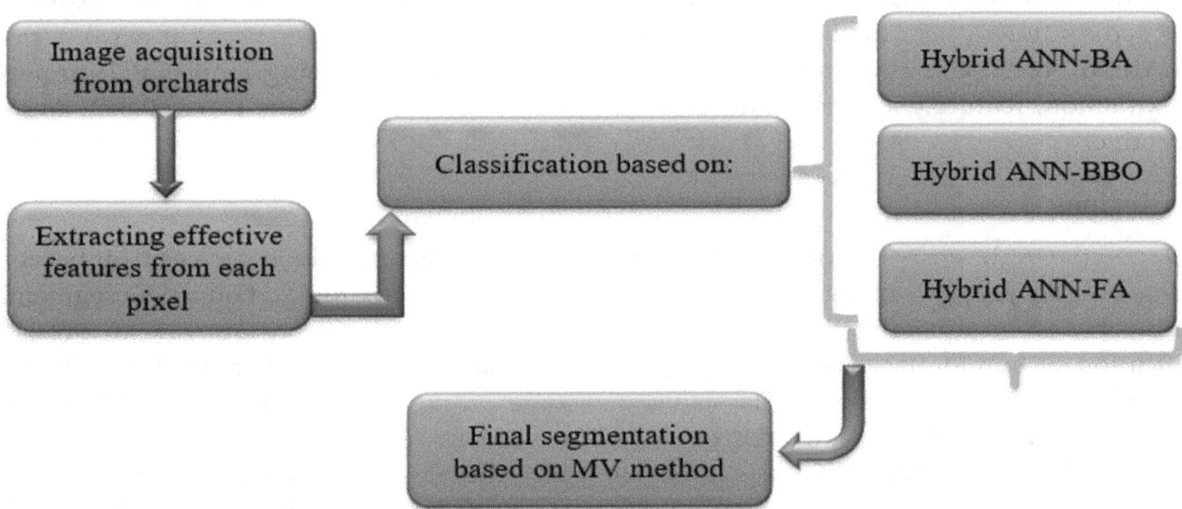

Figure 5. The final flowchart of the different stages of the proposed system in this article.

Finally, Table 14 shows the statistical t-test for surveying significant differences between MV and the other proposed methods. In this table, the MV method is compared with the hybrid ANN-BA, ANN-BBO, and ANN-FA. A statistically significant difference was found between the hybrid ANN-BA and MV method, hybrid ANN-BBO and MV method, and hybrid ANN-FA and MV method. Even if the accuracy of single classification methods was close to the accuracy of the majority voting method, the combination of these models together can limit the overfitting tendencies.

Table 14. Statistical differences between accuracy of majority voting (MV) and other method in 500 iterations.

	t	df	Sig.	Mean Accuracy	95% Confidence Interval	
					Lower	Upper
MV	4420	499	.000	97.20	97.16	97.24
ANN-FA	4739	499	.000	96.91	96.87	96.95
ANN-BBO	4174	499	.000	96.46	96.41	96.51
ANN-BA	3813	499	.000	96.47	96.42	96.52

3.7. Comparison of the Proposed Method with Other Methods Used for Segmentation

The comparison of the results of the proposed method with the results of methods used by other researchers greatly contributes to the importance of the present proposed method. Table 15 compares the correct classification rate of the proposed method with other studies. As can be seen, the proposed method had a higher CCR than the other methods.

Table 15. Comparison of different studies in the field of segmentation with the proposed method in this article.

Method	Number of Samples	Correct Classification Rate (%)
Proposed in this study	7200	98.01
Sabzi et al. [32]	210,752	96.80
Aquino et al. [20]	152	95.72
Hernández-Hernández et al. [19]	182	97
Miao et al. [18]	380	84.58

After the survey of the performance of proposed algorithm, the time consumption of the algorithm to calculate at different stages of the algorithm was estimated. The implementation hardware was a laptop computer with an Intel Core i3 processor CFI, 330 M at 2.13 GHz, 4 GB of RAM-4 GB, and MATLAB 2015b. The average time consumed by the Central Processing Unit (CPU) to perform pixel extraction, feature extraction, and classification for each frame were 0.163, 0.312, and 0.236 s, respectively.

4. Conclusions

Due to the high sensitivity of agricultural robots to various operations such as fruit harvesting, spraying, etc., the proper training of its software is crucial. This study focused on a new segmentation algorithm based on a combination of video processing and majority voting rule to provide an identification system for harvester robots of plums in orchard conditions. Three hybrid ANNs were used to provide the voting process. These hybrids were ANN-BA, ANN-BBO, and ANN-FA. Video processing was based on the extracted color features of each pixel, which led to highly efficient predictions when compared with object analysis. Since there were different types of objects in the orchard, the use of different pixels related to different objects in each frame extracted from different videos captured in different light intensities throughout the day ensures that the identification system was properly trained to detect the plum fruits on the trees.

Although it was not possible to use all of the extracted features of the identification system due to the time consumption, however, the greater the number of extracted features, the more the effective features were selected. Among the 33 extracted features, the hybrid ANN-HS selected four optimal features as inputs of classifiers that included the third channel of HSL, the second channel of LCH, the first channel of L*a*b*, and the first channel of the HIS color space. The single classification methods, namely hybrid ANN-BA, hybrid ANN-BBO, and hybrid ANN-FA had accuracies of 97.86%, 97.59%, and 97.77%, respectively. The majority voting method performed classification with an accuracy over 98% and was more accurate than the other classification methods.

Author Contributions: Conceptualization, S.S. and R.P.; Methodology, S.S., R.P. and D.K.; Validation, S.S., R.P., and D.K.; Formal analysis, S.S.; Investigation, S.S.; Writing—original draft preparation, S.S., T.P., and R.P.; Writing—review and editing, T.P.; Visualization, T.P.; Supervision, T.P, R.P., and D.K. All authors have read and agreed to the published version of the manuscript.

Acknowledgments: This paper was supported by the Research Center on Electronics, Optoelectronics, and Telecommunications (CEOT) project NIBAP—Núcleo de Investigação em Biotecnologia e Agricultura de Precisão FEDER ALG-01-0247-FEDER-037303.

References

1. Bechar, A.; Vigneault, C. Agricultural robots for field operations: Concepts and components. *Biosyst. Eng.* **2016**, *149*, 94–111. [CrossRef]
2. Cavaco, A.M.; Pires, R.; Antunes, M.D.; Panagopoulos, T.; Brázio, A.; Afonso, A.M.; Silva, L.; Lucas, R.M.; Cadeiras, B.; Cruz, S.P.; et al. Validation of short wave near infrared calibration models for the quality and ripening of 'Newhall' orange on tree across years and orchards. *Postharvest Biol. Technol.* **2018**, *141*, 86–97. [CrossRef]
3. Mehta, S.S.; Burks, T.F. Multi-camera fruit localization in robotic harvesting. *IFAC-PapersOnLine* **2016**, *49*, 90–95. [CrossRef]
4. Blasco, J.; Aleixos, N.; Cubero, S.; Gomez-Sanchis, J.; Molto, E. Automatic sorting of Satsuma (Citrus unshiu) segments using computer vision and morphological features. *Comput. Electron. Agric.* **2009**, *66*, 1–8. [CrossRef]
5. Garrido-Novell, C.; Perez-Marin, D.; Amigo, J.M.; Fernández-Novales, J.; Guerrero, J.E.; GarridoVaro, A. Grading and color evolution of apples using RGB and hyperspectral imaging vision cameras. *J. Food Eng.* **2012**, *113*, 281–288. [CrossRef]
6. Mendoza, F.; Dejmek, P.; Aguilera, J.M. Calibrated color measurements of agricultural foods using image analysis. *Postharvest Biol. Technol.* **2006**, *41*, 285–295. [CrossRef]
7. Naderi-Boldaji, M.; Fattahi, R.; Ghasemi-Varnamkhasti, M.; Tabatabaeefar, A.; Jannatizadeh, A. Models for predicting the mass of apricot fruits by geometrical attributes (cv. Shams, Nakhjavan, and Jahangiri). *Sci. Hortic.* **2008**, *118*, 293–298. [CrossRef]
8. Sabzi, S.; Abbaspour-Gilandeh, Y.; Javadikia, H. Machine vision system for the automatic segmentation of plants under different lighting conditions. *Biosyst. Eng.* **2017**, *161*, 157–173. [CrossRef]
9. Hu, J.; Li, D.; Duan, Q.; Han, Y.; Chen, G.; Si, X. Fish species classification by color, texture and multi-class support vector machine using computer vision. *Comput. Electron. Agric.* **2012**, *88*, 133–140. [CrossRef]
10. Valenzuela, C.; Aguilera, J.M. Aerated apple leathers: Effect of microstructure on drying and mechanical properties. *Dry. Technol.* **2013**, *31*, 1951–1959. [CrossRef]
11. Lin, S.; Xinchao, M.; Jiucheng, X.; Yun, T. An Image Segmentation method using an active contour model based on improved SPF and LIF. *Appl. Sci.* **2019**, *8*, 2576. [CrossRef]
12. Silwal, A.; Davidson, J.R.; Karkee, M.; Mo, C.; Zhang, Q.; Lewis, K. Design, integration, and field evaluation of a robotic apple harvester. *J. Field Robot.* **2017**, *34*, 1140–1159. [CrossRef]
13. Cui, Y.; Su, S.; Wang, X.; Tian, Y.; Li, P.; Zhang, F. Recognition and feature extraction of kiwifruit in natural environment based on machine vision. *Trans. Chin. Soc. Agric. Mach.* **2013**, *44*, 247–252.
14. Bechar, A.; Vigneault, C. Agricultural robots for field operations. Part 2: Operations and systems. *Biosyst. Eng.* **2017**, *153*, 110–128. [CrossRef]
15. Vidoni, R.; Bietresato, M.; Gasparetto, A.; Mazzetto, F. Evaluation and stability comparison of different vehicle configurations for robotic agricultural operations on sideslopes. *Biosyst. Eng.* **2015**, *129*, 197–211. [CrossRef]
16. Wang, C.; Yunchao, T.; Xiangjun, Z.; Lufeng, L.; Xiong, C. Recognition and matching of clustered mature litchi fruits using binocular charge-coupled device (CCD) color cameras. *Sensors* **2017**, *17*, 2564.

17. Sabzi, S.; Abbaspour-Gilandeh, Y.; Hernandez, J.; Azadshahraki, F.; Karimzadeh, R. The use of the combination of texture, color and intensity transformation features for segmentation in the outdoors with emphasis on video processing. *Agriculture* **2019**, *9*, 104. [CrossRef]
18. Miao, R.H.; Tang, J.L.; Chen, X.Q. Classification of farmland images based on color features. *J. Vis. Commun. Image Represent.* **2015**, *29*, 138–146. [CrossRef]
19. Hernández-Hernández, J.L.; García-Mateos, G.; González-Esquiva, J.M.; Escarabajal-Henarejos, D.; Ruiz-Canales, A.; Molina-Martínez, J.M. Optimal color space selection method for plant/soil segmentation in agriculture. *Comput. Electron. Agric.* **2016**, *122*, 124–132. [CrossRef]
20. Aquino, A.; Diago, M.P.; Millán, B.; Tardáguila, J. A new methodology for estimating the grapevine-berry number per cluster using image analysis. *Biosyst. Eng.* **2017**, *156*, 80–95. [CrossRef]
21. Kang, H.; Chen, C. Fast implementation of real-time fruit detection in apple orchards using deep learning. *Comput. Electron. Agric.* **2020**, *168*, 105108. [CrossRef]
22. Lin, G.; Tang, Y.; Zou, X.; Xiong, J.; Li, J. Guava detection and pose estimation using a low-cost RGB-D sensor in the field. *Sensors* **2019**, *19*, 428. [CrossRef] [PubMed]
23. Sulistyo, S.B.; Woo, W.L.; Dlay, S.S. Regularized neural networks fusion and genetic algorithm based on-field nitrogen status estimation of wheat plants. *IEEE Trans. Ind. Informa.* **2016**, *13*, 103–114. [CrossRef]
24. Sulistyo, S.B.; Woo, W.L.; Dlay, S.S.; Gao, B. Building a globally optimized computational intelligence image processing algorithm for on-site nitrogen status analysis in plants. *IEEE Intell. Syst.* **2018**, *33*, 15–26. [CrossRef]
25. Sulistyo, B.S.; Wu, D.; Woo, W.L.; Dlay, S.S.; Gao, B. Computational deep intelligence vision sensing for nutrient content estimation in agricultural automation. *IEEE Trans. Autom. Sci. Eng.* **2018**, *15*, 1243–1257. [CrossRef]
26. Lee, K.S.; Geem, Z.W. A new meta-heuristic algorithm for continuous engineering optimization: Harmony search theory and practice. *Comput. Methods Appl. Mech. Eng.* **2005**, *194*, 3902–3933. [CrossRef]
27. Pham, D.; Ghanbarzadeh, A.; Koc, E.; Otri, S.; Rahim, S.; Zaidi, M. The bees algorithm—A novel tool for complex optimisation problems. In Proceedings of the 2nd Virtual International Conference on Intelligent Production Machines and Systems (IPROMS 2006), Cardiff, UK, 3–14 July 2006.
28. Hussein, W.A.; Sahran, S.; Abdullah, S.N.H.S. A fast scheme for multilevel thresholding based on a modified bees algorithm. *Knowl. Based Syst.* **2016**, *101*, 114–134. [CrossRef]
29. Simon, D. Biogeography based optimization. *IEEE Trans. Evolut. Comput.* **2008**, *12*, 702–713. [CrossRef]
30. Yang, X.S. Firefly algorithms for multimodal optimization. In *Stochastic Algorithms: Foundations and Applications*; Watanabe, O., Zeugmann, T., Eds.; Lecture Notes in Computer Science; Springer: Berlin, Germany, 2009; Volume 5792, pp. 169–178. [CrossRef]
31. Guijarro, M.; Riomoros, I.; Pajares, G.; Zitinski, P. Discrete wavelets transform for improving greenness image segmentation in agricultural images. *Comput. Electron. Agric.* **2015**, *118*, 396–407. [CrossRef]
32. Sabzi, S.; Abbaspour-Gilandeh, Y.; García-Mateos, G.; Ruiz-Canales, A.; Molina-Martínez, J.M. Segmentation of apples in aerial images under sixteen different lighting conditions using color and texture for optimal irrigation. *Water* **2018**, *10*, 1634. [CrossRef]

5

Effect of Missing Vines on Total Leaf Area Determined by NDVI Calculated from Sentinel Satellite Data: Progressive Vine Removal Experiments

Sergio Vélez [1,*], Enrique Barajas [1], José Antonio Rubio [1], Rubén Vacas [1] and Carlos Poblete-Echeverría [2,*]

[1] Instituto Tecnológico Agrario de Castilla y León (ITACyL), Unidad de Cultivos Leñosos y Hortícolas, 47071 Valladolid, Spain; bartolen@itacyl.es (E.B.); rubcanjo@itacyl.es (J.A.R.); ita-vacizqru@itacyl.es (R.V.)
[2] Department of Viticulture and Oenology, Faculty of AgriSciences, Stellenbosch University, Private Bag X1, Matieland 7602, South Africa
[*] Correspondence: velmarse@itacyl.es (S.V.); cpe@sun.ac.za (C.P.-E.)

Featured Application: Sentinel-2 images were sensitive to change in the vegetation contained in the pixel. The reduction in the NDVI values was proportional to the reduction in the vegetation, following a linear relationship. The quantitative relationship obtained in this study is valuable since a vineyard, once established, generally loses grapevines each year due to diseases, abiotic stress, etc., so it is important to consider the effect of the missing vines in order to have a correct estimation of the vineyard vigour.

Abstract: Remote Sensing (RS) allows the estimation of some important vineyard parameters. There are several platforms for obtaining RS information. In this context, Sentinel satellites are a valuable tool for RS since they provide free and regular images of the earth's surface. However, several problems regarding the low-resolution of the imagery arise when using this technology, such as handling mixed pixels that include vegetation, soil and shadows. Under this condition, the Normalized Difference Vegetation Index (NDVI) value in a particular pixel is an indicator of the amount of vegetation (canopy area) rather than the NDVI from the canopy (as a vigour expression), but its reliability varies depending on several factors, such as the presence of mixed pixels or the effect of missing vines (a vineyard, once established, generally loses grapevines each year due to diseases, abiotic stress, etc.). In this study, a vine removal simulation (greenhouse experiment) and an actual vine removal (field experiment) were carried out. In the field experiment, the position of the Sentinel-2 pixels was marked using high-precision GPS. Controlled removal of vines from a block of cv. Cabernet Sauvignon was done in four steps. The removal of the vines was done during the summer of 2019, matching with the start of the maximum vegetative growth. The Total Leaf Area (TLA) of each pixel was calculated using destructive field measurements. The operations were planned to have two satellite images available between each removal step. As a result, a strong linear relationship ($R^2 = 0.986$ and $R^2 = 0.72$) was obtained between the TLA and NDVI reductions, which quantitatively indicates the effect of the missing vines on the NDVI values.

Keywords: total leaf area; mixed pixels; Cabernet Sauvignon; NDVI; Normalized Difference Vegetation Index; precision viticulture

1. Introduction

Remote sensing (RS) is a tool that allows information on distant objects to be obtained quickly and accurately [1]. A practical way to use remote sensing in viticulture is by using vegetation indices (VIs)

and its potential relies on their ability to estimate grape quality and yield using spectral information [2]. The VIs are algebraic combinations designed to highlight the contrast of plant vigour and its properties (e.g., canopy biomass, absorbed radiation, chlorophyll content). These indices are based on the fact that healthy plants show a high Near-InfraRed (NIR) reflectance and very low red reflectance [3,4]. The Normalized Difference Vegetation Index (NDVI) [5] has proved to be a useful indicator of the status of the vineyard with several applications, such as for sub-block management [6–12] and estimating the leaf area index, (LAI and can correlate with certain parameters such as total anthocyanins, total phenols, soil moisture, clay and sand content, berry pH, soluble solids, vine size and yield components [13–15]. NDVI has also been useful for establishing a correlation between Photosynthetically Active Biomass (PAB) and total phenolics and colour [16], assessing the water status spatial variability within the vineyard [17] and monitoring quality characteristics in table grapes [18]. Furthermore, within-field NDVI patterns are quite stable between seasons [19].

In RS, a key parameter to choose is the platform on which the sensor is mounted. At present, unmanned aerial vehicle (UAV) platforms have been extensively used for studying and exploring vineyards [20–25]. In general, UAV offers the possibility to obtain high-resolution multispectral imagery, however, the benefit of the high resolution is restricted by some UAV limitations, such as stability on windy days/areas, as well as piloting capabilities and global navigation satellite system/inertial navigation system (GNNS/INA) quality [26,27]. Also, regulations established in most countries might be a problem for properly developing the capabilities of UAVs [28]. Another issue is the cost of each operation, and above a certain scale size, an image taken by satellite may be more convenient than others [29]. In this context, satellites can be used for several applications, for example, mapping vineyard plant and soil water status [30], harvest prediction [31] and to analyse the spatial heterogeneity in the evapotranspiration [32]. Modern image satellite analysis allows the combining of information from different sensors mounted on different satellites in order to improve spatial resolution [33,34], even in the presence of clouds [35], although it should be noted that not all sensors provide the same information. There are several satellites which are used to obtain spatial information and they can be divided into two main groups depending on the cost of the images: free-to-use satellites and paid satellites. Regarding free-to-use satellites, Landsat and Sentinel satellites can be very useful and they have been used for applications as disparate as detecting motions before a landslide [36], ice flow measurements and the quantification of seasonal ice velocities [37], to assess the bloom dynamics of almond orchards [38] and to classify vineyards according to their vigour [39,40]. Some authors [41] have discussed the differences in the information collected from the Sentinel-2A MSI sensor and the Landsat-8 OLI sensor. They found that the MSI surface reflectance was greater than the OLI surface reflectance in almost all bands and that the MSI surface NDVI was greater than the OLI surface NDVI. In this sense, Sentinel-2 satellites (Copernicus Project of the European Space Agency) can be particularly useful due to their free status and the relative ease of access to their web platform (https://scihub.copernicus.eu/). In addition, Sentinel-2 imagery has a spatial resolution of 10 m on the pixel side and a temporal resolution of 10 days; 5 days if we combine the images from the two existing satellites currently in the constellation. Sentinel-2 provides multiple bands from which to obtain information, including the Near-InfraRed (NIR) and the Red, which allow the calculation of the NDVI [42].

Remotely sensed images can be classified into two groups [43] according to their spatial resolution: (i) low-spatial-resolution imagery, in which the majority of pixels contain reflectance information from the grapevines and the inter-row space, and (ii) high-spatial-resolution imagery, in which the majority of pixels contain information only from grapevines or only from inter-row space. Therefore, when using low-resolution imagery, the NDVI value of the pixel is an indicator of the amount of vegetation (canopy area) rather than the NDVI from the canopy of the vines (pure value without the influence of the background). In this context, the main limitation of Sentinel-2 is that the spatial resolution and within-block information could not be accurate in the case of small blocks or blocks with complex borders [44]. This is a widespread problem in satellite imagery because within a vineyard pixel there are plants, soil and shadows, which influence the correct calculation of the coefficients of the crop [45]. More precisely, the NDVI obtained by the satellite and the LAI measured with a photographic ground-truth method can be related [46] and the images from the Worldview-2 satellite, with a resolution of 0.5 to 2 m^2, can be used to indicate that the amount of vegetation contained in a pixel varies according to its size. Therefore, with 0.5 m^2 it will be possible to find pixels with 100% vegetation, however, with 2m^2 there will only be mixed pixels of vegetation and soil. Instead of Worldview-2, Sentinel-2 can be used, but this will lead to a much greater problem, since, instead of a 2m^2 resolution, Sentinel-2 will have a resolution of 100m^2, so all the pixels will be mixed and will contain several plants as well as soil and shadows. It is important to note that soils have a great influence on the calculation of the NDVI [47] and that the average NDVI values for vines can be up to 3 times larger than the average NDVI values for pure soil [48]. Some authors [49] have established that pixels with an NDVI lower than 0.6 should be removed because they are not vegetation. Shadows are mainly influenced by the distance between plants and rows, but also by the characteristics of the plants [50]. Thus, in the same lighting, more vegetation will imply more shadow, an important factor since shaded pixels have low reflectance and modify the values we would expect if there were only vegetation and soil. Several authors have tried to solve this problem, trying to correct the shading in UAV and satellite images [51–55]. Additionally, NDVI can be greatly influenced by viticulture practices (e.g., canopy management and irrigation), so these practices should be considered [56].

Considering the inherent mixed-pixel characteristic of Sentinel-2 imagery in vineyards and the effect of missing vines (a vineyard, once established, generally loses grapevines each year due to diseases, abiotic stress, etc.), the objective of the present study was to analyse the effect of missing vines on mixed pixels using the NDVI as a reference index (NDVI = (NIR − Red)/(NIR + Red)). Our approach was to work with real measurements of vegetation reductions at pixel level evaluated by Sentinel-2 satellites, in order to understand the relationship between the vegetation contained in a pixel and the information captured by the satellite. To this end, two related experiments were performed: (i) a simulation under control conditions (greenhouse experiment) and (ii) a field experiment implementing a progressive vine removal protocol in four steps to check the sensitivity of the satellite images to the loss of vegetative mass within the study area.

2. Materials and Methods

2.1. Greenhouse Experiment

To test the concept of the NDVI reduction, a vine removal simulation was done under control conditions in a greenhouse using pot-grown cv. Cabernet Sauvignon grapevines (Department of Viticulture and Oenology, Stellenbosch University, South Africa). In order to carry out the progressive removal simulation, 12 one-year-old vines were selected and maintained in a greenhouse under natural light at 26 °C and 65% humidity.

The vines were located in two rows to simulate a Vertical Shoot Positioned (VSP) trellis system with a distance between rows of 55 cm and a distance between vines of 27.5 cm. Three random vines were removed each time until all vines in the simulated pixel area were removed.

A multispectral camera (Sequoia, Parrot SA, Paris, France) was used to capture images in each step of the removal simulation. The Sequoia camera has four 1.2-megapixel monochrome sensors which collect global shutter imagery along four discrete spectral bands: Green (550 nm), Red (660 nm), Red-Edge (735 nm) and Near-Infrared (NIR) (790 nm); a standard RGB camera and a sunshine sensor that continuously captures the light conditions in the same spectral bands as the multispectral sensor. The pipeline of the image analysis is shown in Figure 1. Since the satellite images are mixed pixels, a single pixel was simulated to encompass both the soil and plants.

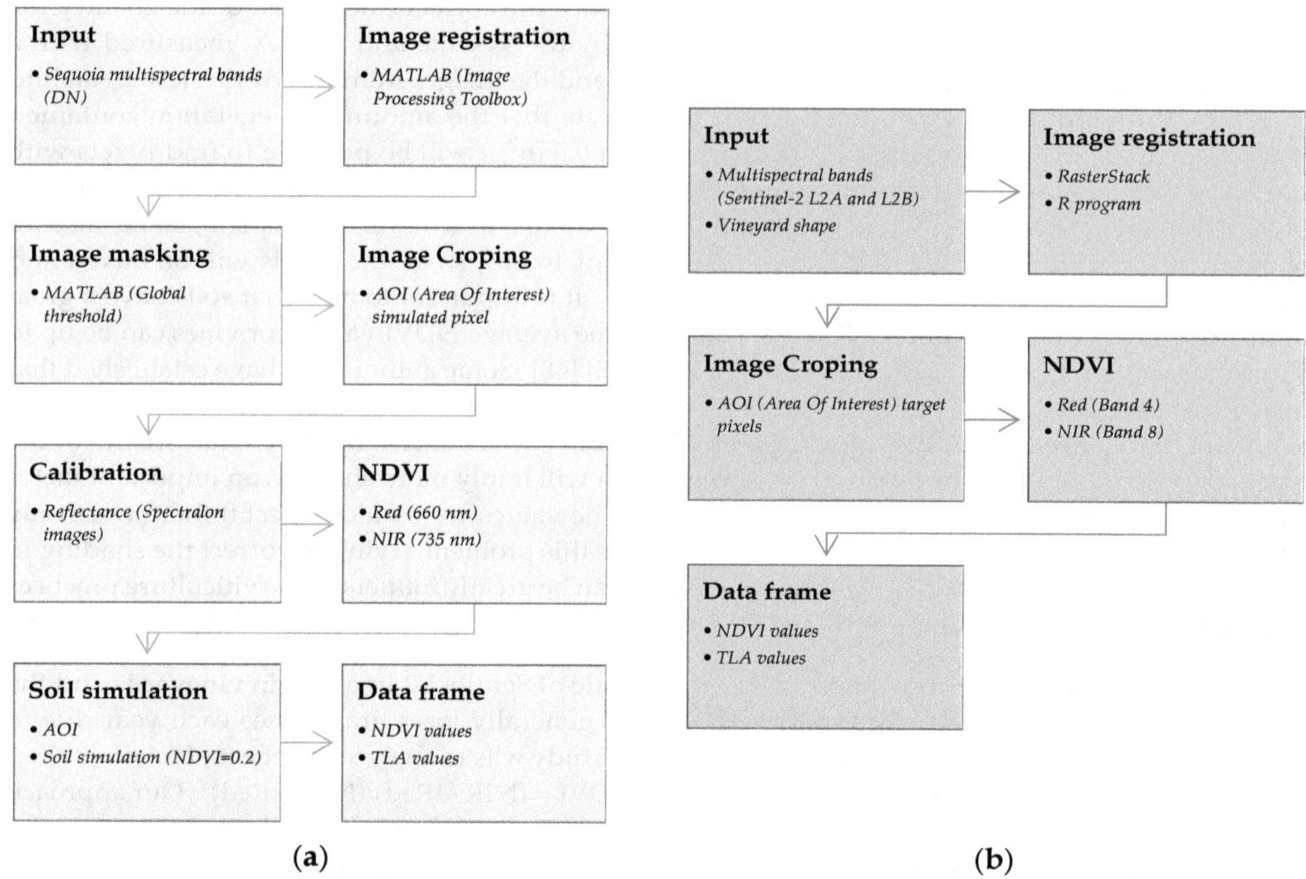

Figure 1. Pipeline of (**a**) greenhouse image analysis (**b**) field experiment image analysis. NDVI: Normalized Difference Vegetation Index. TLA: Total Leaf Area.

2.2. Field Experiment

Subsequently, a large-scale field experiment was carried out on a vineyard (cv. Cabernet Sauvignon), located in Zamadueñas Estate (coordinates: 41.7013° N, 4.7088° W, Valladolid, Spain), belonging to the Agricultural Technology Institute of Castilla y León (ITACyL). The vineyard was trellis-trained, with a bilateral Royat cordon pruning system, with eight spurs per plant and two buds per thumb, with 2.2 × 1.2 m row and plant-spacing, respectively, and NE-SO orientation. The soil was kept free of weeds and any other element that could affect the NDVI [56]. The vineyard was not irrigated, although in previous years it had been irrigated. The accumulated rainfall from 1 January to 31 July 2019 was 133.12 mm.

During the months of June and July 2019, a progressive vine removal experiment was developed in four phases (Figure 2). Three Sentinel-2 pixels (10 × 10 m) were selected within the vineyard, with 38 grapevines inside each pixel. In each phase, a quarter of the vegetative mass of each pixel was eliminated and the last to be removed equated to the elimination of the remnant vegetation. Each grapevine was cut in the lower-middle part of the trunk (Figure 3a,b) and all the material was extracted from the vineyard. A GPS Triumph-2 JAVAD GNSS model with centimetre accuracy was used to mark the pixels in the field to ensure that the grapevines within the studied area (Figure 3c) were removed. The GPS TRIUMPH-2 (JAVAD GNSS INC, San Jose, CA, USA) has 216 channels of dual-frequency GPS and GLONASS and can connect via Bluetooth and WiFi to a mobile phone to access the local GNSS Reference Station Network. A work schedule was established in order to obtain two Sentinel-2 images between each removal, one for each satellite.

Figure 2. Schematic representation of the four-step vine removal implemented in the field experiment.

The dates of the removal of the vines were 28 June, 8 July, 19 July and 29 July 2019. The experiment was carried out at this specific time due to the proximity of veraison, since at this phenological stage the relationship between leaf area and NDVI is greater [41]. To calculate the Total Leaf Area (TLA), the area of each leaf of the removed plants was measured in the laboratory (Figure 3d) using Easy Leaf Area application [57]. Easy Leaf Area measures leaf area non-destructively, calculated automatically from green leaf and red scale areas. This procedure was performed each time the vines were removed.

Regarding the spatial information, free-cloud atmospherically corrected images were downloaded (between 11 June 2019 and 20 August 2019) from the European Space Agency (ESA) Copernicus Project website. NDVI was calculated for each step using a customized code in R v.3.6.2. (Figure 1b) from the Sentinel-2 satellite images corresponding to the T30TUM tile. Sentinel 2A and 2B were used in combination to minimize variations in sensors, satellite orbit, pixel misregistration, clouds and radiometry. Images corresponding to the two satellites were downloaded and the values of the available free-cloud images were averaged between each vine removal.

All image and data analyses were carried out using customized codes written in an R statistical program (version 3.6X, R Foundation for Statistical Computing (R Core Team 2019), https://www.R-project.org/, Vienna, Austria) and MATLAB (Version R2019b, The MathWorks Inc., http://www.mathworks.com, Natick, MA, USA). A *t*-test for independent samples was performed in R for the statistical comparison of each removal step.

(a) (b) (c) (d)

Figure 3. (**a**) Removal of the vines; (**b**) Detail of the removed vines; (**c**) Marked pixels in the vineyard and (**d**) Example of the leaf area measurements using the Easy Leaf Area application.

3. Results

3.1. Greenhouse Experiment

The greenhouse simulation (Figure 4) showed a clear relationship ($R^2 = 0.986$) between the reduction of NDVI and TLA. The NDVI values are the simulated mixed pixel values. When the vines were removed from the simulated mixed pixel (around 2 m^2), the NDVI values decreased linearly until reaching the base soil values (defined in this case as NDVI = 0.20). The slope of the linear equation is 0.3 (y = 0.3x), so for each 20% of reduction in the TLA, the reduction in NDVI is around 6%.

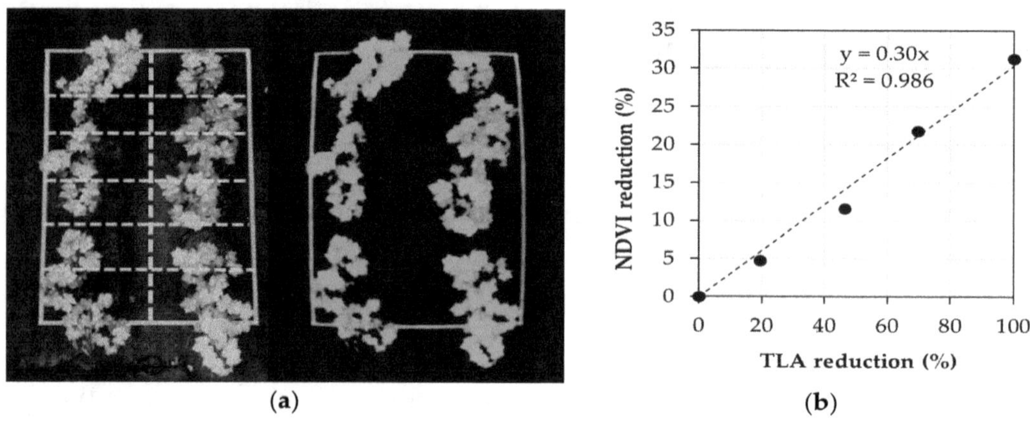

(**a**) (**b**)

Figure 4. (**a**) Example of the mask used to isolate the vines; (**b**) The relationship between them. Normalized Difference Vegetation Index (NDVI) reduction (%) and Total Leaf Area (TLA) reduction (%) as pixel-based.

3.2. Field Experiment

A reduction in the NDVI value is observed after each vine removal until it reaches the base soil values (NDVI values between 0.17 to 0.19), except in pixel 1 (Figure 5). The starting NDVI values, which correspond to the maximum vine cover, were 0.251, 0.321 and 0.306, for pixels 1, 2 and 3, respectively. In Figure 6a all values from all of the pixels from the dataset were used and a linear relationship was found between the reduction in NDVI and TLA ($R^2 = 0.72$) with a slope of 0.32 (y = 0.3104x), showing that for each 20% of reduction in the TLA, the NDVI is reduced by around 7%, similar to the greenhouse experiment result. Moreover, if the dataset is disaggregated by pixel (Figure 6b), the determination coefficients are 0.92, 0.68 and 0.99 for pixels 1, 2 and 3, respectively.

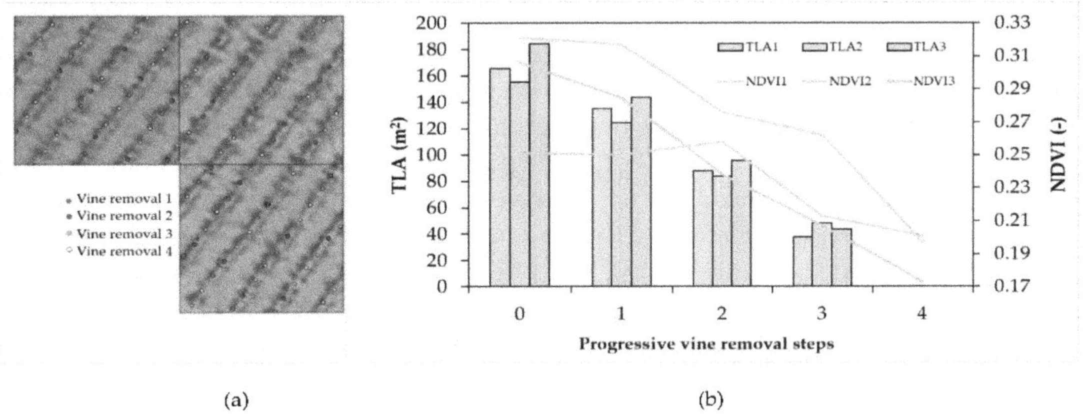

Figure 5. (**a**) Detail of the vine removal by phase. (**b**) TLA (m^2) and NDVI values in each pixel.

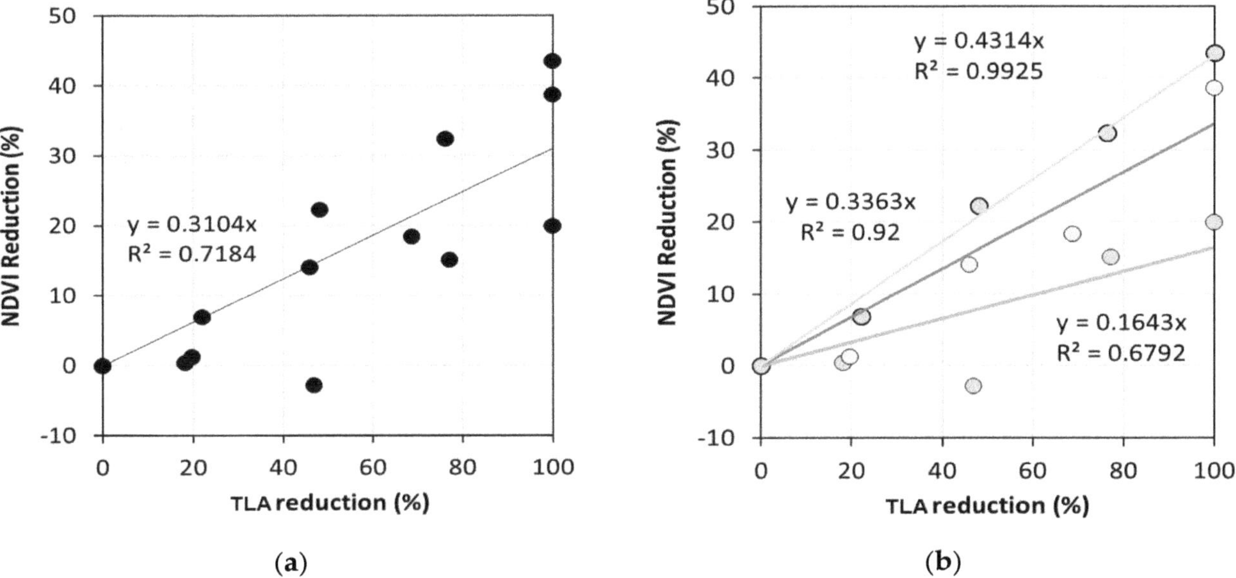

Figure 6. Relationship between NDVI (%) and TLA reduction (%) (**a**) in all pixels (**b**) per pixel.

The *p*-values for the comparison of each pair of removal steps are shown in Table 1. As expected, all removal steps had significant differences in TLA. However, concerning NDVI, only a few pairs of treatments had significant differences and in no case between two consecutive steps, which indicate a low sensitivity of the index to small changes in the vegetation amount.

Table 1. *p*-values from the *t*-test comparison of Normalized Difference Vegetation Index (NDVI) and total leaf area (TLA) in the field experiment.

		TLA				
		Step0	**Step1**	**Step2**	**Step3**	**Step4**
NDVI	Step0		0.018	0.003	0.001	0.001
	Step1	0.396		0.002	0.001	0.001
	Step2	0.120	0.148		0.001	0.001
	Step3	0.039	0.045	0.113		0.002
	Step4	0.013	0.011	0.005	0.079	

Grey colour denotes *p*-values 0.05. Diagonal values were excluded and marked in black.

4. Discussion

This study has approached the mixed pixel effect in two different ways: a laboratory approach (greenhouse) and a full-scale approach with a field experiment. Two experiments presented consistent results, showing the same pattern of NDVI reduction. The gradual reduction in NDVI showed in Figure 5b is expected because the leaf area within the pixel is reduced and NDVI is sensitive to changes in the vegetation [15]. There is an inconsistency in the pixel 1 trend, probably because the vegetation of the adjacent vines was moved and therefore the horizontal leaf area exposed to the satellite increased, although it could also be due to a variation in the reflectivity collected by the Sentinel in that specific image, since, if an analysis is performed using the image of 6/7/2019 corresponding to the Sentinel-2B satellite, an increase in NDVI can be observed in all the pixels of the tile. The influence of humidity in the soil is disregarded [58] since there was no significant rainfall in the period.

In the field experiment, initial NDVI values were between 0.25 and 0.32 with final values between 0.17 and 0.19. The starting NDVI values correspond to the maximum vine cover and indicate a different level of vigour in the selected pixels (Figure 5b). The final NDVI values correspond to bare soil and are consistent with the results of other authors [47]. It has been observed that the reduction in the NDVI value is proportional to the loss of TLA, finding that in the greenhouse experiment and the field experiment the slopes of the regression line were very similar. Therefore, if the vegetation within the pixel is reduced by 20%, the NDVI will be reduced proportionally by 6–7%. This also indicates that, considering the components of a mixed pixel [46], the vegetation and the associated shade effect 30–35%, so the remaining value is being influenced by the soil. These results are similar to the values reported by other authors [48], which indicate that the effect of the soil can be up to 3 times that of the vegetation.

If the field dataset is disaggregated per pixel, the difference between pixels is clearly observed (Figure 6b). This could be due to differences in vigour within the vineyard, since the vigour of the vineyard is related to the NDVI values [39,40]. Another important aspect to consider is the orientation of the row regarding the position of the Sentinel-2 pixels. The orientation combined with the space between rows creates an irregular grid effect. Therefore, if the orientation is not perfectly aligned with the pixel position, the number of vines per pixel is irregular. In our case, this effect was clear, with values of 38, 37 and 39 vines for the pixels 1, 2 and 3, respectively.

The trend lines presented in the results have very high R^2 values, however, these lines were just made in order to observe the trend and to highlight the relationship that has been shown, not to establish a model, since they are strongly influenced by the extreme values (0% and 100% of reduction), and it would be desirable to have a greater amount of data to develop a model. It is also important to note that there may be misregistration of pixels, which are improved regularly by updating the Processing Baseline, so in future studies the algorithm might be ameliorated, allowing more accuracy in the process [59].

Looking at the results of the *t*-test, although the TLA was significantly reduced between each removal step, the differences in the NDVI were not significant until the amount of vegetation was reduced with two removal steps. For example, there are differences between the first and the last removal, and between the second and the last, but there is no difference between the third and the fourth. This indicates that, although the NDVI is affected by the reduction of TLA, it is not overly sensitive to small reductions in the vegetation amount.

5. Conclusions

In this study, a relationship between the Total Leaf Area (TLA) and the Normalized Difference Vegetation Index (NDVI) was developed to analyse the effect of missing vines on NDVI values at pixel level (10 × 10 m). This study has demonstrated that it is possible to estimate quantitatively the impact that the decrease in vegetation in a vineyard has on the NDVI values. Our results show that it is possible to use the NDVI calculated from the Sentinel-2 images to identify the change in the vegetation in the pixel. Furthermore, it is worth noting that the reduction in the NDVI values is proportional to the reduction in the vegetation, following a linear relationship. The quantitative relationship obtained in this study is valuable since a vineyard, once established, generally loses grapevines each year due to diseases, abiotic stress, etc., so it is worth analysing the effect of the missing vines in order to have a clear understanding of the vineyard vigour.

The field experiment was conducted in a vineyard with a vertical trellis and this system has become a standard in today's viticulture. Moreover, in this work, the results of the two experiments are very similar, since the greenhouse experiment simulated the same row and plant distance as the field experiment, so the results obtained in this study could be used as a reference for vineyards with similar trellis characteristics (distance between rows and vines). However, it might be worthwhile to check whether this result can be extrapolated to other trellis systems or vineyards with different canopy or soil management practices, analysing the influence of the elimination of vegetation in vineyards with different soils and different characteristics. Further research would be desirable in this direction.

In further studies, it might be interesting to explore the possibility of removing a specific area of vines within a vineyard to calibrate the entire vineyard and use this technique to calibrate the background adjustment factor (L) of vegetation indexes such as SAVI, or even develop new indexes that take into account parameters related to the canopy, such as the influence of shadows or the linear meters of vertical trellis contained in each pixel.

Although the results are clear and promising, the limitations of this study should be considered due to the complexity and effort involved in an operation of this type in a vineyard. Operations on a larger scale are desirable in order to cover a greater number of pixels and vines to obtain more robust results covering different vineyard conditions.

Author Contributions: Conceptualization: S.V., E.B., J.A.R., C.P.-E.; methodology: S.V., E.B., J.A.R., C.P.-E., R.V.; Data acquisition: S.V., E.B., R.V., C.P.-E.; Data processing: S.V., E.B., R.V., C.P.-E.; Data analysis: S.V., E.B., C.P.-E.; resources, S.V., E.B., J.A.R., C.P.-E.; writing—original draft: S.V.; writing—review and editing: E.B., C.P.-E.; supervision: E.B., J.A.R., C.P.-E.; project administration: E.B., J.A.R.; funding acquisition: E.B., J.A.R. All authors have read and agreed to the published version of the manuscript.

Acknowledgments: We thank the staff of Viticulture for their cooperation in the vineyard operations.

References

1. Krishna, K.R. *Push Button Agriculture: Robotics, Drones, Satellite-Guided Soil and Crop Management*; Apple Academic Press: Oxfordshire, UK, 2016.
2. Hall, A.; Lamb, D.W.; Holzapfel, B.; Louis, J. Optical remote sensing applications in viticulture—A review. *Aust. J. Grape Wine Res.* **2002**, *8*, 36–47. [CrossRef]
3. Proffitt, A.P.B. *Precision Viticulture: A New Era in Vineyard Management and Wine Production*; Winetitles Pty Ltd.: Broadview, SA, Australia, 2006.
4. Bachmann, F.; Herbst, R.; Gebbers, R.; Hafner, V.V. Micro UAV Based Georeferenced Orthophoto Generation in VIS + NIR for Precision Agriculture. *Int. Arch. Photogramm. Remote Sens. Spatial Inf. Sci.* **2013**, *XL-1/W2*, 11–16. [CrossRef]
5. Rouse, J.W., Jr.; Haas, R.H.; Schell, J.A.; Deering, D.W. Monitoring vegetation systems in the Great Plains with ERTS. In Proceedings of the Third ERTS Symposium, NASA SP-351 1. U.S., Government Printing Office, Washington, DC, USA, 10–14 December 1973; pp. 309–317.
6. Johnson, L.F.; Bosch, D.F.; Williams, D.C.; Lobitz, B.M. Remote sensing of vineyard management zones: Implications for wine quality. *Appl. Eng. Agric.* **2001**, *17*, 557–560. [CrossRef]
7. Tagarakis, A.; Liakos, V.; Fountas, S.; Koundouras, S.; Gemtos, T.A. Management zones delineation using fuzzy clustering techniques in grapevines. *Precis. Agric.* **2013**, *14*, 18–39. [CrossRef]

8. Martinez-Casasnovas, J.A.; Agelet-Fernandez, J.; Arno, J.; Ramos, M.C. Analysis of vineyard differential management zones and relation to vine development, grape maturity and quality. *Span. J. Agric. Res.* **2012**, *10*, 326. [CrossRef]
9. Santesteban, L.G.; Guillaume, S.; Royo, J.B.; Tisseyre, B. Are precision agriculture tools and methods relevant at the whole-vineyard scale? *Precis. Agric.* **2013**, *14*, 2–17. [CrossRef]
10. Santesteban, L.G.; Urretavizcaya, I.; Miranda, C.; Garcia, A.; Royo, J.B. Agronomic significance of the zones defined within vineyards early in the season using NDVI and fruit load information. In Proceedings of the Precision Agriculture '13: Papers Presented at the 9th European Conference on Precision Agriculture, Lleida, Catalonia, Spain, 7–11 July 2013; ISBN 978-90-8686-224-5 .
11. Urretavizcaya, I.; Miranda, C.; Royo, J.B.; Santesteban, L.G. Within-vineyard zone delineation in an area with diversity of training systems and plant spacing using parameters of vegetative growth and crop load. In Proceedings of the Precision Agriculture '15: Papers Presented at the 10th European Conference on Precision Agriculture, Volcani Center, Rishon LeTsiyon, Israel, 12–16 July 2015; ISBN 978-90-8686-267-2.
12. Cancela, J.J.; Fandiño, M.; Rey, B.J.; Dafonte, J.; González, X.P. Discrimination of irrigation water management effects in pergola trellis system vineyards using a vegetation and soil index. *Agric. Water Manag.* **2017**, *183*, 70–77. [CrossRef]
13. Johnson, L.F. Temporal stability of an NDVI-LAI relationship in a Napa Valley vineyard. *Aust. J. Grape Wine Res.* **2003**, *9*, 96–101. [CrossRef]
14. Towers, P.C.; Strever, A.; Poblete-Echeverría, C. Comparison of Vegetation Indices for Leaf Area Index Estimation in Vertical Shoot Positioned Vine Canopies with and without Grenbiule Hail-Protection Netting. *Remote Sens.* **2019**, *11*, 1073. [CrossRef]
15. Ledderhof, D.; Brown, R.; Reynolds, A.; Jollineau, M. Using remote sensing to understand Pinot noir vineyard variability in Ontario. *Can. J. Plant Sci.* **2016**, *96*, 89–108. [CrossRef]
16. Lamb, D.W.; Weedon, M.M.; Bramley, R.G.V. Using remote sensing to predict grape phenolics and colour at harvest in a Cabernet Sauvignon vineyard: Timing observations against vine phenology and optimising image resolution. *Aust. J. Grape Wine Res.* **2008**, *10*, 46–54. [CrossRef]
17. Baluja, J.; Diago, M.P.; Balda, P.; Zorer, R.; Meggio, F.; Morales, F.; Tardaguila, J. Assessment of vineyard water status variability by thermal and multispectral imagery using an unmanned aerial vehicle (UAV). *Irrig. Sci.* **2012**, *30*, 511–522. [CrossRef]
18. Anastasiou, E.; Balafoutis, A.; Darra, N.; Psiroukis, V.; Biniari, A.; Xanthopoulos, G.; Fountas, S. Satellite and Proximal Sensing to Estimate the Yield and Quality of Table Grapes. *Agriculture* **2018**, *8*, 94. [CrossRef]
19. Kazmierski, M.; Glémas, P.; Rousseau, J.; Tisseyre, B. Temporal stability of within-field patterns of NDVI in non irrigated Mediterranean vineyards. *OENO One* **2011**, *45*, 61. [CrossRef]
20. Matese, A.; Di Gennaro, S.F.; Berton, A. Assessment of a canopy height model (CHM) in a vineyard using UAV-based multispectral imaging. *Int. J. Remote Sens.* **2017**, *38*, 2150–2160. [CrossRef]
21. Santesteban, L.G.; Di Gennaro, S.F.; Herrero-Langreo, A.; Miranda, C.; Royo, J.B.; Matese, A. High-resolution UAV-based thermal imaging to estimate the instantaneous and seasonal variability of plant water status within a vineyard. *Agr. Wat. Manag.* **2017**, *183*, 49–59. [CrossRef]
22. Weiss, M.; Baret, F. Using 3D Point Clouds Derived from UAV RGB Imagery to Describe Vineyard 3D Macro-Structure. *Remote Sens.* **2017**, *9*, 111. [CrossRef]
23. Mathews, A.; Jensen, J. Visualizing and Quantifying Vineyard Canopy LAI Using an Unmanned Aerial Vehicle (UAV) Collected High Density Structure from Motion Point Cloud. *Remote Sens.* **2013**, *5*, 2164–2183. [CrossRef]
24. Zarco-Tejada, P.J.; Guillén-Climent, M.L.; Hernández-Clemente, R.; Catalina, A.; González, M.R.; Martín, P. Estimating leaf carotenoid content in vineyards using high resolution hyperspectral imagery acquired from an unmanned aerial vehicle (UAV). *Agric. Forest Meteorol.* **2013**, *171–172*, 281–294. [CrossRef]
25. Poblete-Echeverría, C.; Olmedo, G.; Ingram, B.; Bardeen, M. Detection and Segmentation of Vine Canopy in Ultra-High Spatial Resolution RGB Imagery Obtained from Unmanned Aerial Vehicle (UAV): A Case Study in a Commercial Vineyard. *Remote Sens.* **2017**, *9*, 268. [CrossRef]
26. Maes, W.H.; Steppe, K. Perspectives for Remote Sensing with Unmanned Aerial Vehicles in Precision Agriculture. *Trends Plant Sci.* **2019**, *24*, 152–164. [CrossRef] [PubMed]
27. Nex, F.; Remondino, F. UAV for 3D mapping applications: A review. *Appl. Geomat.* **2014**, *6*, 1–15. [CrossRef]

28. Stöcker, C.; Bennett, R.; Nex, F.; Gerke, M.; Zevenbergen, J. Review of the Current State of UAV Regulations. *Remote Sens.* **2017**, *9*, 459. [CrossRef]
29. Matese, A.; Toscano, P.; Di Gennaro, S.; Genesio, L.; Vaccari, F.; Primicerio, J.; Belli, C.; Zaldei, A.; Bianconi, R.; Gioli, B. Intercomparison of UAV, Aircraft and Satellite Remote Sensing Platforms for Precision Viticulture. *Remote Sens.* **2015**, *7*, 2971–2990. [CrossRef]
30. Borgogno-Mondino, E.; Novello, V.; Lessio, A.; Tarricone, L.; de Palma, L. Intra-vineyard variability description through satellite-derived spectral indices as related to soil and vine water status. *Acta Hortic.* **2018**, *1197*, 59–68. [CrossRef]
31. Sun, L.; Gao, F.; Anderson, M.C.; Kustas, W.P.; Alsina, M.M.; Sanchez, L.; Sams, B.; McKee, L.; Dulaney, W.; White, W.A.; et al. Daily Mapping of 30 m LAI and NDVI for Grape Yield Prediction in California Vineyards. *Remote Sens.* **2017**, *9*, 317. [CrossRef]
32. Knipper; Kustas; Anderson; Alsina; Hain; Alfieri; Prueger; Gao; McKee; Sanchez Using High-Spatiotemporal Thermal Satellite ET Retrievals for Operational Water Use and Stress Monitoring in a California Vineyard. *Remote Sens.* **2019**, *11*, 2124. [CrossRef]
33. Chang, J.; Shoshany, M. Mediterranean shrublands biomass estimation using Sentinel-1 and Sentinel-2. In Proceedings of the 2016 IEEE International Geoscience and Remote Sensing Symposium (IGARSS), Beijing, China, 10 July 2016; pp. 5300–5303. [CrossRef]
34. Stumpf, A.; Michéa, D.; Malet, J.-P. Improved Co-Registration of Sentinel-2 and Landsat-8 Imagery for Earth Surface Motion Measurements. *Remote Sens.* **2018**, *10*, 160. [CrossRef]
35. Navarro, A.; Rolim, J.; Miguel, I.; Catalão, J.; Silva, J.; Painho, M.; Vekerdy, Z. Crop Monitoring Based on SPOT-5 Take-5 and Sentinel-1A Data for the Estimation of Crop Water Requirements. *Remote Sens.* **2016**, *8*, 525. [CrossRef]
36. Lacroix, P.; Bièvre, G.; Pathier, E.; Kniess, U.; Jongmans, D. Use of Sentinel-2 images for the detection of precursory motions before landslide failures. *Remote Sens. Environ.* **2018**, *215*, 507–516. [CrossRef]
37. Kääb, A.; Winsvold, S.; Altena, B.; Nuth, C.; Nagler, T.; Wuite, J. Glacier Remote Sensing Using Sentinel-2. *Part I: Radiometric and Geometric Performance, and Application to Ice Velocity. Remote Sens.* **2016**, *8*, 598. [CrossRef]
38. Chen, B.; Jin, Y.; Brown, P. An enhanced bloom index for quantifying floral phenology using multi-scale remote sensing observations. *ISPRS J. Photogramm. Remote Sens.* **2019**, *156*, 108–120. [CrossRef]
39. Vélez, S.; Rubio, J.A.; Andrés, M.I.; Barajas, E. Agronomic classification between vineyards ('Verdejo') using NDVI and Sentinel-2 and evaluation of their wines. *Vitis J. Grapevine Res.* **2019**, *58*, 33–38. [CrossRef]
40. Di Gennaro, S.F.; Dainelli, R.; Palliotti, A.; Toscano, P.; Matese, A. Sentinel-2 Validation for Spatial Variability Assessment in Overhead Trellis System Viticulture Versus UAV and Agronomic Data. *Remote Sens.* **2019**, *11*, 2573. [CrossRef]
41. Zhang, H.K.; Roy, D.P.; Yan, L.; Li, Z.; Huang, H.; Vermote, E.; Skakun, S.; Roger, J.-C. Characterization of Sentinel-2A and Landsat-8 top of atmosphere, surface, and nadir BRDF adjusted reflectance and NDVI differences. *Remote Sens. Environ.* **2018**, *215*, 482–494. [CrossRef]
42. European Space Agency (ESA). *SENTINEL-2 User Handbook*; ESA Standard Document: 2015; Available online: https://sentinel.esa.int/documents/247904/685211/Sentinel-2_User_Handbook (accessed on 25 March 2020).
43. Hall, A.; Louis, J.P.; Lamb, D.W. Low-resolution remotely sensed images of winegrape vineyards map spatial variability in planimetric canopy area instead of leaf area index. *Aust. J. Grape Wine Res.* **2008**, *14*, 9–17. [CrossRef]
44. Devaux, N.; Crestey, T.; Leroux, C.; Tisseyre, B. Potential of Sentinel-2 satellite images to monitor vine fields grown at a territorial scale. *OENO One* **2019**, *53*. [CrossRef]
45. Johnson, L.; Scholasch, T. Remote Sensing of Shaded Area in Vineyards. *Horttech* **2005**, *15*, 859–863. [CrossRef]
46. Fuentes, S.; Poblete-Echeverría, C.; Ortega-Farias, S.; Tyerman, S.; De Bei, R. Automated estimation of leaf area index from grapevine canopies using cover photography, video and computational analysis methods: New automated canopy vigour monitoring tool. *Aust. J. Grape Wine Res.* **2014**, *20*, 465–473. [CrossRef]
47. Montandon, L.; Small, E. The impact of soil reflectance on the quantification of the green vegetation fraction from NDVI. *Remote Sens. Environ.* **2008**, *112*, 1835–1845. [CrossRef]
48. Poblete-Echeverría, C.; Acevedo-Opazo, C.; Ortega-Farías, S.; Valdés-Gómez, H.; Nuñez, R. Study of NDVI spatial variability over a Merlot vineyard-plot in Maule Region using a hand held Spectroradiometer. In Proceedings of the 8th Fruit, Nut, and Veg Prod Eng Symp FRUTIC, Concepción, Chile, 5–9 January 2009; pp. 182–189.

49. Hall, A.; Lamb, D.W.; Holzapfel, B.P.; Louis, J.P. Within-season temporal variation in correlations between vineyard canopy and winegrape composition and yield. *Precis. Agric.* **2011**, *12*, 103–117. [CrossRef]
50. Prichard, T.; Hanson, B.; Schwankl, L.; Verdegaal, P.; Smith, R. *Deficit Irrigation of Quality Winegrapes Using Micro-Irrigation Techniques*; UC Coop Extension, Dept. of LAWR. UC Davis, 2004; Available online: http://cesanluisobispo.ucdavis.edu/files/89518.pdf (accessed on 22 March 2020).
51. Wang, Q.J.; Tian, Q.J.; Lin, Q.Z.; Li, M.X.; Wang, L.M. An improved algorithm for shadow restoration of high spatial resolution imagery. In Proceedings of the Proc. SPIE 7123, Remote Sensing of the Environment: 16th National Symposium on Remote Sensing of China, Beijing, China, 7–10 September 2007; p. 71230D. [CrossRef]
52. Zhang, Z.; Chen, F. A shadow processing method of high spatial resolution remote sensing image. In Proceedings of the 2010 3rd International Congress on Image and Signal Processing, Yantai, China, 16–18 October 2010; pp. 816–820. [CrossRef]
53. Aboutalebi, M.; Torres-Rua, A.F.; McKee, M.; Kustas, W.; Nieto, H.; Coopmans, C. Behavior of vegetation/soil indices in shaded and sunlit pixels and evaluation of different shadow compensation methods using UAV high-resolution imagery over vineyards. In Proceedings of the Autonomous Air and Ground Sensing Systems for Agricultural Optimization and Phenotyping III., Orlando, FL, USA, 18–19 April 2018; Thomasson, J.A., McKee, M., Moorhead, R.J., Eds.; SPIE: Orlando, FL, USA, 2018; p. 6. [CrossRef]
54. Wu, J.; Bauer, M. Evaluating the Effects of Shadow Detection on QuickBird Image Classification and Spectroradiometric Restoration. *Remote Sens.* **2013**, *5*, 4450–4469. [CrossRef]
55. Ma, H.; Qin, Q.; Shen, X. Shadow Segmentation and Compensation in High Resolution Satellite Images. In Proceedings of the IGARSS 2008-2008 IEEE International Geoscience and Remote Sensing Symposium, Boston, MA, USA, 8–11 July 2008; pp. II-1036–II-1039. [CrossRef]
56. Fountas, S.; Anastasiou, E.; Balafoutis, A.; Koundouras, S.; Theoharis, S.; Theodorou, N. The influence of vine variety and vineyard management on the effectiveness of canopy sensors to predict winegrape yield and quality. In Proceedings of the International Conference of Agricultural Engineering, Zurich, Switzerland, 6–10 July 2014.
57. Easlon, H.M.; Bloom, A. Easy Leaf Area: Automated Digital Image Analysis for Rapid and Accurate Measurement of Leaf Area. *Apps Plant Sci.* **2014**, *2*, 1400033. [CrossRef] [PubMed]
58. Lobell, D.B.; Asner, G.P. Moisture Effects on Soil Reflectance. *Soil Sci. Soc. Am. J.* **2002**, *66*, 6. [CrossRef]
59. Yan, L.; Roy, D.P.; Li, Z.; Zhang, H.K.; Huang, H. Sentinel-2A multi-temporal misregistration characterization and an orbit-based sub-pixel registration methodology. *Remote Sens. Environ.* **2018**, *215*, 495–506. [CrossRef]

6

Systematic Mapping Study on Remote Sensing in Agriculture

José Alberto García-Berná [1], Sofia Ouhbi [2], Brahim Benmouna [1], Ginés García-Mateos [1,*], José Luis Fernández-Alemán [1] and José Miguel Molina-Martínez [3]

1 Department of Computer Science and Systems, University of Murcia, 30100 Murcia, Spain; josealberto.garcia1@um.es (J.A.G.-B.); brahim.benmouna@um.es (B.B.); aleman@um.es (J.L.F.-A.)
2 Department of Computer Science and Software Engineering, CIT, United Arab Emirates University, Al Ain 15551, UAE; sofia.ouhbi@uaeu.ac.ae
3 Food Engineering and Agricultural Equipment Department, Technical University of Cartagena, 30203 Cartagena, Spain; josem.molina@upct.es
* Correspondence: ginesgm@um.es

Abstract: The area of remote sensing techniques in agriculture has reached a significant degree of development and maturity, with numerous journals, conferences, and organizations specialized in it. Moreover, many review papers are available in the literature. The present work describes a literature review that adopts the form of a systematic mapping study, following a formal methodology. Eight mapping questions were defined, analyzing the main types of research, techniques, platforms, topics, and spectral information. A predefined search string was applied in the Scopus database, obtaining 1590 candidate papers. Afterwards, the most relevant 106 papers were selected, considering those with more than six citations per year. These are analyzed in more detail, answering the mapping questions for each paper. In this way, the current trends and new opportunities are discovered. As a result, increasing interest in the area has been observed since 2000; the most frequently addressed problems are those related to parameter estimation, growth vigor, and water usage, using classification techniques, that are mostly applied on RGB and hyperspectral images, captured from drones and satellites. A general recommendation that emerges from this study is to build on existing resources, such as agricultural image datasets, public satellite imagery, and deep learning toolkits.

Keywords: remote images; systematic mapping study; agriculture; applications

1. Introduction

Nowadays, precision agriculture (PA) has become an essential component for modern agricultural businesses and production management. Thanks to the technological improvements, it has played an increasingly important role in agricultural production around the world by helping farmers in increasing crop yield, reducing costs and environmental impacts, and managing their land more efficiently. PA involves the integration of different areas such as geographic information systems (GIS), global positioning systems (GPS), and remote sensing (RS) technology [1]; decision support systems could also be added to this equation.

In general, GIS are computer systems that are used for storing, managing, analyzing, and displaying geospatial data [2]. In agriculture, they enable farmers and managers to handle data obtained from satellites and other types of sensors through georeferenced databases. Several research works have addressed PA problems from the perspective of GIS to reduce the environmental impact of agriculture, in applications such as disaster risk reduction [3], land use change monitoring and modeling [4], climate change detection [5], subsurface tile drains area detection [6], and identification of wetland areas [7].

GPS is closely related to GIS and RS, being used as input for both systems, i.e., GPS offers precise positioning of geospatial data and the collection of data in the field [8]. Some works have addresses PA problems from this point of view, such as solving weed management issues [9–11], but usually in conjunction with other technologies.

RS has been considered by some authors as the most cost-efficient technique for monitoring and analyzing large areas in the agricultural domain [12]. It can be considered as a part of the Earth observation domain, used for capturing and analyzing information about crops and soil features acquired from sensors mounted on different types of platforms such as satellites, aircraft, or ground-based equipment. Thus, the technologies related with remote sensing in agriculture (RSA) include hardware design of the cameras and capturing vehicles, communication technologies used to transfer the images [13], and the necessary tools of image processing, computer vision and machine learning to analyze the images and additional information available [14]. The obtained information is later used in agricultural decision support systems [15]. As the number of tasks and activities involved in the efficient use of these technologies can be overwhelming (from study design to quality assurance), efforts have been done to harmonize these tasks and provide general recommendations [16].

The existing applications of RSA include almost all tasks of the cultivation process [17]: estimation of cropland parameters; drought stress and use of water resources; pathogen and disease detection; weed detection; monitoring nutrient status and growth vigor of the plants; and yield estimation. These applications are affected by a set of parameters specific for each sensor type [12]:

- Type of platform where the sensor is mounted: in-field systems, ground vehicles, aircraft or satellites.
- Wavelengths of the electromagnetic spectrum that are captured; most frequently, they include visible, infrared, ultraviolet and microwaves.
- Number and width of the spectral bands captured: panchromatic (a single wide band), multispectral (a small number of broad bands), and hyperspectral (many narrow bands).
- Spatial resolution, measured in meters per pixel, which can be roughly classified in high (less than 1 mm for in-field cameras), medium, and low (around 1 km in some satellites and bands).
- Temporal resolution, i.e., capture frequency of the system, which can range from real-time (in-field cameras) to several weeks (in some satellites).
- Radiometric resolution, i.e., the number of bits per pixel and band (typically 8, 12, or 16 bits), and the source of energy (passive sensors or active sensors).

Airborne remote sensing is mostly realized with unmanned aerial vehicles (UAV), but also with manned aircraft. UAVs are generally low-cost, light, and low-speed planes that are well suited for remote sensing data collection [18]. UAVs are normally equipped with sensors, and have been used in many problems such as mapping weeds [19,20], monitoring the vegetation growth and yield estimation [21–23], managing water and irrigation [24,25], detecting diseases and monitoring plant health [26,27], crop spraying [28,29], and field phenotyping of the temperature of the canopy using thermal images [30]. In any case, the hardware capabilities depend on parameters such as weight, payload, range of flight, configuration, and cost [31]. Different kinds of UAVs have been used in last decades in PA applications, such as fixed wing drones [32], single rotors [33], quad rotors (or quadcopters) [34], hexa rotors (or hexacopters) [35], and octo rotors (or octocopters) [36]. Normally, the larger number of rotors involves better maneuverability, greater payload, and ease of use. However, they require a greater use of energy and, therefore, have less autonomy.

An alternative to drones is the use of satellites, which have gained popularity in RSA research thanks to projects such as MODIS [37], Landsat series [38], Gaofen-1/2 [39], ATLAS [40], and many others. Although they are considerably much more expensive, many of them are controlled by public or private institutions that provide free access to the obtained images. These systems have a large coverage, lower spatial and temporal resolution than UAVs, and normally each satellite includes many different capturing devices. Additionally, ground-based sensing devices have also been used in PA for certain applications and research studies [13,41,42], for example, mounted on mobile vehicles or static sensor networks.

The sensors most frequently embedded on RSA platforms are RGB cameras, multispectral and hyperspectral cameras, thermal cameras, Light Detection and Ranging (LiDAR), and Synthetic Aperture Radar (SAR) [43]. Multispectral cameras are useful to estimate parameters as chlorophyll content, leaf area index (LAI), leaf water content, and normalized difference vegetation index (NDVI), while thermal images are applied to study water stress in the plants. RGB cameras can be combined with LiDAR to obtain digital terrain/surface models (DTM/DSM) of the area being monitored [44]. SAR systems have the advantage that their quality is independent of light and weather conditions. The most basic applications of agricultural SAR remote sensing are crop identification and mapping [45], crop-type classification [46,47], and crop recognition [48].

In addition to hardware, the other major component of remote sensing systems is software. Image processing and computer vision have proven to be effective tools for analysis in PA applications, including photogrammetry techniques, vegetation indices, and machine learning as the most common areas in RSA. Photogrammetry consists of computing 3-dimensional digital terrain models [49–51] and orthophotos [52,53]. Other systems are based on vegetation indices, that are then used to classify the land cover or the crop type, such as obtaining the crop growing pattern [54,55], managing environmental issues [56,57], and estimating crop yield [55,58].

However, the area in which most research can be classified is machine learning. It is extensively used in PA in order to provide smart solutions for the tasks of interest. Unsupervised and supervised methods have been successfully applied, such as classification, clustering, and regression models [59]. For example, in [60], regression models are used to estimate vegetation indices, and in [24], it is used to predict crop water status. Classification techniques are the other major category, which have been used for weed detection [61,62], identification and quantification of the leaf area [63], disease detection [26,64], and identification of rapeseed [65]. Some of the most common classification techniques are listed as follows.

- Artificial Neural Networks (ANN). ANN models have shown great potential in various RS applications in PA. For example, Hassan-Esfahani et al. [66] used an ANN to compute surface soil moisture. Poblete et al. [67] developed an ANN system to predict vine water status. In [68], the authors used ANNs to separate maize plants from weeds.
- Random Forest algorithm (RF). It is an ensemble classification model that consists of a set of randomized decision trees. It has been used in [35] to estimate biomass and the amount of nitrogen in plants. In [36], RFs are applied to estimate the content of nitrogen in the leaf of the plants.
- Support vector machines (SVM), naïve Bayes classifier, and k-mean clustering. These methods have also been applied in different areas of agricultural machine learning systems. Sannakki et al. [69] proposed a SVM classifier to detect diseases in pomegranate leaves at an early stage. Mokhtar et al. [70] presented a SVM-based technique for detecting diseases in tomato leaves. k-Nearest neighbors algorithm (kNN) was used in [71] to classify large agricultural land cover types. A system to discriminate weeds from crops using naïve Bayesian classifiers is presented in [72]. Moreover, in [73], Mondal et al. proposed a naïve Bayes classifier to detect gourd leaf diseases using color and texture features.
- Deep Learning (DL). The use of DL in agriculture is a recent and promising alternative to traditional methods [74,75]. It has been used in several applications in the domain of PA. For example, a fully convolutional neural network for mapping weed is used in [76]. Castro et al. [77] used a CNN model for the classification of crops using multitemporal optical and SAR data. Mortensen et al. [78] addressed the problem of segmenting mixed crops applying CNN methods. dos Santos Ferreira et al. [20] proposed a deep learning-based CNN algorithm to classify weeds from grass and broadleaf. Moreover, Kussul et al. [79] dealt with the crop mapping problem using a multi-level DL network.

This paper describes a systematic mapping study in the area of remote sensing in agriculture. Many recent and interesting review papers can be found in the literature regarding RSA research [43,45,74,80–83]. However, the present paper is the first to adopt the form of a systematic mapping. These studies are characterized by following a formalized methodology, whose objective is to find the current trends in techniques, problems, applications, publication channels, etc., obtaining recommendations for researchers and practitioners in this area. The rest of the paper is organized as follows. In Section 2, the steps of the methodology used are explained. Then, Section 3 presents the quantitative results of the study. The main findings, suggestions, and limitations are discussed in Section 4. Finally, the conclusions and future perspectives are drawn in Section 5.

2. Research Methodology

The bibliographic review carried out in the present work has taken the form of a systematic mapping study [84], with the purpose of providing an overview of the field of remote sensing in agriculture (RSA) to identify the quantity and channels of the papers published, the type of research that is currently being done, and the results available in the literature. Systematic mapping studies follow a well-established methodology [85], consisting of the following main steps; (i) study planning by determining the mapping questions, the source databases, and the search string; (ii) searching for the relevant papers in the predefined databases; (iii) defining a classification scheme of the papers; (iv) mapping the selected papers; and (v) extracting the main findings, implications, and limitations of the study. All these steps are described in the following sections.

2.1. Formulation of the Mapping Questions

After analyzing the most interesting aspects to extract from the papers, a total of eight mapping questions (MQs) were defined. These questions help to perform the subsequent search and analysis processes. The first four questions (MQ1–4) extract general information about the publication channels, the frequency of the approaches, the research types and the empirical validation of the RSA studies. The rest of the questions (MQ5–8) are related to more specific aspects of RSA, such as the techniques used, the devices for image capturing, the problems addressed, and the type of spectral information considered. All these MQs were formulated to cover the key factors that comprise the field of RSA. Table 1 presents these MQs with the rationale that motivate their importance.

Table 1. Mapping questions defined in the present review.

ID	Mapping Question	Rationale
MQ1	What publication channels are the main targets for RSA?	Identifying where RSA research can be found, and the most adequate publication channels for future works
MQ2	How has the frequency of approaches related to RSA changed over time?	Identifying publication trends over time related to RSA
MQ3	What are the main research types of RSA studies?	Exploring different types of research existing in the literature about RSA
MQ4	Are RSA studies empirically validated?	Discovering if research works on RSA has been validated with empirical methods
MQ5	What types of techniques were reported in RSA research?	Detecting the most important types of computer vision and machine learning techniques reported in the existing RSA literature
MQ6	What are the platforms used to capture the images for RSA?	Exposing the main types of devices employed to obtain the images in RSA
MQ7	What are the research topics addressed by RSA?	Studying what are the most prominent topics currently tackled in RSA research
MQ8	What are the different types of spectral information used?	Analyzing what types of images are the most frequently used in RSA research

2.2. *Definition of the Search Strategy*

After analysing different bibliographic databases, the search was done in Scopus (https://www.scopus.com/). This database indexes an important number of journals and conferences with a certain level of rigor [86], many of them coinciding with those of the other databases. The search was done in December 2019.

Another key factor in the bibliographic search is the definition of the search string used in the database. Scopus allows to define a complex search string with Boolean operators and wildcards. This string takes a form similar to a sentence with subject–adjective–verb–complement, where all the main possibilities are considered for each component. Thus, it was formulated as follows.

TITLE ((sensing **OR** *sensor** **OR** *imaging* **OR** *imagery* **OR** *image**) **AND** *(remote* **OR** *satellit** **OR** *SAR* **OR** *UAV* **OR** *airborne* **OR** *hyperspectral* **OR** *thermal* **OR** *infrared* **OR** *"hyperspectral")* **AND** *(detect** **OR** *management* **OR** *monitor** **OR** *estimat** **OR** *classification* **OR** *recognition* **OR** *diagnosis* **OR** *identif*)* **AND** *(agricultur** **OR** *plant* **OR** *crop** **OR** *cultivar** **OR** *plague* **OR** *canopy* **OR** *leaves* **OR** *infestation))* .

Observe that this string is applied on the title of the paper rather than the abstract or the content, as this is more specific and produces less false-positives. The combination of the four groups of words with an *AND* requires that at least one word of each group appears in the title. The first group corresponds to the *subject*, including terms related to the scope of images and capture devices: sensing, sensor, imaging, imagery, and image. The second group is the *adjective*, and it is used to refine the previous set introducing the property of being remote. It contains the words remote, satellite, SAR, UAV, airborne, hyperspectral, thermal, infrared, and hyperspectral. Although the last four terms do not necessarily involve "remote", these types of spectral information are more common in remote sensing applications. The third group is the *verb*, so the terms correspond to the actions being performed with the images. These terms are the main tasks of RSA applications: detection, management, monitoring, estimation, classification, recognition, diagnosis, and identification. Finally, the fourth group is the *complement*, which indicates some property of the task. This allows to remove research works in remote sensing that are not related to agriculture. The terms in this group are agriculture, plant, crop, cultivar, plague, canopy, leaves, and infestation. The final search string was refined in a trial-and-error process, observing that the papers found are in the area of RSA, and no relevant papers are lost. For example, the terms "plague" and "infestation" were included after observing that some papers did not include other terms in the complement.

2.3. *Study Selection*

The following task after defining the search string is to establish the inclusion and exclusion criteria. Inclusion criteria are the conditions that should be met by the selected papers, whereas the exclusion criteria indicate what candidate papers should be removed from the review. Inclusion criteria (IC) were limited to the search string (IC1), and the papers should be written in English (IC2). On the other hand, the papers that meet one or more of these exclusion criteria (EC) were discarded:

- **EC1**. Editorial papers, papers about colloquium and international meetings, and summer school papers.
- **EC2**. Papers that have a citation ratio of less than 6 citations per year.

In EC1, editorial papers, papers about colloquium, international meetings, and summer school papers were not considered as the material provided in these manuscripts may not be of sufficient relevance and novelty. EC2 was based on the idea of selecting the most highly-cited publications. In addition, the impact of the literature on RSA was also kept in mind. For this purpose, a citation ratio with the number of citations divided by the numbers of years was employed. This ratio defines an objective criterion that allows to order the papers according to their relevance in the literature, taking into account that recent papers can have less citations than older papers.

The PRISMA methodology [87] was followed in the selection of the papers, providing a formal protocol for the accuracy and impartiality in the search of the titles in Scopus. Figure 1 shows the steps occurred during the study selection.

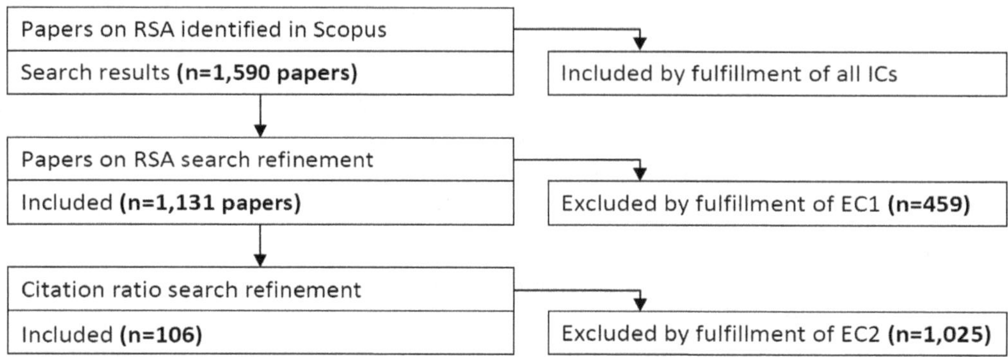

Figure 1. PRISMA flow chart resulting in the present mapping study.

2.4. Data Extraction Strategy

The data extraction strategy refers to the way in which each question should be answered for each selected paper. This step requires a previous classification of the possible answers to each MQ and some indications to extract this information from the papers. The extraction strategy developed for the present study was as follows.

- **MQ1**. To answer this question, the publication source and channel for each paper should be identified. The channel can be classified in journals, books, and conferences. The source refers to the name of the corresponding journal, book, or conference.
- **MQ2**. In order to draw conclusions about the publication trends, articles should be classified per publication year. Therefore, this question extracts the year of each paper.
- **MQ3**. Research works can be of different types, for example, a paper can propose new methods and techniques, it can evaluate existing solutions in a new application, or it can describe a specific experience that could be useful for other researchers. According to the authors of [88], the types of research can be classified into the following categories.
    - Evaluation Research. In this case, the research consists of the evaluation of an approach in RSA. This class also includes identifying new problems in RSA.
    - Solution Proposal. Research works which involve proposing a new solution for an existing problem in RSA. The proposed approach must be new, or it can be relevant modification of some existing method. An extensive experimentation is not required.
    - Experience Papers. These articles describe the personal experience of the authors. The paper explains what has been done and how it has been done in practice, and the obtained results.
    - Other. Other types of research can include, for example, reviews, opinion papers, etc.

  It is also possible to find some papers that can be classified into different categories, for example, an article can propose a new technique and perform an extensive experimental validation.
- **MQ4**. Most of the research works are expected to have an empirical validation of the theoretical advances and proposals. These experimentation can be done in different ways. According to the authors of [89], the empirical research types can be classified into the following.
    - Case study. It is an empirical inquiry that investigates a phenomenon in its real-life context. One or many case studies can be described.
    - Survey. A survey is a method for collecting quantitative information related to aspects in RSA research, for example, through a questionnaire.
    - Experiment. This case refers to an empirical method applied under controlled conditions to observe its effects and the results of certain processes or treatments.

  - Data-based experiments. This is a different case from the previous category, as the research does not involve new experiments, but the data available from previous experiments is used. It can be either a public or private database.
  - Other. Other types can include meta-analysis, history-based evaluation, etc. It is also possible that some papers do not report any empirical validation.

- **MQ5**. Another interesting aspect to analyze is the type of techniques that are used in the papers, that is, the computer vision or machine learning tasks that are addressed in [14]. Many different classifications can be found in the literature. Following the work in [82], in the present review, the techniques are classified as follows.

  - Image preprocessing and segmentation. Although they are different problems, the two are closely related since the input are images and the output are also images. Besides, they are typically the first steps of many computer vision systems. Image preprocessing includes the techniques whose purpose is to improve the quality of the images captured [90], e.g., to remove noise, enhance image contrast, correct geometric deformations, or remove artifacts. Image segmentation consists in separating image regions in different categories [78], e.g., separating plants and background, or detecting the regions of a crop of interest. Segmentation can be considered a result by itself, or it can be the input for further processing.
  - Feature extraction. Most frequently, after segmenting the regions of interest in the images, a set of features are extracted from them, although it can also be applied to the entire image. Feature extractors are a set of techniques to obtain relevant and high-level data from the images. The most usual types of features in RSA are color, texture, shape, and spectral features [91]. In many cases, the features are not explicitly predefined by the human experts, but they are given by a machine learning algorithm [75]. The extracted features can be used later for computing parameters of interest from the images, such as the water stress of the plants, or the crop yield.
  - Similarity measures and maximum likelihood. Most empirical research has been dedicated to find effective similarity measures on the extracted features. Then, the similarity values can be used in a maximum likelihood approach [92]. This can be used, for example, to predict the evolution of a certain crop from other previously observed cases with similar characteristics.
  - Classification systems. Given an image, or an image region, classification consists of determining the most likely class among a predefined set of classes [32,39,40]. For example, it can be used to classify a segmented region of plants in crop or weed, it can be used to classify a plot in dry land or irrigated, or to classify a fruit in unripe/ripe/overripe. Common classifiers used in RSA include support vector machines (SVM) [69,70], decision trees (DT), and artificial neural networks (ANN) [52], although they can also be used in the other problems.
  - Recognition systems. The purpose of a recognition system is to find the specific identity of the object of the given class. For example, a segmentation step can be used to separate an image in plant/background; then, a classifier is applied to find if a plant region is a tree, a grass or a weed; finally, the recognition step would determine the specific type of tree, grass, or weed [77]. Obviously, a recognizer should not be prepared to deal with all the instances from all the classes, but only for those species of interest that have been trained.
  - Other machine learning algorithms. In this category we include additional applications of machine learning algorithms [14]. These can include regression algorithms (e.g., for estimating the crop evapotranspiration), decision support systems (e.g., for deciding the fertirrigation schedules), or methods to automatize different processes (e.g., harvesting or fumigation machines).

A complete computer vision system in agriculture should include many (if not all) of these techniques. Therefore, the papers have been classified according to the area where the most important contributions are done, although they could be classified into different categories.

- **MQ6**. The platforms typically used to capture the images in agriculture are highly diversified [93]. They can be classified according to different criteria, such as the type of information captured (spectral or depth maps), the spatial, spectral and photometric resolution, or the type of cameras. However, they are most commonly classified considering the type of vehicles or devices in which they are mounted [80]. The main categories are listed as follows.
  - Satellite imagery. They are characterized for offering images of very large areas, with lower temporal resolution compared to the other platforms [94,95]. The high cost of this kind of device places them beyond the reach of farmers, being controlled by governmental or international institutions. However, in many cases, these organizations provide free access to the obtained satellite images for research purposes. Another characteristic of satellites is that most of them are equipped with multispectral or hyperspectral cameras [96].
  - Drones, UAVs, and manned aircraft. The use of these types of devices in agriculture has experienced a huge growth in the last decade [18]. In general, an aircraft is any vehicle which is able to fly. When they include a human pilot, they are referred as manned aircraft, while the term Unmanned Aerial Vehicle (UAV) is used when the vehicle can fly remotely (controlled by a human) or autonomously (without human control) [81]. The term drone is normally used as a synonym of UAV; however, it can also be used for other types of aquatic or land vehicles. Thus, all UAVs are drones, but not all drones are UAVs. The use of the term Unmanned Aerial System (UAS) is also frequent [97], which refers not only to the flying vehicle, but also to the ground control, communication units, support systems, etc. Compared to manned aircraft, UAVs are normally less expensive, less invasive, and safer tools, so they can be used in sensitive areas such as the polar regions [98]. The most common type of operation is the so-called visual line of sight (VLOS), where the pilot can directly see the UAV at all times; however, some systems are prepared to operate beyond visual line of sight (BVLOS) [99] allowing to cover larger extensions.
  - Other types of vehicles. In many cases, remote capture systems can be incorporated into the existing farm machinery [41], such as trucks, tractors, combine harvesters, etc. In this case, the images are typically used in real-time during the agricultural processes of plowing, irrigation, planting, weeding, or harvesting, more than for out-of-line analysis. We also include in this category other types of autonomous vehicles that can not be considered as UAVs, such as aerial balloons.
  - In-field installations. Remote image capture systems in agriculture also include field installations of fixed cameras. They can be considered *remote* in the sense that they are used and controlled remotely, not in the capture distance. They are usually based on inexpensive cameras communicating wirelessly, which are able to perform a real-time monitoring of the crops [13]. In counterpart, they have lower resolution than the other modalities, they only capture a small portion of the plots, and normally only RGB images are used. In some cases, they can be integrated into a wider Wireless Sensor Network (WSN) installed in the farms; these include other types of sensors (thermometers, barometers, lysimeters, etc.) that are out of the scope of the present review.

- **MQ7**. To date, a large number of different problems have been addressed with the RSA techniques listed above [83]. However, this fact does not limit the possibility that other new topics and areas of application will appear in the future. According to the recent reviews [17,100], the main applications of interest can be classified as follows.
  - Agricultural parameters estimation. In this case, remote images are used to estimate parameters of large plots that would be difficult or expensive to be obtained using in-field methods. These parameters of interest can include crops or cropland parameters [45], for example, the height of the plants, the leaf area index (LAI), the percentage of green cover (PGC), the total biomass, the depth of the roots, or the surface roughness can be estimated.

  - Drought stress, irrigation, and water productivity. Due to the great importance of water in agriculture, this category includes all applications related to water and irrigation (although some of them could also be understood as parameter estimation) [91,101]. Optimization of water resources is an essential aspect of global sustainability due to the great water shortage in many regions. A key parameter is water balance, which measures the water incomes and outcomes, including the crop evapotranspiration (ET).
  - Nutrient status. Nutrient efficiency and avoiding nutrient losses are other topics that have received much attention in the literature of RSA. The proper use of nutrients can also be aimed at reducing pollution of the environment. It is particularly relevant the use of nitrogen (N), which has proved to affect the leaf and plant reflectance signatures [17].
  - Growth vigor. Monitoring plant vigor during the different stages of growth is another of the principal applications of RSA [23]. It can be based on different parameters such as the growth of the plant height, the total biomass, and the PGC. We distinguish this category from the parameter estimation in that these works perform a temporal analysis of the images.
  - Detection of pathogens, diseases, and insect pests. Early detection of these problems can help reduce losses. Precision agriculture systems are able to reduce pesticide use by performing site-specific spraying [102]. Thus, the effectiveness of these systems is related with the obtained quality, yield and sustainability of the crops.
  - Weed detection. The appearance of weeds is another problem that can appear during the cultivation process, leading to a reduction in the water and nutrients available for the crops of interest [103]. As weeds are also plants, the distinction between crops and weeds must be done using color, texture, shape, or spectral features.
  - Yield prediction. Regarding the last stages of the cultivation process, remote sensing images have been used to predict the yield before the actual harvesting [58]. These systems are usually based on regression models using parameters extracted from the images, although the most precise methods use accumulated temporal information and crop growth models.
  - Automatic crop harvesting. Intelligent harvesting machinery and picking robots have emerged in the last years as a feasible alternative to traditional harvesting methods [80], although the first experimental systems for automatic harvesting using machine vision date back to the 1980s.

- **MQ8**. Computer vision systems in agriculture are not exclusively based on the use of visible light; a wide range of the electromagnetic spectrum has shown to be effective in different RSA applications, normally is frequencies lower than the visible wavelengths. Several reviews have analyzed the suitability of spectral information in different RSA problems [17,80,104]. The main types can be classified as follows.

  - RGB (visible spectrum). The visible spectrum corresponds to the wavelengths between 380 and 740 nm, which are visible by the human eye [105]. RGB cameras do not capture a complete spectrum of these wavelengths, but only three bands corresponding to red, green, and blue color. The main advantage of this category is the high availability, high spatial resolution, and low cost of the cameras, with respect to the other types of sensors. For these reasons, it is the predominant class in computer vision in general.
  - Red edge spectrum. This class corresponds to a small part of the visible spectrum, located at the end of the lowest frequencies, approximately from 670 to 740 nm. It is particularly important in agriculture [104], as the chlorophyll contained in vegetation reflects most of these wavelengths, while it absorbs a great part of the rest of the visible spectrum. Therefore, several vegetation indices have been defined based on the relationship between the reflection of red edge and red.
  - Near-infrared (NIR) and Vis-NIR. NIR includes the part of the infrared spectrum nearest to the visible region, approximately from 740 to 1500 nm. This class is also characterized by a high reflectance by the plants. The normalized difference vegetation index (NDVI) [23] is based on NIR and red bands, and is a very common parameter to study the amount

and healthiness of vegetation. Consequently, most works include NIR and visible bands, being a typical range from 400 nm to 1500 nm; this is usually called visible-NIR or Vis-NIR.

- Short-wave infrared. The term infrared refers to a broad slice of the electromagnetic spectrum ranging from 740 nm to 1 mm [93]. It is subdivided in near, short, mid, high, and far infrared, from lowest to highest wavelength. Short-wave infrared is located approximately from 1.5 to 3 µm. This range is characterized by a high absorption from the water, so it is specially interesting for moisture analysis.
- Long-wave infrared. This range corresponds to 8–15 µm. It is also called thermal infrared [106], as it contains the wavelengths of the thermal emission of the objects. It is widely used in studies about soil moisture, crop evapotranspiration and water balance, which can be estimated from the relative temperatures [107].
- Synthetic aperture radar (SAR). Unlike the previous passive sensing methods, SAR is an active sensing technique [45]. This means that the capture device emits some kind of radiation and receives the echo; normally, microwave radiations in different bands are used. This type of radar is called *synthetic aperture* because it takes advantage of the motion of the satellite or aircraft to simulate a large antenna, thus providing higher resolution images. Polarization properties of the waves are also used to provide more information of the land. The captured images are unaffected by the clouds, and it can be used in night-time operation. Although passive microwave capture is also possible, it is less used in RSA.
- Light Detection and Ranging (LiDAR). This method also belongs to the category of active remote sensing, usually mounted on satellites and aircraft. In this case, the radiation is emitted by a laser beam, and the echo time is measured to calculate the distance to the objective. Unlike the other methods, which obtain radiation/absorption images in different wavelengths, the data obtained are depth images [108]. This type of images are also called digital elevation models (DEM). They can be used, for example, to estimate the height and volume of the plants.

In addition, two other related terms are multispectral and hyperspectral images. These categories do not correspond to specific wavelengths, but to the number of channels that are captured.

- Multispectral images (broad band). When the number of channels captured for each pixel is small, usually between three and 10 channels, we call them multispectral images [18]. Each channel corresponds to a broad range of the spectrum, which can have a descriptive name. For example, an RGB image can be understood as a multispectral image with three channels. In the review, this category has been used only when the paper cannot be classified in the previous classes. For example, satellite Lansat-8 (https://www.usgs.gov/land-resources/nli/landsat/landsat-8) is able to capture 11 different bands (although not all of them with the same spatial resolution).
- Hyperspectral images (narrow band). These images are characterized by having a large number of channels, which can be some hundreds or even thousands [18]. For example, Hyperion imaging spectrometer is able to capture 224 bands with 10 nm wavelength intervals [109]. This high number of channels allows obtaining the spectral signature of the observed objects, in order to analyze their chemical composition. However, most computer vision techniques are designed for images with few channels. Specific methods should be applied when the spatial resolution of the images is small but the number of channels is very large.

2.5. Synthesis Procedure

After defining the mapping questions of interest, selecting the candidate papers, and performing the data extraction, the last step of the systematic mapping study is to synthesize the results. For each MQ, the papers are classified into the corresponding category (or categories, if more than one is applicable), and the results are presented in charts. Afterwards, these results are discussed using a variety of evaluation approaches. Finally, a narrative summary draws the main findings of the mapping study.

3. Results of the Systematic Mapping Study

As shown in Figure 1, 1590 candidate papers were first obtained by applying the search string in the Scopus database. From these, 1131 publications were selected after the application of exclusion criterion EC1. However, due to this large number, the more restrictive criterion EC2 was also applied; recall that this second criterion requires an average of 6 citations per year, so it is expected to extract the most relevant works. Finally, a total of 106 studies were selected and analysed to answer the MQs. The results obtained in the classification are presented in the following subsections.

3.1. MQ1. What Publication Channels Are the Main Targets for RSA?

This question refers both to the type of channel and the name of the publication. Figure 2 shows that almost all the selected papers were published in scientific journals, except for two conference papers and one book. The names of the journals with more than one publication are shown in Table 2. It is interesting to observe that all these journals are indexed in the Journal Citation Reports, being most of them in quartiles Q1 and Q2.

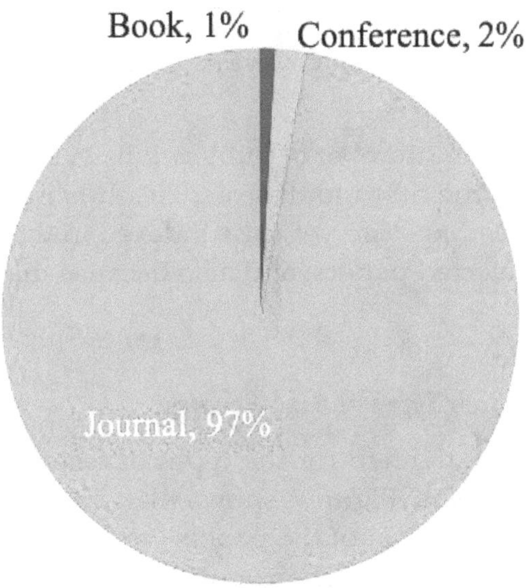

Figure 2. Publication channels of the selected studies.

Table 2. Publication sources with more than one selected paper.

Journal Name	Total
Remote Sensing	20
Remote Sensing of Environment	15
Journal of Experimental Botany	6
International Journal of Applied Earth Observation and Geoinformation	6
Computers and Electronics in Agriculture	5
Agricultural Water Management	4
Agricultural and Forest Meteorology	3
ISPRS Journal of Photogrammetry and Remote Sensing	3
Journal of Stored Products Research	2
Ecological Modelling	2
International Journal of Digital Earth	2

3.2. MQ2. How Has the Frequency of Approaches Related to RSA Changed over Time?

For this mapping question, it is interesting to consider both the set of 1131 candidate papers after applying EC1, and the final set of 106 papers after applying EC2. Figure 3 presents the number of articles published per year until 2019. This figure shows that there has been an important increase in the number of publications in RSA field in the last decade. Since 2000, this growth has followed a linear trend. Although the first papers date back to the 1970s, no paper meets the strict EC2 criterion until 1997; from 2002 onwards, there are always selected papers.

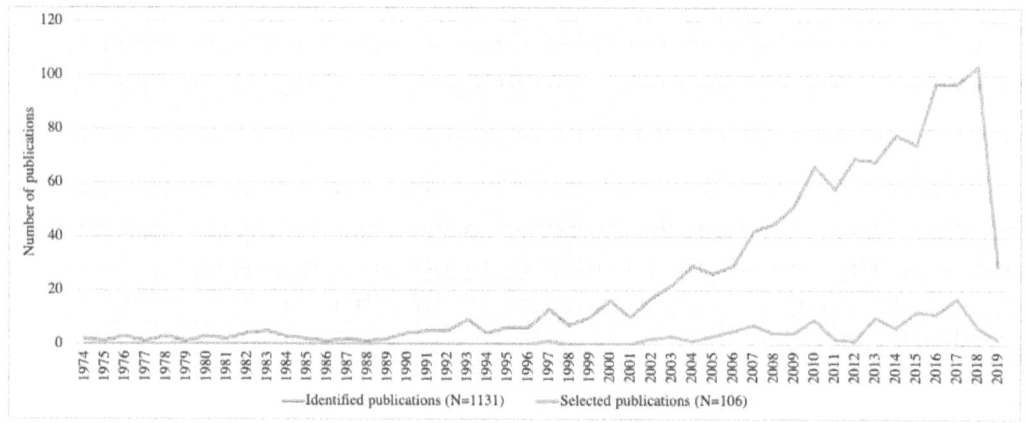

Figure 3. Publication trends throughout the years for the candidate and selected papers.

There is an evident decrease in the number of publications in 2019. However, this is a consequence that the study was carried out in the first months of 2020. It is possible that many publications at the end of 2019, particularly proceedings, are yet to be indexed in the database used. The same reason applies to the small number of selected papers, and also because they have not had time to receive a sufficient number of citations.

3.3. MQ3. *What Are the Main Research Types of RSA Studies?*

Four standard categories were defined for the types of research: evaluation research, solution proposal, experience papers, and others. Figure 4 shows that only three of these types were identified in the highly cited papers about RSA. Most of the papers were evaluation research (57%), and almost one-third of selected publications were solution proposals (28%). Reviews represented the remaining 15% of the selected papers.

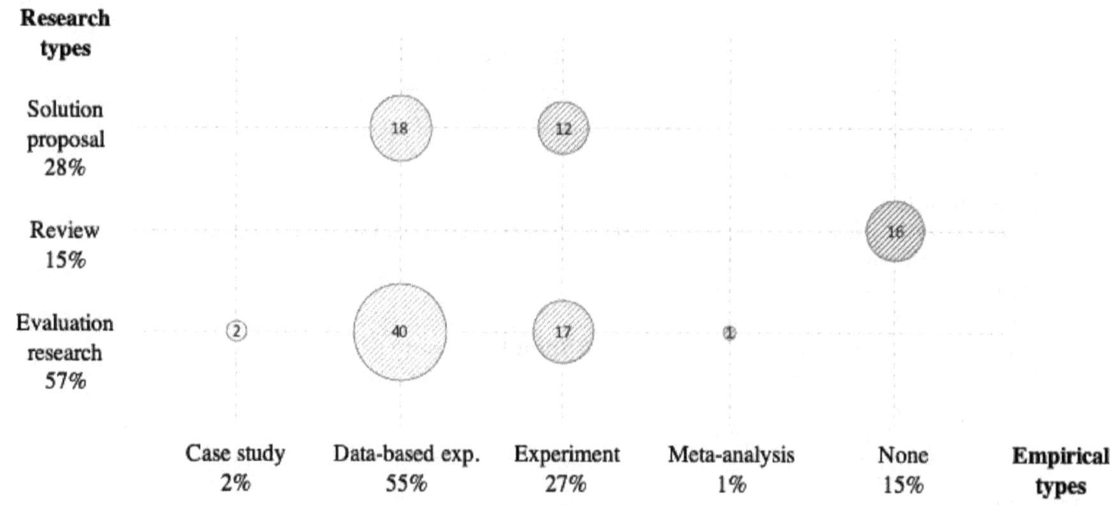

Figure 4. Research types and empirical validation types of the selected papers.

It can be surprising the large proportion of review papers found, which can be explained by the large number of citations that they receive.

3.4. MQ4. Are RSA Studies Empirically Validated?

This question is closely related with the previous one, as both give an overview of how research is done. For this reason, Figure 4 shows the relationship between research types and empirical validation. Except for the review papers (which do not require validation), all the selected works were empirically evaluated. Most of the papers were evaluated through experiments, particularly data-based experiments. One paper explicitly stated using meta-analysis approach in its evaluation. Moreover, only 2% of the selected papers conducted case studies. These results demonstrate the importance of creating complete, verified and public available remote image databases for RSA research.

3.5. MQ5. What Types of Techniques Were Reported in RSA Research?

The most frequent types of techniques identified in the selected papers are presented in Table 3. More than half of the papers (54/106) focused on classification systems. It has to be noted that the type does not necessarily correspond to the final result of a proposed system. For example, a classification system can be used to classify crops and weeds, and the final result is presented as a method for detecting weeds, or it can perform a binary classification plant/soil, in order to perform an estimation of the crop coefficient. Therefore, the content of the papers were analyzed in detail to extract their main contributions.

Table 3. Classification of the types of techniques used in the selected papers.

Techniques	**Ref.**	**Total**
Classification systems	[101,104,106,108,110–159]	54
Feature extraction	[41,91,107,108,111,155,160–169]	16
Similarity and maximum likelihood	[92,117,118,170–180]	14
Preprocessing and segmentation	[108,112–115,181–186]	11
Recognition systems	[96,187–196]	11
Other machine learning algorithms	[79,95,110,116,197–203]	11

The second most frequent computer vision task is feature extraction (16/106), which can be used for a subsequent classification, estimation, recognition, or monitoring process. The rest of categories, similarity measures and maximum likelihood, image processing and segmentation, recognition systems and other machine learning algorithms, present a very similar number of papers. Besides, a total of 10 papers were classified in more than one technique [108,110–118], considering their main contributions.

3.6. MQ6. What Are the Platforms Used to Capture the Images for RSA?

Figure 5 depicts the most frequent types of systems used in RSA to capture the images. In some cases, different capture devices are used, so the total number of systems is higher than the number of papers. Moreover, the capture process should not necessarily be done by the authors, as the research could be based on existing datasets.

Concerning the obtained results, it is remarkable the small number of research works that are based on in-field low cost cameras. Although most research has been done in this area, this may be due to the fact that, in some contexts, they are not considered to be included in the remote sensing category. On the other hand, the most frequent type of platform used in the research are satellites, followed by drones and manned aircraft, and finally other types of vehicles.

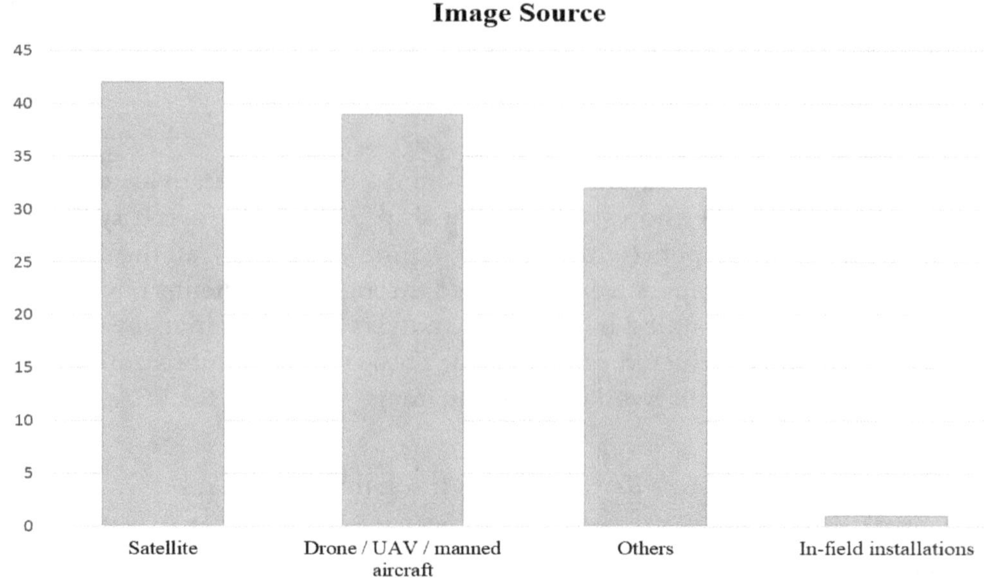

Figure 5. Frequency of the main types of capture platforms for the research in RSA.

3.7. MQ7. What Are the Research Topics by RSA?

Again, this mapping question may be subject to different interpretations, since a paper can address different topics or it can be in the borderline between some of them. Thus, a careful inspection of the literature was done to classify the papers in the most adequate category. As a result, the main problems detected are shown in Table 4. In the case of automatic crop harvesting, no papers were found in the present mapping study, possibly because they do not consider the use of remote images.

Table 4. Classification of the main types of research topics addressed in remote sensing in agriculture (RSA) papers.

Research Topic	Ref.	Total
Agricultural parameters extraction	[79,91,95,96,104,108,110,111,114,120,122,125,126,129,136,142,147, 148,150,153,154,157,161,175,178,181–184,188,191,196,201,202]	34
Growth vigor	[91,92,114–117,132,139,142,144,145,152,158,159,161,171,174,176, 177,184,189,190,194,195,197,199]	26
Drought stress, irrigation and water productivity	[91,101,106,107,119,124,127,128,130,131,135,138,151,155,162,163, 165–167,172,180]	21
Detection of pathogens, diseases and insect pests	[133,141,143,146,155,156,173,187,192,200,203]	11
Yield prediction	[91,134,140,151,152,159,179,186]	8
Weed detection	[113,137,149,169,170]	5
Nutrient status	[118,121,151,198]	4
Automatic crop harvesting	-	0

The results indicate that the different categories are not equally distributed. The topics that received more attention are the estimation of agricultural parameters, the analysis of crop vigor, and the problems related to water usage. These represent more than 77% of the papers. The works related to detection of pathogens, diseases, and insect pests are about 10% of the total. Moreover, at the other end, the classes with relatively fewer publications are yield prediction, weed detection, and the analysis of nutrient status. Therefore, these types of problems represent a good opportunity to advance in RSA research.

3.8. MQ8. What Are the Different Types of Spectral Information Used?

This last mapping question refers to the type of spectral information of the images used in the research. As described, a wide range of the electromagnetic spectrum has proved to be useful in RSA. Each class can be suitable for some specific problem, or it can be applied to different tasks. The classification of the papers is presented in Table 5. As the labels of multispectral and hyperspectral data are not incompatible with the rest of categories, some papers are classified into more than one class. In addition, many works use different types of images, so they can also be classified in different classes. For example, in [204] three types of images (RGB, multispectral, and thermal images) are compared for the problem of high-throughput plant phenotyping, using UAV and in-field cameras.

Table 5. Classification of types of spectral information used in RSA research.

Spectral Information	Ref.	Total
RGB (visible spectrum)	[79,91,95,104,110,111,113,115–118,121,122,125,126,130,132, 140,142,145,147,148,150–152,160–162,164,165,169,171,176, 178–180,182–184,186–188,190,192,193,199]	46
Hyperspectral (narrow band)	[91,96,104,108,112,118,123,127,134,137,141,143,144,146, 153,155–157,168,170,187,194–196,198,200,201,203]	28
Long-wave infrared (thermal)	[41,101,106,107,124,131,135,136,138,141,151,155,164–168, 172,177,189,193,197]	22
Near infrared (NIR)	[116–118,121,130,133,134,142,145,151,158,181,186,193]	14
Multispectral (broad band)	[114,129,139,146,149,154,159,163,171,173,197,200,201]	13
Red edge spectrum	[101,104,119,128,132,142,158,160,181,185,188,191]	12
Synthetic aperture radar (SAR)	[79,104,120,121,126,183,202]	7
Light detection and ranging (LiDAR)	[92,108,148,174]	4
Short-wave infrared	[117,121]	2

According to these results, standard RGB images (i.e., the visible spectrum) are clearly the most frequent type of images employed (46/106). Hyperspectral images are also found very frequently (28/106), in many cases mounted on UAVs or acquired from satellites. Apart from the visible spectrum, the following bands most used in the research are thermal and near infrared. In the opposite side, LiDAR and short wavelength infrared are to less commonly used.

4. Discussion

4.1. Main Findings and Implications for Researchers and Practitioners

The ultimate purpose of the study is to gain an in-depth understanding of the current state of research in remote sensing in agriculture, in order to give suggestions about future lines of research and finding new possibilities and application areas. This is achieved by an analysis and discussion of the results presented in the previous section. The major findings that can be extracted are the following.

- The main publication channels of the selected papers are journals, at a great distance from books and conferences. This is caused by the introduction of the strict exclusion criterion EC2 of six citations per year. Publications in journals are known to be cited more than those in conferences. Although conferences are important publication venues for computer science researchers [205], the research community tends to prefer publishing in journals due to the tenure and promotion guidelines in many institutions which only consider publications in high-impact factor journals [206]. However, the role of conferences as a means of spreading new ideas, showing ongoing research and connecting researchers should not be dismissed.
- The research field of RSA has gained an increasing interest since the beginning of the millennium. This can be explained by the new technologies that appeared in this period (cameras, satellites, and UAVs) in addition to the improvements in telecommunications and data transmission. The decrease of publications observed in 2019 is a collateral effect of the review procedure

and the minimum required number of citations. Thus, the increasing interest in RSA is expected to continue in the near future, favoring the appearance of new journals and conferences more specialized in the different areas of RSA.

- Most of the selected papers are evaluation research using data sets. Solution proposals represent almost one-third of the selected papers, which indicates that the field has reached a certain maturity and researchers are more interested in evaluating existing technologies rather than proposing new ones. This is also supported by the large number of reviews identified. On the other hand, this highlights the importance of creating public and comprehensive datasets where the results of different authors can be compared. It would be recommended that this effort be carried out by existing institutions and associations, rather than particular research groups. One example of these public resources is EuroSAT [40], a public dataset of 27,000 labeled and georeferenced images from Sentinel-2 satellite useful for the classification of land usages; the DeepSat Airborne Dataset [94], with 500,000 image patches in Vis-NIR range; or the Copernicus Programme, which offers satellite and in-situ images for land monitoring (https://land.copernicus.eu/).
- The majority of the selected empirically evaluated publications were conducted through experiments. Only two case studies have been identified in the selected papers, which means that it is difficult to perform this type of studies in RSA, as normally the research is done under uncontrolled settings. The small number of meta-analysis papers that were found indicates that there is an interesting opportunity to apply this type of statistical analysis whose purpose is to combine the results of multiple previous scientific works to assess these results and derive conclusions.
- The computer vision task most frequently found in the selected papers is classification: given a image patch or region, classify it into a predefined set of classes of interest. This is an expected observation, as it is one of the most studied machine learning problems, it has a simple a clear definition, and its results can be used in different applications. Decision trees, support vector machines, classical neural networks, k-nearest neighbors. and Bayes classifiers are among the most frequent techniques. However, deep learning methods are gaining popularity [20,74,76–79], proving to overcome other techniques in many domains. It is recommended that it should be applied when it is really of interest and not as a fad. An interesting alternative could be the use of ensemble classification systems that have not been widely used in RSA research. Other problems that have been identified in the selected papers include feature extraction, maximum likelihood, image preprocessing and segmentation, and recognition systems. As a general recommendation, we advice to make use of free tools and libraries for machine learning and computer vision, taking advantage of the great effort done by the free software community, for example using Python with tools such as the scientific programming environment Scikit-Learn (https://scikit-learn.org/stable/) and deep learning ecosystem PyTorch (https://pytorch.org/).
- The main types of platforms employed in RSA to capture the images are satellites, UAVs, and manned aircraft. However, in-field cameras and ground vehicles have not been widely used in RSA. The reason these images are used instead of the ground-based ones could be explained by the fact that they provide a broader view of the land. In addition, the resolution of the cameras permits to go from a global perspective to a more detailed view of a specific area. The fusion of satellite and UAV imagery [207] is an emerging field that would be very useful to harness the power of both capture systems. It is interesting to observe that many satellite imagery is freely available for research purposes, so this would be a convenient source for beginners. Among the most cited satellites in the selected papers, NASA's Landsat missions (https://landsat.gsfc.nasa.gov), ESA's Sentinel missions (https://sentinel.esa.int/), and ESA's Envisat satellite (https://www.esa.int/Applications/Observing_the_Earth/Envisat/Mission_overview) can be mentioned. Other satellites that are referred in several papers include Proba-1/2, Spot, QuickBird, Ikonos, TerraSAR-X, and Radarsat-2. In the domain of UAVs, some research teams are specifically dedicated to the hardware development of capture systems that can be applied for different task,

such as the system PhenoFly (https://kp.ethz.ch/infrastructure/uav-phenofly.html), which has been used in many publications.

- The main research topics addressed by the RSA community were growth vigor, cropland parameter extraction, and water usage. From a general perspective, these problems have the potential to impact on the sustainability of agriculture. As a matter of fact, the better usage of the water together with an adequate knowledge on the cropland, the better growth vigor, crop quality, and efficiency is achieved. Indeed, sustainability is one of the main goals of precision agriculture, in line with the United Nations' Sustainable Development Goals (SDG2: End hunger, achieve food security and improved nutrition and promote sustainable agriculture. SDG14. Protect, restore and promote sustainable use of terrestrial ecosystems, sustainably manage forests, combat desertification, and halt and reverse land degradation and halt biodiversity loss.). Sustainability was not a topic specifically addressed in the present mapping study, although all the problems are in some way related to it. The detection of pathogens, diseases, and insect pests attempts to reduce the amount of pesticides and insecticides; weed detection allows the use of site-specific spraying of herbicides, and nutrient analysis is related to the optimal use of fertilizers. On the other hand, yield prediction and automatic crop harvesting seek to optimize the productivity of farming. It cannot be discarded that new problems and applications appear in the future with the advance in technology, such as those involved in a completely automated cropping cycle.
- Concerning the types of the images used, standard RGB images continue to be the most frequently used image type. This can be explained by the low cost and high availability of RGB cameras, and the fact that they are the main source for computer vision in general. In this sense, RSA research is commonly observed as a sub-domain of computer vision and image processing. New methodologies should be developed more specific to the agricultural domain, for example considering the spectral, temporal, and phenotypic dimensions. In many works, NIR channel is added to RGB tuples, forming a 4-valued tuple for each pixel. Visible spectrum allows to validate the results from simple inspection, but important information may be lost or not detected with this type of images. For example, the temperature of the objects cannot be measured with Vis-NIR images, but this could be useful to estimate the water status of the plants. In these applications, hyperspectral images and thermal infrared were found in a second level of usage, and on a third group near infrared, red edge spectrum and multispectral images. When the spectral bands of interest can be known a priori, multispectral images can be more interesting than hyperspectral, focusing on the wavelengths of interest. In this way, an interesting domain of research in RSA that deserves much more work is the determination of the optimal spectral bands for each problem [208]. Ideally, new cameras could be made that are specific to the selected and reduced number of wavelengths for each problem.

4.2. Limitations of the Mapping Study

The last step in the execution of a systematic mapping study is the evaluation of the limitations and weak points of the study itself. This analysis has to consider all the steps of the process. Several possible limitations have been recognized for the present study:

- Using only Scopus as a source of publications. Other relevant publications that are not indexed in Scopus could have impacted the final results. However, our focus was to provide an overview on the most highly cited papers. For this reason, we chose to focus on Scopus as it is one of the largest databases available. Moreover, it has one of the most complete searching methods allowing Boolean combinations and wildcards.
- Some missing terms in the search string might have impacted the results. In order to reduce this limitation, we formed the search string to include a broad range of terms of interest to this study.

Besides, the search string was refine in several trials, by observing some papers that were not initially included. Therefore, we consider this threat is low.

- Other classification criteria not present in this study might have provided interesting views on the selected papers. The eight mapping questions included provide interesting findings to researchers and practitioners, although other questions could be also useful.

- The exclusion criterion of six cites per year could have rejected some interesting papers. This criterion was added to select only the most relevant works in the literature. This has shown to be very restrictive both for recent papers (only two papers from 2019 received more than six citations) and for older publications (e.g., the papers before year 2000 require more than 120 citations). However, as the purpose of the mapping study is to analyze the trends of the most relevant works, this is not a threat to the validity of the study.

5. Conclusions

Precision agriculture is a very active area of research with a significant impact on the improvement of global sustainability and the optimization of natural resources. It is based on information and communication technologies to achieve its goals, being remote image capture systems one of the main branches. This includes the development of cameras and capture devices, the remote communications of the images, image processing and computer vision tasks, and machine learning methods to automate the farming decisions.

The present systematic mapping study has presented a quantitative and qualitative analysis of the state-of-the-art in this rapidly evolving area. Only since the year 2000, more than 1400 journal and conference papers were found, and this trend is expected to continue in the future. A selection of the 106 most highly cited papers has been done to obtain an in-depth view of the state of the research. The archetype paper is a journal manuscript describing a classification problem on a dataset of satellite or UAV imagery, using existing computer vision and machine learning techniques, possibly with minor adaptations, applied in problems of parameter estimation, growth vigor, and water usage. Standard RGB and hyperspectral images are the most frequently found, although many works use different modalities.

Current trends are towards the popularization of the use of UAVs and the increasing availability of satellite imagery. However, we believe that a solution should integrate in-field cameras and airborne images in order to achieve high spatial and temporal resolution to cover large areas and reduce operational costs. Deep neural networks are also a very marked tendency as they can obtain excellent results in tasks of classification, segmentation, feature extraction, recognition, and analysis of time series. Finally, the integration should also be referred to the development of more holistic approaches that consider all the aspects involved in the cultivation cycle, and not just the problems in isolation.

Author Contributions: Conceptualization, J.A.G.-B. and S.O.; methodology, J.A.G.-B. and S.O.; validation, B.B., G.G.-M., and J.M.M.-M.; formal analysis, J.A.G.-B., S.O., and B.B.; investigation, J.A.G.-B. and S.O.; data curation, J.A.G.-B., S.O., G.G.-M., and J.L.F.-A.; writing—original draft preparation, J.A.G.-B., S.O., B.B., and G.G.-M.; writing—review and editing, J.A.G.-B., S.O., B.B., G.G.-M., J.L.F.-A., and J.M.M.-M.; visualization, J.A.G.-B. and S.O.; supervision, G.G.-M., J.L.F.-A., and J.M.M.-M.; project administration, J.L.F.-A.; funding acquisition, G.G.-M., J.L.F.-A., and J.M.M.-M. All authors have read and agreed to the published version of the manuscript.

References

1. Brisco, B.; Brown, R.; Hirose, T.; McNairn, H.; Staenz, K. Precision agriculture and the role of remote sensing: A review. *Can. J. Remote Sens.* **1998**, *24*, 315–327. [CrossRef]

2. Chang, K.T. *Introduction to Geographic Information Systems*; McGraw-Hill Higher Education: Boston, FL, USA, 2006.

3. Der Sarkissian, R.; Zaninetti, J.M.; Abdallah, C. The use of geospatial information as support for Disaster Risk Reduction; contextualization to Baalbek-Hermel Governorate/Lebanon. *Appl. Geogr.* **2019**, *111*, 102075. [CrossRef]

4. Chaikaew, P. Land Use Change Monitoring and Modelling using GIS and Remote Sensing Data for Watershed Scale in Thailand. In *Land Use-Assessing the Past, Envisioning the Future*; IntechOpen: London, UK, 2019.

5. Bai, Y.; Kaneko, I.; Kobayashi, H.; Kurihara, K.; Takayabu, I.; Sasaki, H.; Murata, A. A Geographic Information System (GIS)-based approach to adaptation to regional climate change: A case study of Okutama-machi, Tokyo, Japan. *Mitig. Adapt. Strateg. Glob. Chang.* **2014**, *19*, 589–614. [CrossRef]

6. Gökkaya, K.; Budhathoki, M.; Christopher, S.F.; Hanrahan, B.R.; Tank, J.L. Subsurface tile drained area detection using GIS and remote sensing in an agricultural watershed. *Ecol. Eng.* **2017**, *108*, 370–379. [CrossRef]

7. Wu, Q. GIS and remote sensing applications in wetland mapping and monitoring. In *Comprehensive Geographic Information Systems*; Elsevier: Amsterdam, The Netherlands, 2018; pp. 140–157.

8. Sanga, B.; Mohanty, D.; Singh, A.; Singh, R. *Nexgen Technologies for Mining and Fuel Industries*; Allied Publishers: New Delhi, India, 2017.

9. Rocha, F.; Oliveira Neto, A.; Bottega, E.; Guerra, N.; Rocha, R.; Vilar, C. Weed mapping using techniques of precision agriculture. *Planta Daninha* **2015**, *33*, 157–164. [CrossRef]

10. Slaughter, D.C.; Pérez Ruiz, M.; Fathallah, F.; Upadhyaya, S.; Gliever, C.J.; Miller, B. GPS-based intra-row weed control system: Performance and labor savings. In *Automation Technology for Off-Road Equipment*; ASABE: St. Joseph, MO, USA, 2012.

11. De Castro, A.I.; Torres-Sánchez, J.; Peña, J.M.; Jiménez-Brenes, F.M.; Csillik, O.; López-Granados, F. An automatic random forest-OBIA algorithm for early weed mapping between and within crop rows using UAV imagery. *Remote Sens.* **2018**, *10*, 285. [CrossRef]

12. Khanal, S.; Fulton, J.; Shearer, S. An overview of current and potential applications of thermal remote sensing in precision agriculture. *Comput. Electron. Agric.* **2017**, *139*, 22–32. [CrossRef]

13. Mateo-Aroca, A.; García-Mateos, G.; Ruiz-Canales, A.; Molina-García-Pardo, J.M.; Molina-Martínez, J.M. Remote Image Capture System to Improve Aerial Supervision for Precision Irrigation in Agriculture. *Water* **2019**, *11*, 255. [CrossRef]

14. Liakos, K.G.; Busato, P.; Moshou, D.; Pearson, S.; Bochtis, D. Machine learning in agriculture: A review. *Sensors* **2018**, *18*, 2674. [CrossRef]

15. Lindblom, J.; Lundström, C.; Ljung, M.; Jonsson, A. Promoting sustainable intensification in precision agriculture: review of decision support systems development and strategies. *Precis. Agric.* **2017**, *18*, 309–331. [CrossRef]

16. Tmušić, G.; Manfreda, S.; Aasen, H.; James, M.R.; Gonçalves, G.; Ben-Dor, E.; Brook, A.; Polinova, M.; Arranz, J.J.; Mészáros, J.; et al. Current Practices in UAS-based Environmental Monitoring. *Remote Sens.* **2020**, *12*, 1001. [CrossRef]

17. Maes, W.H.; Steppe, K. Perspectives for remote sensing with unmanned aerial vehicles in precision agriculture. *Trends Plant Sci.* **2019**, *24*, 152–164. [CrossRef] [PubMed]

18. Adão, T.; Hruška, J.; Pádua, L.; Bessa, J.; Peres, E.; Morais, R.; Sousa, J.J. Hyperspectral imaging: A review on UAV-based sensors, data processing and applications for agriculture and forestry. *Remote Sens.* **2017**, *9*, 1110. [CrossRef]

19. Bah, M.D.; Hafiane, A.; Canals, R. Weeds detection in UAV imagery using SLIC and the hough transform. In Proceedings of the 2017 Seventh International Conference on Image Processing Theory, Tools and Applications (IPTA), Montreal, QC, Canada, 28 November–1 December 2017; pp. 1–6.

20. dos Santos Ferreira, A.; Freitas, D.M.; da Silva, G.G.; Pistori, H.; Folhes, M.T. Weed detection in soybean crops using ConvNets. *Comput. Electron. Agric.* **2017**, *143*, 314–324. [CrossRef]

21. Jung, J.; Maeda, M.; Chang, A.; Landivar, J.; Yeom, J.; McGinty, J. Unmanned aerial system assisted framework for the selection of high yielding cotton genotypes. *Comput. Electron. Agric.* **2018**, *152*, 74–81. [CrossRef]

22. Han, L.; Yang, G.; Yang, H.; Xu, B.; Li, Z.; Yang, X. Clustering field-based maize phenotyping of plant-height growth and canopy spectral dynamics using a UAV remote-sensing approach. *Front. Plant Sci.* **2018**, *9*, 1638. [CrossRef]
23. Wahab, I.; Hall, O.; Jirström, M. Remote sensing of yields: Application of uav imagery-derived ndvi for estimating maize vigor and yields in complex farming systems in sub-saharan africa. *Drones* **2018**, *2*, 28. [CrossRef]
24. Quebrajo, L.; Perez-Ruiz, M.; Pérez-Urrestarazu, L.; Martínez, G.; Egea, G. Linking thermal imaging and soil remote sensing to enhance irrigation management of sugar beet. *Biosyst. Eng.* **2018**, *165*, 77–87. [CrossRef]
25. Albornoz, C.; Giraldo, L.F. Trajectory design for efficient crop irrigation with a UAV. In Proceedings of the 2017 IEEE 3rd Colombian Conference on Automatic Control (CCAC), Cartagena, Colombia, 18–20 October 2017; pp. 1–6.
26. Kerkech, M.; Hafiane, A.; Canals, R. Deep leaning approach with colorimetric spaces and vegetation indices for vine diseases detection in UAV images. *Comput. Electron. Agric.* **2018**, *155*, 237–243. [CrossRef]
27. Montero, D.; Rueda, C. Detection of palm oil bud rot employing artificial vision. In Proceedings of the IOP Conference Series: Materials Science and Engineering, Constanta, Romania, 13–16 June 2018; Volume 437, p. 012004.
28. Xue, X.; Lan, Y.; Sun, Z.; Chang, C.; Hoffmann, W.C. Develop an unmanned aerial vehicle based automatic aerial spraying system. *Comput. Electron. Agric.* **2016**, *128*, 58–66. [CrossRef]
29. Garre, P.; Harish, A. Autonomous agricultural pesticide spraying uav. In Proceedings of the IOP Conference Series: Materials Science and Engineering, Constanta, Romania, 13–16 June 2018; Volume 455, p. 012030.
30. Perich, G.; Hund, A.; Anderegg, J.; Roth, L.; Boer, M.P.; Walter, A.; Liebisch, F.; Aasen, H. Assessment of multi-image UAV based high-throughput field phenotyping of canopy temperature. *Front. Plant Sci.* **2020**, *11*, 150. [CrossRef] [PubMed]
31. Maurya, P. Hardware Implementation of a Flight Control System for an Unmanned Aerial Vehicle. 2015. Available online: http://www.cse.iitk.ac.in/users/moona/students/Y2258.pdf (accessed on 14 March 2020).
32. Park, J.K.; Park, J.H. Crops classification using imagery of unmanned aerial vehicle (UAV). *J. Korean Soc. Agric. Eng.* **2015**, *57*, 91–97.
33. Punjani, A.; Abbeel, P. Deep learning helicopter dynamics models. In Proceedings of the 2015 IEEE International Conference on Robotics and Automation (ICRA), Seattle, WA, USA, 26–30 May 2015; pp. 3223–3230.
34. Peña Barragán, J.M.; Kelly, M.; Castro, A.I.d.; López Granados, F. Object-based approach for crop row characterization in UAV images for site-specific weed management. In Proceedings of the 4th GEOBIA, Rio de Janeiro, Brazil, 7–9 May 2012; pp. 426–430.
35. Näsi, R.; Viljanen, N.; Kaivosoja, J.; Alhonoja, K.; Hakala, T.; Markelin, L.; Honkavaara, E. Estimating biomass and nitrogen amount of barley and grass using UAV and aircraft based spectral and photogrammetric 3D features. *Remote Sens.* **2018**, *10*, 1082. [CrossRef]
36. Zheng, H.; Li, W.; Jiang, J.; Liu, Y.; Cheng, T.; Tian, Y.; Zhu, Y.; Cao, W.; Zhang, Y.; Yao, X. A comparative assessment of different modeling algorithms for estimating leaf nitrogen content in winter wheat using multispectral images from an unmanned aerial vehicle. *Remote Sens.* **2018**, *10*, 2026. [CrossRef]
37. Wang, D.C.; Zhang, G.L.; Zhao, M.S.; Pan, X.Z.; Zhao, Y.G.; Li, D.C.; Macmillan, B. Retrieval and mapping of soil texture based on land surface diurnal temperature range data from MODIS. *PLoS ONE* **2015**, *10*, 1–14. [CrossRef]
38. Shafian, S.; Maas, S.J. Index of soil moisture using raw Landsat image digital count data in Texas high plains. *Remote Sens.* **2015**, *7*, 2352–2372. [CrossRef]
39. Tong, X.Y.; Xia, G.S.; Lu, Q.; Shen, H.; Li, S.; You, S.; Zhang, L. Land-cover classification with high-resolution remote sensing images using transferable deep models. *Remote Sens. Environ.* **2020**, *237*, 111322. [CrossRef]
40. Helber, P.; Bischke, B.; Dengel, A.; Borth, D. Eurosat: A novel dataset and deep learning benchmark for land use and land cover classification. *IEEE J. Sel. Top. Appl. Earth Obs. Remote Sens.* **2019**, *12*, 2217–2226. [CrossRef]
41. Maes, W.; Steppe, K. Estimating evapotranspiration and drought stress with ground-based thermal remote sensing in agriculture: A review. *J. Exp. Bot.* **2012**, *63*, 4671–4712. [CrossRef]
42. Soliman, A.; Heck, R.J.; Brenning, A.; Brown, R.; Miller, S. Remote sensing of soil moisture in vineyards using airborne and ground-based thermal inertia data. *Remote Sens.* **2013**, *5*, 3729–3748. [CrossRef]

43. Daponte, P.; De Vito, L.; Glielmo, L.; Iannelli, L.; Liuzza, D.; Picariello, F.; Silano, G. A review on the use of drones for precision agriculture. In Proceedings of the IOP Conference Series: Earth and Environmental Science, Bogor, Indonesia, 10–11 September 2019; Volume 275, p. 012022.
44. Daponte, P.; De Vito, L.; Mazzilli, G.; Picariello, F.; Rapuano, S. A height measurement uncertainty model for archaeological surveys by aerial photogrammetry. *Measurement* **2017**, *98*, 192–198. [CrossRef]
45. Liu, C.A.; Chen, Z.X.; Yun, S.; Chen, J.S.; Hasi, T.; Pan, H.Z. Research advances of SAR remote sensing for agriculture applications: A review. *J. Integr. Agric.* **2019**, *18*, 506–525. [CrossRef]
46. Tan, C.P.; Ewe, H.T.; Chuah, H.T. Agricultural crop-type classification of multi-polarization SAR images using a hybrid entropy decomposition and support vector machine technique. *Int. J. Remote Sens.* **2011**, *32*, 7057–7071. [CrossRef]
47. Kussul, N.; Lemoine, G.; Gallego, F.J.; Skakun, S.V.; Lavreniuk, M.; Shelestov, A.Y. Parcel-based crop classification in Ukraine using Landsat-8 data and Sentinel-1A data. *IEEE J. Sel. Top. Appl. Earth Obs. Remote Sens.* **2016**, *9*, 2500–2508. [CrossRef]
48. Stankiewicz, K.A. The efficiency of crop recognition on ENVISAT ASAR images in two growing seasons. *IEEE Trans. Geosci. Remote Sens.* **2006**, *44*, 806–814. [CrossRef]
49. Ziliani, M.G.; Parkes, S.D.; Hoteit, I.; McCabe, M.F. Intra-season crop height variability at commercial farm scales using a fixed-wing UAV. *Remote Sens.* **2018**, *10*, 2007. [CrossRef]
50. Torres-Sánchez, J.; de Castro, A.I.; Pena, J.M.; Jiménez-Brenes, F.M.; Arquero, O.; Lovera, M.; López-Granados, F. Mapping the 3D structure of almond trees using UAV acquired photogrammetric point clouds and object-based image analysis. *Biosyst. Eng.* **2018**, *176*, 172–184. [CrossRef]
51. De Castro, A.I.; Jiménez-Brenes, F.M.; Torres-Sánchez, J.; Peña, J.M.; Borra-Serrano, I.; López-Granados, F. 3-D characterization of vineyards using a novel UAV imagery-based OBIA procedure for precision viticulture applications. *Remote Sens.* **2018**, *10*, 584. [CrossRef]
52. Sa, I.; Popović, M.; Khanna, R.; Chen, Z.; Lottes, P.; Liebisch, F.; Nieto, J.; Stachniss, C.; Walter, A.; Siegwart, R. Weedmap: A large-scale semantic weed mapping framework using aerial multispectral imaging and deep neural network for precision farming. *Remote Sens.* **2018**, *10*, 1423. [CrossRef]
53. Marino, S.; Alvino, A. Detection of homogeneous wheat areas using multitemporal UAS images and ground truth data analyzed by cluster analysis. *Eur. J. Remote Sens.* **2018**, *51*, 266–275. [CrossRef]
54. de Souza, C.H.W.; Mercante, E.; Johann, J.A.; Lamparelli, R.A.C.; Uribe-Opazo, M.A. Mapping and discrimination of soya bean and corn crops using spectro-temporal profiles of vegetation indices. *Int. J. Remote Sens.* **2015**, *36*, 1809–1824. [CrossRef]
55. Zheng, Y.; Zhang, M.; Zhang, X.; Zeng, H.; Wu, B. Mapping winter wheat biomass and yield using time series data blended from PROBA-V 100-and 300-m S1 products. *Remote Sens.* **2016**, *8*, 824. [CrossRef]
56. Gouveia, C.; Trigo, R.; Beguería, S.; Vicente-Serrano, S.M. Drought impacts on vegetation activity in the Mediterranean region: An assessment using remote sensing data and multi-scale drought indicators. *Glob. Planet. Chang.* **2017**, *151*, 15–27. [CrossRef]
57. Zhang, C.; Ren, H.; Qin, Q.; Ersoy, O.K. A new narrow band vegetation index for characterizing the degree of vegetation stress due to copper: The copper stress vegetation index (CSVI). *Remote Sens. Lett.* **2017**, *8*, 576–585. [CrossRef]
58. Rembold, F.; Atzberger, C.; Savin, I.; Rojas, O. Using low resolution satellite imagery for yield prediction and yield anomaly detection. *Remote Sens.* **2013**, *5*, 1704–1733. [CrossRef]
59. Mavridou, E.; Vrochidou, E.; Papakostas, G.A.; Pachidis, T.; Kaburlasos, V.G. Machine Vision Systems in Precision Agriculture for Crop Farming. *J. Imaging* **2019**, *5*, 89. [CrossRef]
60. Khan, Z.; Rahimi-Eichi, V.; Haefele, S.; Garnett, T.; Miklavcic, S.J. Estimation of vegetation indices for high-throughput phenotyping of wheat using aerial imaging. *Plant Methods* **2018**, *14*, 20. [CrossRef]
61. Bah, M.D.; Hafiane, A.; Canals, R. Deep learning with unsupervised data labeling for weed detection in line crops in UAV images. *Remote Sens.* **2018**, *10*, 1690. [CrossRef]
62. Bah, M.D.; Dericquebourg, E.; Hafiane, A.; Canals, R. Deep learning based classification system for identifying weeds using high-resolution UAV imagery. In Proceedings of the Science and Information Conference, Las Vegas, NV, USA, 25–26 April 2018; pp 176–187. pp. 176–187.
63. Kruse, O.M.O.; Prats-Montalbán, J.M.; Indahl, U.G.; Kvaal, K.; Ferrer, A.; Futsaether, C.M. Pixel classification methods for identifying and quantifying leaf surface injury from digital images. *Comput. Electron. Agric.* **2014**, *108*, 155–165. [CrossRef]

64. Albetis, J.; Jacquin, A.; Goulard, M.; Poilvé, H.; Rousseau, J.; Clenet, H.; Dedieu, G.; Duthoit, S. On the potentiality of UAV multispectral imagery to detect Flavescence dorée and Grapevine Trunk Diseases. *Remote Sens.* **2019**, *11*, 23. [CrossRef]
65. Kurtulmuş, F.; Ünal, H. Discriminating rapeseed varieties using computer vision and machine learning. *Expert Syst. Appl.* **2015**, *42*, 1880–1891. [CrossRef]
66. Hassan-Esfahani, L.; Torres-Rua, A.; Jensen, A.; McKee, M. Assessment of surface soil moisture using high-resolution multi-spectral imagery and artificial neural networks. *Remote Sens.* **2015**, *7*, 2627–2646. [CrossRef]
67. Poblete, T.; Ortega-Farías, S.; Moreno, M.A.; Bardeen, M. Artificial neural network to predict vine water status spatial variability using multispectral information obtained from an unmanned aerial vehicle (UAV). *Sensors* **2017**, *17*, 2488. [CrossRef] [PubMed]
68. Jeon, H.Y.; Tian, L.F.; Zhu, H. Robust crop and weed segmentation under uncontrolled outdoor illumination. *Sensors* **2011**, *11*, 6270–6283. [CrossRef] [PubMed]
69. Sannakki, S.S.; Rajpurohit, V.S.; Nargund, V. SVM-DSD: SVM Based diagnostic system for the detection of pomegranate leaf diseases. In Proceedings of International Conference on Advances in Computing, Mumbai, India, 18–19 January 2013; pp. 715–720.
70. Mokhtar, U.; El Bendary, N.; Hassenian, A.E.; Emary, E.; Mahmoud, M.A.; Hefny, H.; Tolba, M.F. SVM-based detection of tomato leaves diseases. In *Intelligent Systems' 2014*; Springer: Berlin/Heidelberg, Germany, 2015; pp. 641–652.
71. Dingle Robertson, L.; King, D.J. Comparison of pixel-and object-based classification in land cover change mapping. *Int. J. Remote Sens.* **2011**, *32*, 1505–1529. [CrossRef]
72. De Rainville, F.M.; Durand, A.; Fortin, F.A.; Tanguy, K.; Maldague, X.; Panneton, B.; Simard, M.J. Bayesian classification and unsupervised learning for isolating weeds in row crops. *Pattern Anal. Appl.* **2014**, *17*, 401–414. [CrossRef]
73. Mondal, D.; Kole, D.K.; Roy, K. Gradation of yellow mosaic virus disease of okra and bitter gourd based on entropy based binning and Naive Bayes classifier after identification of leaves. *Comput. Electron. Agric.* **2017**, *142*, 485–493. [CrossRef]
74. Tsouros, D.C.; Bibi, S.; Sarigiannidis, P.G. A review on UAV-based applications for precision agriculture. *Information* **2019**, *10*, 349. [CrossRef]
75. Kamilaris, A.; Prenafeta-Boldú, F.X. Deep learning in agriculture: A survey. *Comput. Electron. Agric.* **2018**, *147*, 70–90. [CrossRef]
76. Huang, H.; Deng, J.; Lan, Y.; Yang, A.; Deng, X.; Zhang, L. A fully convolutional network for weed mapping of unmanned aerial vehicle (UAV) imagery. *PLoS ONE* **2018**, *13*, e0196302. [CrossRef]
77. Castro, J.D.B.; Feitoza, R.Q.; La Rosa, L.C.; Diaz, P.M.A.; Sanches, I.D.A. A Comparative analysis of deep learning techniques for sub-tropical crop types recognition from multitemporal optical/SAR image sequences. In Proceedings of the 2017 30th SIBGRAPI Conference on Graphics, Patterns and Images (SIBGRAPI), Niteroi, Brazil, 17–20 October 2017; pp. 382–389.
78. Mortensen, A.K.; Dyrmann, M.; Karstoft, H.; Jørgensen, R.N.; Gislum, R. Semantic segmentation of mixed crops using deep convolutional neural network. In Proceedings of the International Conference of Agricultural Engineering (CIGR), Aarhus, Denmark, 26–29 June 2016.
79. Kussul, N.; Lavreniuk, M.; Skakun, S.; Shelestov, A. Deep learning classification of land cover and crop types using remote sensing data. *IEEE Geosci. Remote Sens. Lett.* **2017**, *14*, 778–782. [CrossRef]
80. Tian, H.; Wang, T.; Liu, Y.; Qiao, X.; Li, Y. Computer Vision Technology in Agricultural Automation—A review. *Inf. Process. Agric.* **2019**, *7*, 1–19. [CrossRef]
81. Mogili, U.R.; Deepak, B. Review on application of drone systems in precision agriculture. *Procedia Comput. Sci.* **2018**, *133*, 502–509. [CrossRef]
82. Tripathi, M.K.; Maktedar, D.D. A role of computer vision in fruits and vegetables among various horticulture products of agriculture fields: A survey. *Information Processing in Agriculture* **2019**. [CrossRef]
83. Weiss, M.; Jacob, F.; Duveiller, G. Remote sensing for agricultural applications: A meta-review. *Remote Sens. Environ.* **2020**, *236*, 111402. [CrossRef]
84. Ouhbi, S.; Idri, A.; Fernández-Alemán, J.L.; Toval, A. Predicting software product quality: A systematic mapping study. *Comput. Sist.* **2015**, *19*, 547–562. [CrossRef]

85. Petersen, K.; Vakkalanka, S.; Kuzniarz, L. Guidelines for conducting systematic mapping studies in software engineering: An update. *Inf. Softw. Technol.* **2015**, *64*, 1–18. [CrossRef]
86. Mongeon, P.; Paul-Hus, A. The journal coverage of Web of Science and Scopus: A comparative analysis. *Scientometrics* **2016**, *106*, 213–228. [CrossRef]
87. Moher, D.; Liberati, A.; Tetzlaff, J.; Altman, D.G. Preferred reporting items for systematic reviews and meta-analyses: the PRISMA statement. *Ann. Intern. Med.* **2009**, *151*, 264–269. [CrossRef]
88. Ouhbi, S.; Idri, A.; Fernández-Alemán, J.L.; Toval, A. Requirements engineering education: A systematic mapping study. *Requir. Eng.* **2015**, *20*, 119–138. [CrossRef]
89. Ouhbi, S.; Idri, A.; Aleman, J.L.F.; Toval, A. Evaluating software product quality: A systematic mapping study. In Proceedings of the 2014 Joint Conference of the International Workshop on Software Measurement and the International Conference on Software Process and Product Measurement, Rotterdam, The Netherlands, 6–8 October 2014; pp. 141–151.
90. Minu, S.; Shetty, A.; Gopal, B. Review of preprocessing techniques used in soil property prediction from hyperspectral data. *Cogent Geosci.* **2016**, *2*, 1145878. [CrossRef]
91. Atzberger, C. Advances in remote sensing of agriculture: Context description, existing operational monitoring systems and major information needs. *Remote Sens.* **2013**, *5*, 949–981. [CrossRef]
92. Zhao, K.; García, M.; Liu, S.; Guo, Q.; Chen, G.; Zhang, X.; Zhou, Y.; Meng, X. Terrestrial lidar remote sensing of forests: Maximum likelihood estimates of canopy profile, leaf area index, and leaf angle distribution. *Agric. For. Meteorol.* **2015**, *209*, 100–113. [CrossRef]
93. Angelopoulou, T.; Tziolas, N.; Balafoutis, A.; Zalidis, G.; Bochtis, D. Remote sensing techniques for soil organic carbon estimation: A review. *Remote Sens.* **2019**, *11*, 676. [CrossRef]
94. Basu, S.; Ganguly, S.; Mukhopadhyay, S.; DiBiano, R.; Karki, M.; Nemani, R. Deepsat: A learning framework for satellite imagery. In Proceedings of the 23rd SIGSPATIAL International Conference on Advances in Geographic Information Systems, Seattle, WA, USA, 3–6 November 2015; pp. 1–10.
95. Shelestov, A.; Lavreniuk, M.; Kussul, N.; Novikov, A.; Skakun, S. Exploring Google Earth Engine platform for big data processing: Classification of multitemporal satellite imagery for crop mapping. *Front. Earth Sci.* **2017**, *5*, 17. [CrossRef]
96. Stagakis, S.; Markos, N.; Sykioti, O.; Kyparissis, A. Monitoring canopy biophysical and biochemical parameters in ecosystem scale using satellite hyperspectral imagery: An application on a Phlomis fruticosa Mediterranean ecosystem using multiangular CHRIS/PROBA observations. *Remote Sens. Environ.* **2010**, *114*, 977–994. [CrossRef]
97. Zmarz, A. Introduction to the special issue UAS for mapping and monitoring. *Eur. J. Remote Sens.* **2019**, *52*, 1. [CrossRef]
98. Korczak-Abshire, M.; Zmarz, A.; Rodzewicz, M.; Kycko, M.; Karsznia, I.; Chwedorzewska, K.J. Study of fauna population changes on Penguin Island and Turret Point Oasis (King George Island, Antarctica) using an unmanned aerial vehicle. *Polar Biol.* **2019**, *42*, 217–224. [CrossRef]
99. Fang, S.X.; O'Young, S.; Rolland, L. Development of small uas beyond-visual-line-of-sight (bvlos) flight operations: System requirements and procedures. *Drones* **2018**, *2*, 13. [CrossRef]
100. Rehman, T.U.; Mahmud, M.S.; Chang, Y.K.; Jin, J.; Shin, J. Current and future applications of statistical machine learning algorithms for agricultural machine vision systems. *Comput. Electron. Agric.* **2019**, *156*, 585–605. [CrossRef]
101. Alchanatis, V.; Cohen, Y.; Cohen, S.; Moller, M.; Sprinstin, M.; Meron, M.; Tsipris, J.; Saranga, Y.; Sela, E. Evaluation of different approaches for estimating and mapping crop water status in cotton with thermal imaging. *Precis. Agric.* **2010**, *11*, 27–41. [CrossRef]
102. Lamichhane, J.R.; Dachbrodt-Saaydeh, S.; Kudsk, P.; Messéan, A. Toward a reduced reliance on conventional pesticides in European agriculture. *Plant Dis.* **2016**, *100*, 10–24. [CrossRef] [PubMed]
103. Sabzi, S.; Abbaspour-Gilandeh, Y.; García-Mateos, G. A fast and accurate expert system for weed identification in potato crops using metaheuristic algorithms. *Comput. Ind.* **2018**, *98*, 80–89. [CrossRef]
104. Huang, Y.; Chen, Z.X.; Tao, Y.; Huang, X.Z.; Gu, X.F. Agricultural remote sensing big data: Management and applications. *J. Integr. Agric.* **2018**, *17*, 1915–1931. [CrossRef]
105. Hernández-Hernández, J.; García-Mateos, G.; González-Esquiva, J.; Escarabajal-Henarejos, D.; Ruiz-Canales, A.; Molina-Martínez, J.M. Optimal color space selection method for plant/soil segmentation in agriculture. *Comput. Electron. Agric.* **2016**, *122*, 124–132. [CrossRef]

106. Jones, H.G.; Serraj, R.; Loveys, B.R.; Xiong, L.; Wheaton, A.; Price, A.H. Thermal infrared imaging of crop canopies for the remote diagnosis and quantification of plant responses to water stress in the field. *Funct. Plant Biol.* **2009**, *36*, 978–989. [CrossRef]
107. Mangus, D.L.; Sharda, A.; Zhang, N. Development and evaluation of thermal infrared imaging system for high spatial and temporal resolution crop water stress monitoring of corn within a greenhouse. *Comput. Electron. Agric.* **2016**, *121*, 149–159. [CrossRef]
108. Liu, X.; Bo, Y. Object-based crop species classification based on the combination of airborne hyperspectral images and LiDAR data. *Remote Sens.* **2015**, *7*, 922–950. [CrossRef]
109. Pearlman, J.; Carman, S.; Segal, C.; Jarecke, P.; Clancy, P.; Browne, W. Overview of the Hyperion imaging spectrometer for the NASA EO-1 mission. In Proceedings of the IGARSS 2001. Scanning the Present and Resolving the Future. IEEE 2001 International Geoscience and Remote Sensing Symposium (Cat. No. 01CH37217), Piscataway, NJ, USA, 9–13 July 2001; Volume 7, pp. 3036–3038.
110. Poblete-Echeverría, C.; Olmedo, G.; Ingram, B.; Bardeen, M. Detection and segmentation of vine canopy in ultra-high spatial resolution RGB imagery obtained from unmanned aerial vehicle (UAV): A case study in a commercial vineyard. *Remote Sens.* **2017**, *9*, 268. [CrossRef]
111. Lucas, R.; Rowlands, A.; Brown, A.; Keyworth, S.; Bunting, P. Rule-based classification of multitemporal satellite imagery for habitat and agricultural land cover mapping. *ISPRS J. Photogramm. Remote Sens.* **2007**, *62*, 165–185. [CrossRef]
112. Asaari, M.S.M.; Mishra, P.; Mertens, S.; Dhondt, S.; Inzé, D.; Wuyts, N.; Scheunders, P. Close-range hyperspectral image analysis for the early detection of stress responses in individual plants in a high-throughput phenotyping platform. *ISPRS J. Photogramm. Remote Sens.* **2018**, *138*, 121–138. [CrossRef]
113. Müllerová, J.; Brůna, J.; Bartaloš, T.; Dvořák, P.; Vítková, M.; Pyšek, P. Timing is important: Unmanned aircraft vs. satellite imagery in plant invasion monitoring. *Front. Plant Sci.* **2017**, *8*, 887. [CrossRef] [PubMed]
114. Duveiller, G.; Defourny, P. A conceptual framework to define the spatial resolution requirements for agricultural monitoring using remote sensing. *Remote Sens. Environ.* **2010**, *114*, 2637–2650. [CrossRef]
115. Wei, Z.; Han, Y.; Li, M.; Yang, K.; Yang, Y.; Luo, Y.; Ong, S.H. A small UAV based multitemporal image registration for dynamic agricultural terrace monitoring. *Remote Sens.* **2017**, *9*, 904. [CrossRef]
116. Ji, S.; Zhang, C.; Xu, A.; Shi, Y.; Duan, Y. 3D convolutional neural networks for crop classification with multitemporal remote sensing images. *Remote Sens.* **2018**, *10*, 75. [CrossRef]
117. Yang, C.; Everitt, J.H.; Murden, D. Evaluating high resolution SPOT 5 satellite imagery for crop identification. *Comput. Electron. Agric.* **2011**, *75*, 347–354. [CrossRef]
118. Zhang, X.; Liu, F.; He, Y.; Gong, X. Detecting macronutrients content and distribution in oilseed rape leaves based on hyperspectral imaging. *Biosyst. Eng.* **2013**, *115*, 56–65. [CrossRef]
119. Ač, A.; Malenovský, Z.; Olejníčková, J.; Gallé, A.; Rascher, U.; Mohammed, G. Meta-analysis assessing potential of steady-state chlorophyll fluorescence for remote sensing detection of plant water, temperature and nitrogen stress. *Remote Sens. Environ.* **2015**, *168*, 420–436. [CrossRef]
120. Baghdadi, N.; Boyer, N.; Todoroff, P.; El Hajj, M.; Bégué, A. Potential of SAR sensors TerraSAR-X, ASAR/ENVISAT and PALSAR/ALOS for monitoring sugarcane crops on Reunion Island. *Remote Sens. Environ.* **2009**, *113*, 1724–1738. [CrossRef]
121. Baret, F.; Houles, V.; Guérif, M. Quantification of plant stress using remote sensing observations and crop models: The case of nitrogen management. *J. Exp. Bot.* **2007**, *58*, 869–880. [CrossRef]
122. Baret, F.; Buis, S. Estimating canopy characteristics from remote sensing observations: Review of methods and associated problems. In *Advances in Land Remote Sensing*; Springer: Berlin/Heidelberg, Germany, 2008; pp. 173–201.
123. Behmann, J.; Steinrücken, J.; Plümer, L. Detection of early plant stress responses in hyperspectral images. *ISPRS J. Photogramm. Remote Sens.* **2014**, *93*, 98–111. [CrossRef]
124. Bellvert, J.; Marsal, J.; Girona, J.; Gonzalez-Dugo, V.; Fereres, E.; Ustin, S.; Zarco-Tejada, P. Airborne thermal imagery to detect the seasonal evolution of crop water status in peach, nectarine and Saturn peach orchards. *Remote Sens.* **2016**, *8*, 39. [CrossRef]
125. Bendig, J.; Bolten, A.; Bennertz, S.; Broscheit, J.; Eichfuss, S.; Bareth, G. Estimating biomass of barley using crop surface models (CSMs) derived from UAV-based RGB imaging. *Remote Sens.* **2014**, *6*, 10395–10412. [CrossRef]

126. Blaes, X.; Vanhalle, L.; Defourny, P. Efficiency of crop identification based on optical and SAR image time series. *Remote Sens. Environ.* **2005**, *96*, 352–365. [CrossRef]
127. Clevers, J.G.; Kooistra, L.; Schaepman, M.E. Estimating canopy water content using hyperspectral remote sensing data. *Int. J. Appl. Earth Obs. Geoinf.* **2010**, *12*, 119–125. [CrossRef]
128. Er-Raki, S.; Chehbouni, A.; Guemouria, N.; Duchemin, B.; Ezzahar, J.; Hadria, R. Combining FAO-56 model and ground-based remote sensing to estimate water consumptions of wheat crops in a semi-arid region. *Agric. Water Manag.* **2007**, *87*, 41–54. [CrossRef]
129. Garrigues, S.; Allard, D.; Baret, F.; Weiss, M. Influence of landscape spatial heterogeneity on the non-linear estimation of leaf area index from moderate spatial resolution remote sensing data. *Remote Sens. Environ.* **2006**, *105*, 286–298. [CrossRef]
130. Glenn, E.P.; Neale, C.M.; Hunsaker, D.J.; Nagler, P.L. Vegetation index-based crop coefficients to estimate evapotranspiration by remote sensing in agricultural and natural ecosystems. *Hydrol. Process.* **2011**, *25*, 4050–4062. [CrossRef]
131. Gonzalez-Dugo, M.; Neale, C.; Mateos, L.; Kustas, W.; Prueger, J.; Anderson, M.; Li, F. A comparison of operational remote sensing-based models for estimating crop evapotranspiration. *Agric. For. Meteorol.* **2009**, *149*, 1843–1853. [CrossRef]
132. Houborg, R.; Anderson, M.; Daughtry, C. Utility of an image-based canopy reflectance modeling tool for remote estimation of LAI and leaf chlorophyll content at the field scale. *Remote Sens. Environ.* **2009**, *113*, 259–274. [CrossRef]
133. Kaliramesh, S.; Chelladurai, V.; Jayas, D.; Alagusundaram, K.; White, N.; Fields, P. Detection of infestation by Callosobruchus maculatus in mung bean using near-infrared hyperspectral imaging. *J. Stored Prod. Res.* **2013**, *52*, 107–111. [CrossRef]
134. Kong, W.; Zhang, C.; Liu, F.; Nie, P.; He, Y. Rice seed cultivar identification using near-infrared hyperspectral imaging and multivariate data analysis. *Sensors* **2013**, *13*, 8916–8927. [CrossRef] [PubMed]
135. Kullberg, E.G.; DeJonge, K.C.; Chávez, J.L. Evaluation of thermal remote sensing indices to estimate crop evapotranspiration coefficients. *Agric. Water Manag.* **2017**, *179*, 64–73. [CrossRef]
136. Lanorte, A.; De Santis, F.; Nolè, G.; Blanco, I.; Loisi, R.V.; Schettini, E.; Vox, G. Agricultural plastic waste spatial estimation by Landsat 8 satellite images. *Comput. Electron. Agric.* **2017**, *141*, 35–45. [CrossRef]
137. Lawrence, R.L.; Wood, S.D.; Sheley, R.L. Mapping invasive plants using hyperspectral imagery and Breiman Cutler classifications (RandomForest). *Remote Sens. Environ.* **2006**, *100*, 356–362. [CrossRef]
138. Li, H.; Zheng, L.; Lei, Y.; Li, C.; Liu, Z.; Zhang, S. Estimation of water consumption and crop water productivity of winter wheat in North China Plain using remote sensing technology. *Agric. Water Manag.* **2008**, *95*, 1271–1278. [CrossRef]
139. Li, Y.; Zhou, Q.; Zhou, J.; Zhang, G.; Chen, C.; Wang, J. Assimilating remote sensing information into a coupled hydrology-crop growth model to estimate regional maize yield in arid regions. *Ecol. Model.* **2014**, *291*, 15–27. [CrossRef]
140. Lobell, D.B.; Asner, G.P.; Ortiz-Monasterio, J.I.; Benning, T.L. Remote sensing of regional crop production in the Yaqui Valley, Mexico: Estimates and uncertainties. *Agric. Ecosyst. Environ.* **2003**, *94*, 205–220. [CrossRef]
141. López-López, M.; Calderón, R.; González-Dugo, V.; Zarco-Tejada, P.; Fereres, E. Early detection and quantification of almond red leaf blotch using high-resolution hyperspectral and thermal imagery. *Remote Sens.* **2016**, *8*, 276. [CrossRef]
142. Löw, F.; Duveiller, G. Defining the spatial resolution requirements for crop identification using optical remote sensing. *Remote Sens.* **2014**, *6*, 9034–9063. [CrossRef]
143. Lowe, A.; Harrison, N.; French, A.P. Hyperspectral image analysis techniques for the detection and classification of the early onset of plant disease and stress. *Plant Methods* **2017**, *13*, 80. [CrossRef] [PubMed]
144. Mahesh, S.; Jayas, D.; Paliwal, J.; White, N. Hyperspectral imaging to classify and monitor quality of agricultural materials. *J. Stored Prod. Res.* **2015**, *61*, 17–26. [CrossRef]
145. Marshall, M.; Thenkabail, P. Developing in situ non-destructive estimates of crop biomass to address issues of scale in remote sensing. *Remote Sens.* **2015**, *7*, 808–835. [CrossRef]
146. Mehl, P.; Chao, K.; Kim, M.; Chen, Y. Detection of defects on selected apple cultivars using hyperspectral and multispectral image analysis. *Appl. Eng. Agric.* **2002**, *18*, 219.

147. Moran, M.S.; Inoue, Y.; Barnes, E. Opportunities and limitations for image-based remote sensing in precision crop management. *Remote Sens. Environ.* **1997**, *61*, 319–346. [CrossRef]
148. Moudrý, V.; Gdulová, K.; Fogl, M.; Klápště, P.; Urban, R.; Komárek, J.; Moudrá, L.; Štroner, M.; Barták, V.; Solský, M. Comparison of leaf-off and leaf-on combined UAV imagery and airborne LiDAR for assessment of a post-mining site terrain and vegetation structure: Prospects for monitoring hazards and restoration success. *Appl. Geogr.* **2019**, *104*, 32–41. [CrossRef]
149. Müllerová, J.; Pergl, J.; Pyšek, P. Remote sensing as a tool for monitoring plant invasions: Testing the effects of data resolution and image classification approach on the detection of a model plant species Heracleum mantegazzianum (giant hogweed). *Int. J. Appl. Earth Obs. Geoinf.* **2013**, *25*, 55–65. [CrossRef]
150. Pandey, A.; Chowdary, V.; Mal, B. Identification of critical erosion prone areas in the small agricultural watershed using USLE, GIS and remote sensing. *Water Resour. Manag.* **2007**, *21*, 729–746. [CrossRef]
151. Pinter, P.J., Jr.; Hatfield, J.L.; Schepers, J.S.; Barnes, E.M.; Moran, M.S.; Daughtry, C.S.; Upchurch, D.R. Remote sensing for crop management. *Photogramm. Eng. Remote Sens.* **2003**, *69*, 647–664. [CrossRef]
152. Prasad, A.K.; Chai, L.; Singh, R.P.; Kafatos, M. Crop yield estimation model for Iowa using remote sensing and surface parameters. *Int. J. Appl. Earth Obs. Geoinf.* **2006**, *8*, 26–33. [CrossRef]
153. Schlerf, M.; Atzberger, C. Inversion of a forest reflectance model to estimate structural canopy variables from hyperspectral remote sensing data. *Remote Sens. Environ.* **2006**, *100*, 281–294. [CrossRef]
154. Siachalou, S.; Mallinis, G.; Tsakiri-Strati, M. A hidden Markov models approach for crop classification: Linking crop phenology to time series of multi-sensor remote sensing data. *Remote Sens.* **2015**, *7*, 3633–3650. [CrossRef]
155. Thomas, S.; Kuska, M.T.; Bohnenkamp, D.; Brugger, A.; Alisaac, E.; Wahabzada, M.; Behmann, J.; Mahlein, A.K. Benefits of hyperspectral imaging for plant disease detection and plant protection: A technical perspective. *J. Plant Dis. Prot.* **2018**, *125*, 5–20. [CrossRef]
156. Xie, C.; Yang, C.; He, Y. Hyperspectral imaging for classification of healthy and gray mold diseased tomato leaves with different infection severities. *Comput. Electron. Agric.* **2017**, *135*, 154–162. [CrossRef]
157. Yue, J.; Feng, H.; Jin, X.; Yuan, H.; Li, Z.; Zhou, C.; Yang, G.; Tian, Q. A comparison of crop parameters estimation using images from UAV-mounted snapshot hyperspectral sensor and high-definition digital camera. *Remote Sens.* **2018**, *10*, 1138. [CrossRef]
158. Ma, Y.; Wang, S.; Zhang, L.; Hou, Y.; Zhuang, L.; He, Y.; Wang, F. Monitoring winter wheat growth in North China by combining a crop model and remote sensing data. *Int. J. Appl. Earth Obs. Geoinf.* **2008**, *10*, 426–437.
159. Zhao, Y.; Chen, S.; Shen, S. Assimilating remote sensing information with crop model using Ensemble Kalman Filter for improving LAI monitoring and yield estimation. *Ecol. Model.* **2013**, *270*, 30–42. [CrossRef]
160. Anderson, L.O.; Malhi, Y.; Aragão, L.E.; Ladle, R.; Arai, E.; Barbier, N.; Phillips, O. Remote sensing detection of droughts in Amazonian forest canopies. *New Phytol.* **2010**, *187*, 733–750. [CrossRef]
161. Bendig, J.; Bolten, A.; Bareth, G. UAV-based imaging for multitemporal, very high Resolution Crop Surface Models to monitor Crop Growth VariabilityMonitoring des Pflanzenwachstums mit Hilfe multitemporaler und hoch auflösender Oberflächenmodelle von Getreidebeständen auf Basis von Bildern aus UAV-Befliegungen. *Photogramm. Fernerkund. Geoinf.* **2013**, *2013*, 551–562.
162. Calera, A.; Campos, I.; Osann, A.; D'Urso, G.; Menenti, M. Remote sensing for crop water management: From ET modelling to services for the end users. *Sensors* **2017**, *17*, 1104. [CrossRef]
163. Kamble, B.; Kilic, A.; Hubbard, K. Estimating crop coefficients using remote sensing-based vegetation index. *Remote Sens.* **2013**, *5*, 1588–1602. [CrossRef]
164. Leinonen, I.; Jones, H.G. Combining thermal and visible imagery for estimating canopy temperature and identifying plant stress. *J. Exp. Bot.* **2004**, *55*, 1423–1431. [CrossRef]
165. Möller, M.; Alchanatis, V.; Cohen, Y.; Meron, M.; Tsipris, J.; Naor, A.; Ostrovsky, V.; Sprintsin, M.; Cohen, S. Use of thermal and visible imagery for estimating crop water status of irrigated grapevine. *J. Exp. Bot.* **2006**, *58*, 827–838. [CrossRef] [PubMed]
166. Park, S.; Ryu, D.; Fuentes, S.; Chung, H.; Hernández-Montes, E.; O'Connell, M. Adaptive estimation of crop water stress in nectarine and peach orchards using high-resolution imagery from an unmanned aerial vehicle (UAV). *Remote Sens.* **2017**, *9*, 828. [CrossRef]
167. Santesteban, L.; Di Gennaro, S.; Herrero-Langreo, A.; Miranda, C.; Royo, J.; Matese, A. High-resolution UAV-based thermal imaging to estimate the instantaneous and seasonal variability of plant water status within a vineyard. *Agric. Water Manag.* **2017**, *183*, 49–59. [CrossRef]

168. Suárez, L.; Zarco-Tejada, P.J.; Sepulcre-Cantó, G.; Pérez-Priego, O.; Miller, J.; Jiménez-Muñoz, J.; Sobrino, J. Assessing canopy PRI for water stress detection with diurnal airborne imagery. *Remote Sens. Environ.* **2008**, *112*, 560–575. [CrossRef]
169. Torres-Sánchez, J.; López-Granados, F.; Peña, J.M. An automatic object-based method for optimal thresholding in UAV images: Application for vegetation detection in herbaceous crops. *Comput. Electron. Agric.* **2015**, *114*, 43–52. [CrossRef]
170. Capolupo, A.; Kooistra, L.; Berendonk, C.; Boccia, L.; Suomalainen, J. Estimating plant traits of grasslands from UAV-acquired hyperspectral images: A comparison of statistical approaches. *ISPRS Int. J. Geo-Inf.* **2015**, *4*, 2792–2820. [CrossRef]
171. Carreiras, J.M.; Pereira, J.M.; Pereira, J.S. Estimation of tree canopy cover in evergreen oak woodlands using remote sensing. *For. Ecol. Manag.* **2006**, *223*, 45–53. [CrossRef]
172. Cohen, Y.; Alchanatis, V.; Meron, M.; Saranga, Y.; Tsipris, J. Estimation of leaf water potential by thermal imagery and spatial analysis. *J. Exp. Bot.* **2005**, *56*, 1843–1852. [CrossRef]
173. Di Gennaro, S.F.; Battiston, E.; Di Marco, S.; Facini, O.; Matese, A.; Nocentini, M.; Palliotti, A.; Mugnai, L. Unmanned Aerial Vehicle (UAV)-based remote sensing to monitor grapevine leaf stripe disease within a vineyard affected by esca complex. *Phytopathol. Mediterr.* **2016**, *55*, 262–275.
174. Drake, J.B.; Knox, R.G.; Dubayah, R.O.; Clark, D.B.; Condit, R.; Blair, J.B.; Hofton, M. Above-ground biomass estimation in closed canopy neotropical forests using lidar remote sensing: Factors affecting the generality of relationships. *Glob. Ecol. Biogeogr.* **2003**, *12*, 147–159. [CrossRef]
175. Hütt, C.; Koppe, W.; Miao, Y.; Bareth, G. Best accuracy land use/land cover (LULC) classification to derive crop types using multitemporal, multisensor, and multi-polarization SAR satellite images. *Remote Sens.* **2016**, *8*, 684. [CrossRef]
176. Possoch, M.; Bieker, S.; Hoffmeister, D.; Bolten, A.; Schellberg, J.; Bareth, G. Multi-temporal crop surface models combined with the RGB vegetation index from UAV-based images for forage monitoring in grassland. *Int. Arch. Photogramm. Remote Sens. Spat. Inf. Sci.* **2016**, *41*, 991. [CrossRef]
177. Sagan, V.; Maimaitijiang, M.; Sidike, P.; Eblimit, K.; Peterson, K.T.; Hartling, S.; Esposito, F.; Khanal, K.; Newcomb, M.; Pauli, D.; et al. Uav-based high resolution thermal imaging for vegetation monitoring, and plant phenotyping using ici 8640 p, flir vue pro r 640, and thermomap cameras. *Remote Sens.* **2019**, *11*, 330. [CrossRef]
178. Schirrmann, M.; Giebel, A.; Gleiniger, F.; Pflanz, M.; Lentschke, J.; Dammer, K.H. Monitoring agronomic parameters of winter wheat crops with low-cost UAV imagery. *Remote Sens.* **2016**, *8*, 706. [CrossRef]
179. Swain, K.C.; Thomson, S.J.; Jayasuriya, H.P.; et al. Adoption of an unmanned helicopter for low-altitude remote sensing to estimate yield and total biomass of a rice crop. *Trans. ASAE Am. Soc. Agric. Eng.* **2010**, *53*, 21.
180. Yebra, M.; Van Dijk, A.; Leuning, R.; Huete, A.; Guerschman, J.P. Evaluation of optical remote sensing to estimate actual evapotranspiration and canopy conductance. *Remote Sens. Environ.* **2013**, *129*, 250–261. [CrossRef]
181. Agapiou, A.; Alexakis, D.D.; Hadjimitsis, D.G. Spectral sensitivity of ALOS, ASTER, IKONOS, LANDSAT and SPOT satellite imagery intended for the detection of archaeological crop marks. *Int. J. Digit. Earth* **2014**, *7*, 351–372. [CrossRef]
182. Chianucci, F.; Disperati, L.; Guzzi, D.; Bianchini, D.; Nardino, V.; Lastri, C.; Rindinella, A.; Corona, P. Estimation of canopy attributes in beech forests using true colour digital images from a small fixed-wing UAV. *Int. J. Appl. Earth Obs. Geoinf.* **2016**, *47*, 60–68. [CrossRef]
183. Inglada, J.; Vincent, A.; Arias, M.; Marais-Sicre, C. Improved early crop type identification by joint use of high temporal resolution SAR and optical image time series. *Remote Sens.* **2016**, *8*, 362. [CrossRef]
184. Li, W.; Niu, Z.; Chen, H.; Li, D.; Wu, M.; Zhao, W. Remote estimation of canopy height and aboveground biomass of maize using high-resolution stereo images from a low-cost unmanned aerial vehicle system. *Ecol. Indic.* **2016**, *67*, 637–648. [CrossRef]
185. Mintenig, S.; Int-Veen, I.; Löder, M.G.; Primpke, S.; Gerdts, G. Identification of microplastic in effluents of waste water treatment plants using focal plane array-based micro-Fourier-transform infrared imaging. *Water Res.* **2017**, *108*, 365–372. [CrossRef] [PubMed]
186. Wu, B.; Meng, J.; Li, Q.; Yan, N.; Du, X.; Zhang, M. Remote sensing-based global crop monitoring: Experiences with China's CropWatch system. *Int. J. Digit. Earth* **2014**, *7*, 113–137. [CrossRef]

187. Bock, C.; Poole, G.; Parker, P.; Gottwald, T. Plant disease severity estimated visually, by digital photography and image analysis, and by hyperspectral imaging. *Crit. Rev. Plant Sci.* **2010**, *29*, 59–107. [CrossRef]
188. Bovensmann, H.; Buchwitz, M.; Burrows, J.; Reuter, M.; Krings, T.; Gerilowski, K.; Schneising, O.; Heymann, J.; Tretner, A.; Erzinger, J. A remote sensing technique for global monitoring of power plant CO_2 emissions from space and related applications. *Atmos. Meas. Tech.* **2010**, *3*, 781–811. [CrossRef]
189. Chaerle, L.; Leinonen, I.; Jones, H.G.; Van Der Straeten, D. Monitoring and screening plant populations with combined thermal and chlorophyll fluorescence imaging. *J. Exp. Bot.* **2006**, *58*, 773–784. [CrossRef]
190. Jin, X.; Liu, S.; Baret, F.; Hemerlé, M.; Comar, A. Estimates of plant density of wheat crops at emergence from very low altitude UAV imagery. *Remote Sens. Environ.* **2017**, *198*, 105–114. [CrossRef]
191. Lasaponara, R.; Masini, N. Detection of archaeological crop marks by using satellite QuickBird multispectral imagery. *J. Archaeol. Sci.* **2007**, *34*, 214–221. [CrossRef]
192. Prince, G.; Clarkson, J.P.; Rajpoot, N.M.; et al. Automatic detection of diseased tomato plants using thermal and stereo visible light images. *PLoS ONE* **2015**, *10*, e0123262.
193. Rhee, J.; Im, J.; Carbone, G.J. Monitoring agricultural drought for arid and humid regions using multi-sensor remote sensing data. *Remote Sens. Environ.* **2010**, *114*, 2875–2887. [CrossRef]
194. Smith, M.L.; Ollinger, S.V.; Martin, M.E.; Aber, J.D.; Hallett, R.A.; Goodale, C.L. Direct estimation of aboveground forest productivity through hyperspectral remote sensing of canopy nitrogen. *Ecol. Appl.* **2002**, *12*, 1286–1302. [CrossRef]
195. Yue, J.; Yang, G.; Li, C.; Li, Z.; Wang, Y.; Feng, H.; Xu, B. Estimation of winter wheat above-ground biomass using unmanned aerial vehicle-based snapshot hyperspectral sensor and crop height improved models. *Remote Sens.* **2017**, *9*, 708. [CrossRef]
196. Zarco-Tejada, P.J.; Guillén-Climent, M.L.; Hernández-Clemente, R.; Catalina, A.; González, M.; Martín, P. Estimating leaf carotenoid content in vineyards using high resolution hyperspectral imagery acquired from an unmanned aerial vehicle (UAV). *Agric. For. Meteorol.* **2013**, *171*, 281–294. [CrossRef]
197. Elarab, M.; Ticlavilca, A.M.; Torres-Rua, A.F.; Maslova, I.; McKee, M. Estimating chlorophyll with thermal and broadband multispectral high resolution imagery from an unmanned aerial system using relevance vector machines for precision agriculture. *Int. J. Appl. Earth Obs. Geoinf.* **2015**, *43*, 32–42. [CrossRef]
198. Kalacska, M.; Lalonde, M.; Moore, T. Estimation of foliar chlorophyll and nitrogen content in an ombrotrophic bog from hyperspectral data: Scaling from leaf to image. *Remote Sens. Environ.* **2015**, *169*, 270–279. [CrossRef]
199. Moeckel, T.; Dayananda, S.; Nidamanuri, R.; Nautiyal, S.; Hanumaiah, N.; Buerkert, A.; Wachendorf, M. Estimation of vegetable crop parameter by multitemporal UAV-borne images. *Remote Sens.* **2018**, *10*, 805. [CrossRef]
200. Moshou, D.; Bravo, C.; Oberti, R.; West, J.; Bodria, L.; McCartney, A.; Ramon, H. Plant disease detection based on data fusion of hyperspectral and multi-spectral fluorescence imaging using Kohonen maps. *Real-Time Imaging* **2005**, *11*, 75–83. [CrossRef]
201. Rußwurm, M.; Korner, M. Temporal vegetation modelling using long short-term memory networks for crop identification from medium-resolution multi-spectral satellite images. In Proceedings of the IEEE Conference on Computer Vision and Pattern Recognition Workshops, Honolulu, HI, USA, 21–26 July 2017; pp. 11–19.
202. Skakun, S.; Kussul, N.; Shelestov, A.Y.; Lavreniuk, M.; Kussul, O. Efficiency assessment of multitemporal C-band Radarsat-2 intensity and Landsat-8 surface reflectance satellite imagery for crop classification in Ukraine. *IEEE J. Sel. Top. Appl. Earth Obs. Remote Sens.* **2015**, *9*, 3712–3719. [CrossRef]
203. Xie, C.; Shao, Y.; Li, X.; He, Y. Detection of early blight and late blight diseases on tomato leaves using hyperspectral imaging. *Sci. Rep.* **2015**, *5*, 16564. [CrossRef]
204. Gracia-Romero, A.; Kefauver, S.C.; Fernandez-Gallego, J.A.; Vergara-Díaz, O.; Nieto-Taladriz, M.T.; Araus, J.L. UAV and ground image-based phenotyping: A proof of concept with Durum wheat. *Remote Sens.* **2019**, *11*, 1244. [CrossRef]
205. Franceschet, M. The role of conference publications in CS. *Commun. ACM* **2010**, *53*, 129–132. [CrossRef]
206. Vrettas, G.; Sanderson, M. Conferences versus journals in computer science. *J. Assoc. Inf. Sci. Technol.* **2015**, *66*, 2674–2684. [CrossRef]
207. Zou, Y.; Li, G.; Wang, S. The Fusion of Satellite and Unmanned Aerial Vehicle (UAV) Imagery for Improving Classification Performance. In Proceedings of the 2018 IEEE International Conference on Information and Automation (ICIA), Wuyishan, China, 11–13 August 2018; pp. 836–841.

7

Detecting Banana Plantations in the Wet Tropics, Australia, using Aerial Photography and U-Net

Andrew Clark * and Joel McKechnie

Remote Sensing Centre, Queensland Department of Environment and Science, GPO Box 2454, Brisbane, QLD 4001, Australia; joel.mckechnie@hotmail.com

* Correspondence: andrew.clark@des.qld.gov.au

Abstract: Bananas are the world's most popular fruit and an important staple food source. Recent outbreaks of Panama TR4 disease are threatening the global banana industry, which is worth an estimated $8 billion. Current methods to map land uses are time- and resource-intensive and result in delays in the timely release of data. We have used existing land use mapping to train a U-Net neural network to detect banana plantations in the Wet Tropics of Queensland, Australia, using high-resolution aerial photography. Accuracy assessments, based on a stratified random sample of points, revealed the classification achieves a user's accuracy of 98% and a producer's accuracy of 96%. This is more accurate compared to existing (manual) methods, which achieved a user's and producer's accuracy of 86% and 92% respectively. Using a neural network is substantially more efficient than manual methods and can inform a more rapid respond to existing and new biosecurity threats. The method is robust and repeatable and has potential for mapping other commodities and land uses which is the focus of future work.

Keywords: convolutional neural network; U-Net; segmentation; deep learning; land use; banana plantation; Panama TR4; aerial photography

---

1. Introduction

1.1. Panama TR4

Fusarium oxysporum f. sp. *cubense* tropical race 4 (Foc TR4), is a soil-borne fungus that causes Panama TR4, a form of fusarium wilt that eventually kills infected banana plants [1,2]. Since the 1980s, Foc TR4 has been regarded as the most important biosecurity threat to the global banana industry, and an unparalleled botanical epidemic [2], persisting indefinitely in the soil with no effective control method. There is currently no suitable replacement variety for Cavendish that can meet the needs of the market [1,3,4]. The disease can spread anthropogenically and naturally through the transportation of infected plant material, soil, and water [5].

Bananas are the world's most popular fruit and an important staple food [6], with a global industry worth $8 billion annually [7]. The potential impact of Panama TR4 is severe, because Cavendish accounts for approximately 47% of bananas produced globally, predominantly sourced from Asia, Latin America, and Africa [7].

Foc TR4 was first identified in Sumatra, Indonesia, in 1992 [8], and to date, has spread across several continents [1,9,10] including Australia; the Northern Territory in 1997 and the state of Queensland in 2015 [4,8]. The Queensland Government Department of Agriculture and Fisheries (DAF) initiated the Panama TR4 Program in response to the first detection in the Tully River catchment, within the Wet Tropics bioregion. The program successfully controlled and contained the impact of Panama TR4 within a section of the Tully River catchment, however, three additional plantations within this location were infested in 2017, 2018, and 2020 [9,10]. At the time of the outbreak there was no accurate spatial dataset of all banana plantations [11]. The absence of this data jeopardized the banana industry and DAF's abilities to respond rapidly. Approximately 94% of the national banana supply is concentrated in the Wet Tropics, and is worth approximately AUD 480 million annually to the national economy [12]. Therefore, it is essential that the locations and extents of affected, and unaffected, banana plantations are monitored—particularly where vectors are likely to be transported, for example through erosion of contaminated soil, distribution and processing facilities, and machinery and equipment used across multiple plantations.

1.2. Land Use Mapping

The Queensland Government, currently through the Department of Environment and Science, has mapped land use and land use change throughout Queensland since 1999. The Queensland Land Use Mapping Program (QLUMP) maps land use in accordance with the Australian Land Use and Management (ALUM) classification [13]. In 2015, when Panama TR4 was first detected in the Wet Tropics, banana plantations had not been specifically classified, as they did not explicitly appear in the ALUM classification. For DAF and the banana industry to manage the Panama TR4 infestation, Système Pour l'Observation de la Terre (SPOT) 6 imagery was acquired over the Wet Tropics, for QLUMP to manually digitize the extent of banana plantations.

Timely land use mapping is fundamental for responding to biosecurity incidents, and for other applications such as natural disaster response, natural resource management and environmental monitoring [14]. Advances in big data and imagery availability have created an opportunity to develop methods to automatically and efficiently classify land use features over large geographical areas to allow for higher spatial and temporal resolutions and a more detailed classification for commodity level observations.

Using high-resolution imagery, many different land uses can be identified with human vision, including banana plantations. This is a result of human operators combining a number of image properties including colors, textures, pixel proximity, geometric attributes, and contextual information such as related built infrastructure [15]. Spectral information alone cannot successfully distinguish land use features as some land uses can appear spectrally similar [16] and are usually restricted to a single sensor without cross calibration. Using ancillary datasets and decision trees to derive land use is not always an accurate representation of what is on the ground [14].

The greater availability of high-resolution imagery introduces more complexity into the data, requiring more computing power to process the imagery and more detailed classifications. The integration of textural properties through object-based segmentation techniques have significantly improved the classification results for remote sensing applications [15]. For land use mapping, it has been found that using spatial as well as spectral information outperforms pixel-based classifications [17]. However, object-based image analysis approaches still require human input [16,18] and these complex workflows tend to be just as time- and resource-intensive as entirely drawing the land use features manually [19–21].

1.3. Deep-Learning Classifications

Neural networks have been around for many decades (see review by Schmidhuber [22]). However, only since the recent advancements in GPU technology have they been able to be trained with large amounts of data in a reasonable amount of time [23]. Neural networks simulate the processes of the human brain—interconnected neurons which process incoming information [24]. The solution is obtained by nonalgorithmic and unstructured methods, and by the adjustment of weights connecting the neurons in the network [25]. They can adaptively simulate complex and non-linear patterns [24,26] such as those found in high-resolution aerial photography.

Deep learning methods are based on neural networks [22]. These networks consist of many layers, which can transform images into categories through learning of high-level features [27]. Convolutional Neural Networks (CNN) are situated at the fringe between machine learning and computer vision, combining the power of deep learning with contextual image analysis. CNNs have been used in applications such as number-plate reading, facial recognition, and aerial image classification [28,29]. CNNs have gained momentum for image classification since the AlexNet architecture won the ImageNet contest by a wide margin in 2012 [30].

Mnih [31] and Romero et al. [32] found that deep CNNs outperform shallow CNNs (with fewer hidden layers), Support Vector Machines (SVM), Kernel-based Principal Component Analysis (KPCA), and spectral classifications for land use classification in aerial photography, multispectral and hyperspectral imagery. As CNN classifications integrate spatial as well as spectral information, they achieve higher accuracy compared to SVM and Random Forest (RF) classifications [23].

CNNs evaluate large amounts of contextual information over multiple scales that can result in classifications at a lower resolution than the original image. To overcome this, the review article by Ma et al. [23] suggests using U-Net [33] or an ensemble of models trained with different variables or different architectures. U-Net was originally developed for image segmentation problems in biomedical imaging [33] and has been adopted for use with optical earth observation data with overall accuracies exceeding 90% [34–36]. Issues occur with U-Net at the edges of inference areas and vanishing gradient problems where the network becomes difficult to train and has insufficient learning [37]. To overcome this, Sun et al. [37] suggests using an ensemble model approach to counter the edge effects and the use of concatenation operations and activation functions such as rectified linear units (ReLU) to reduce the vanishing gradient problem.

1.4. *Automated Land Use Mapping*

Most studies that have used deep learning to automatically map land use features have a constrained geographical extent and are limited to a standard set of training images, for example the University of California Merced Land Use Dataset [38] and Banja-Luka [39]. Although this is advantageous in benchmarking different methodologies, no studies have operationalized the applications for real-world land use mapping over a large geographical area [23,40].

Previous work in Queensland has focused on land cover. Pringle et al. [41] developed a time-series based method to operationally map summer and winter crops, and Flood et al. [42] used U-Net to map land cover, specifically the extent of woody vegetation cover, to a resolution of 1 m. However, there is an absence of studies that classify land use in Australia using earth observation data with a resolution of less than 10 m which is required for detailed land use mapping [43,44] and the resolution in which CNNs have the most success [23].

The aim of this study is to demonstrate that using a convolutional neural network and high-resolution imagery (<1 m) to automate land use mapping is more rapid and accurate than existing manual methods. This would be of benefit to the on-going response to Panama TR4, future biosecurity incidents, and other events requiring a rapid response (e.g., natural disasters). Additionally, the improved land use data would better inform natural resource planning and monitoring, biodiversity conservation, and the monitoring and modelling of the effects of land management practices on water quality.

2. Materials and Methods

2.1. *Study Area*

The location of this study is within the Wet Tropics and Atherton Tablelands, located approximately 1200 km northwest of Brisbane, Australia (Figure 1). The region of 2.7 million hectares, includes the Wet Tropics World Heritage Area, and is adjacent to the Great Barrier Reef World Heritage Area.

In 2015, QLUMP reported the major secondary land uses within the project area to be: Nature Conservation (37.5%); Grazing (31.52%); Other Minimal Use (8.2%); and Cropping (7.2%). There were 14,533 hectares (0.65%) of banana plantations mapped.

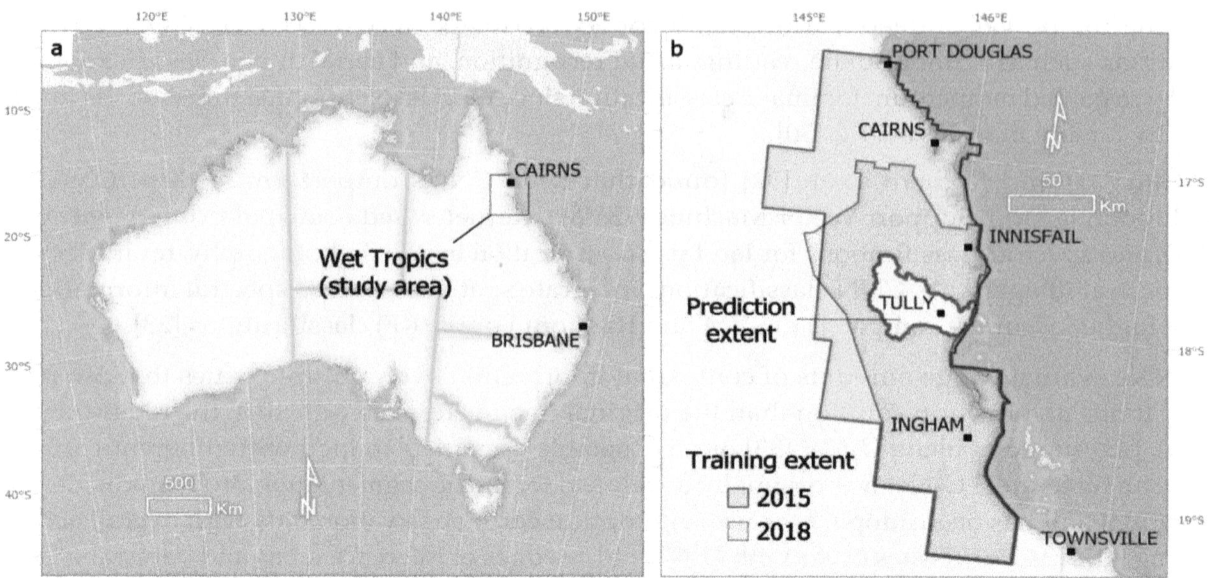

Figure 1. The study area (**a**) and extents of training (**b**). Note the Tully River catchment was excluded from the training and reserved as the extent of the final classification.

2.2. *Remote Sensing Imagery*

Two aerial imagery captures were used for this study (Figure 1). The 2015 data were acquired between 17th July and 14th October 2015, and the 2018 data were acquired between 1st and 27th August 2018, by AEROmetrex. At the time of this study, the full extent of the 2018 Wet Tropics imagery capture was not available so the training was restricted to the middle and southern sections of the region.

The data were captured with a fixed-wing mounted three-band true-color A3 Edge camera, at a spatial resolution of 25 cm and 20 cm for 2015 and 2018 respectively. The data were provided orthorectified by AEROmetrex based on a digital terrain model from LiDAR and stereo aerial imagery. The quality of the imagery is not consistent across the study area with some blurred areas and some discoloration. These artefacts are likely a result of post-processing of the imagery and appear to be located along tile boundaries where stitching the tiles and color balancing was not perfect. Unfortunately, the metadata supplied with the data does not list specific processing details. However the imagery is the best data available for the project area at a resolution suitable for this type of application. The Queensland Government has a large archive of aerial photography and it is likely these data, along with future captures will contain similar artefacts and any model developed will need to be robust enough to account for these inconsistencies.

2.3. *Project Hardware and Software*

Scripts and tools were written using the Python programming language. A combination of NumPy and Geospatial Data Abstraction Library (GDAL) were used to process the imagery and training data and convert them into multi-dimensional arrays, the format required for machine learning data processing.

A combination of Python, Nvidia's CUDA [45], CUDA deep neural network library (cuDNN) [46], Keras [47], and Tensorflow [48] formed the basis for the deep learning part of the project. Because of the volume of data and the number of iterations required to train a model, efficient processing of the data is required.

The Queensland Government's high performance computing (HPC) infrastructure consists of 2256 threads, 8.8 TB of memory and eight Nvidia Tesla V100 GPUs, used to process the training data, train the U-Net model, and to create the model inference (resulting area of banana plantations classified by the model once trained).

2.4. Existing Land Use Data Set

As described in Section 1.2, state-wide land use information is mapped by QLUMP. Land use is mapped to a national standard, according to the ALUM Classification—which has a three-tiered hierarchical structure broadly structured by the potential degree of modification from essentially native land cover [13]. The (six) primary and (32) secondary classes relate to land use, and (159) tertiary classes include commodity and land management practice information (e.g., "Tree fruits" as demonstrated in Figure 2. While tertiary-level information is particularly valuable for many applications, including biosecurity response, it has historically been expensive and impractical to collect, and as a result not consistently recorded.

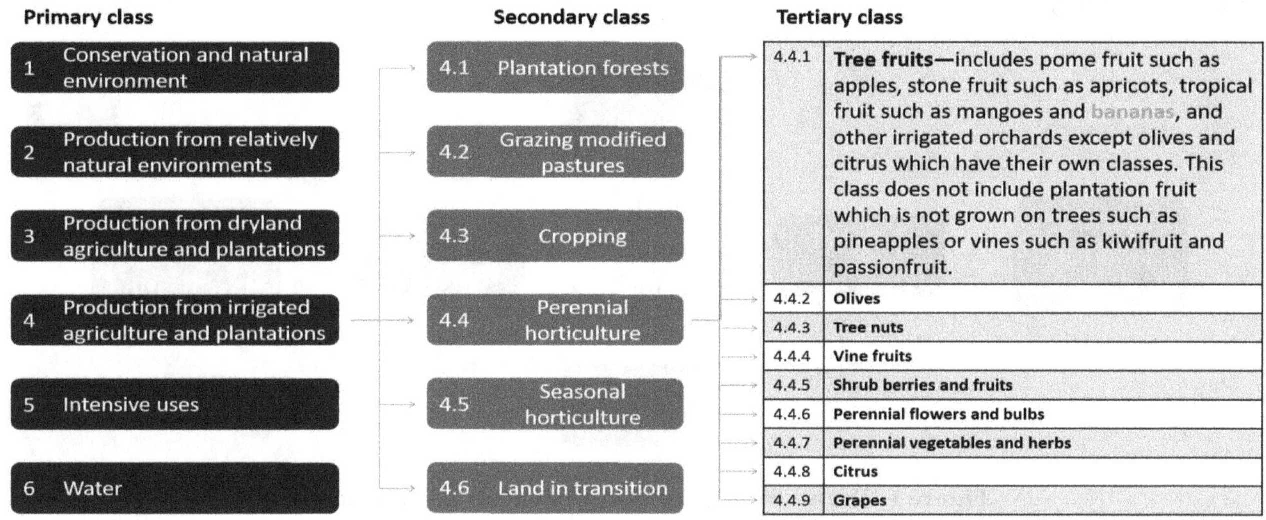

Figure 2. This diagram shows the three-tiered hierarchical structure of the Australian Land Use and Management (ALUM) classification and an extract demonstrating bananas as a commodity within the "Tree fruits" tertiary class, "Perennial horticulture" secondary class, and "Production from irrigated agriculture and plantations" primary class.

The QLUMP methodology has been an accurate, reliable, and cost-effective option since the late 90s—making use of available technology, data, and imagery. Mapping is undertaken primarily at the desktop, combining imagery interpretation and ancillary data to derive land use products. These products are field validated, peer reviewed, and accuracy assessed prior to publishing.

Because of the large area of Queensland (1.85 million squared kilometers), it has not been feasible to update land use information across the entire state at once, therefore updates occur regionally, using natural resource management (NRM) region boundaries. As a result, the currency of data varies from region to region. The most recent data is 2017 (Fitzroy and Burnett Mary NRM regions) and the most dated is 2012 (South East Queensland NRM region). The Wet Tropics NRM region was last updated to 2015. Regional updates occur on an ad hoc basis, dependent on state government priorities—for example the most recent updates were in the Great Barrier Reef catchments to support the Paddock to Reef Monitoring, Modelling and Reporting Program, and the Reef 2050 Water Quality Improvement Plan. A user survey conducted by QLUMP in 2020 indicates that there is a growing need for more current, and higher resolution land use information. The current QLUMP methodology, while proven, requires an intensive amount of manual image interpretation and spatial data analysis. There is a need for a more automated methodology that enables faster publishing of land use information, and CNNs are a possible solution.

2.5. Training Data

The generation of training data was an iterative process (Figure 3). Initially a subset of the existing QLUMP data in the study area was edited to better represent the land use features within the 2015 imagery. This editing was required as the QLUMP data were compiled using lower resolution imagery (SPOT 6 with a resolution of 1.5 m) and mapped land use features at a scale of 1:50,000 (using a minimum mapping unit area of 2 ha and width of 50 m) [13,49]. From these data, image and corresponding mask chips were randomly generated for model training. An initial inference was produced, converted to a polygon using a prediction probability threshold of 50%, and edited to fix any areas of omission and commission errors. This resulted in a more accurate and detailed training dataset compared to the QLUMP data. The image and mask chips were then regenerated for additional training.

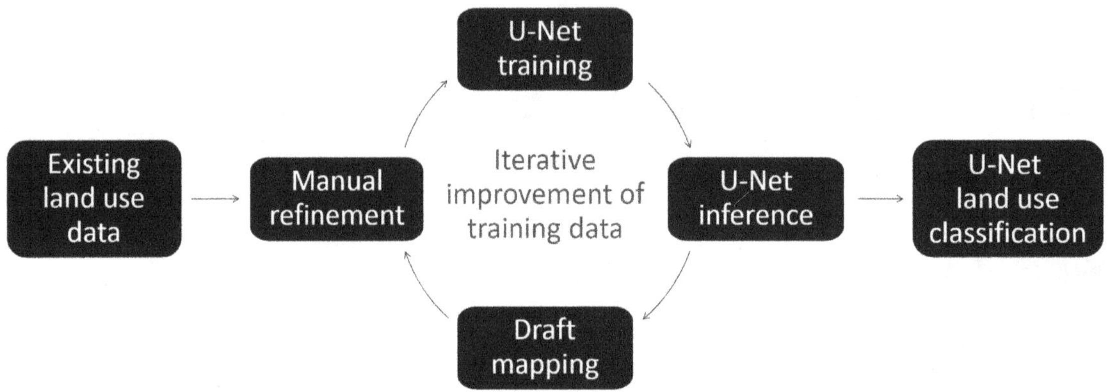

Figure 3. The training–inference–refinement iteration loop.

A total of 91,129 image chips with a size of 256 × 256 pixels were randomly generated from the 2015 and 2018 imagery. Of the total number of chips, 16,560 (18.2%) contained banana plantations, with the remainder located over a range of other land uses. No data were generated which intersected the Tully River catchment (Figure 1b) and this area was not included in the training stage of the study.

2.6. The U-Net Convolutional Neural Network

In this study, we aimed to classify every pixel in the image as banana or non-banana plantation through semantic segmentation using a convolutional neural network. The structure of the CNN follows the U-Net architecture [33] and is shown in Figure 4. It consists of two parts: (i) An encoding stage that downsamples the resolution of the input images; and, (ii) a decoding stage that upsamples and restores the images back to the original resolution. At each level of the encoding stages, two 3 × 3 convolution operations are applied using the rectified linear unit (ReLU) activation and a 2 × 2 max pooling operation to downsample the input images. The first level consists of the original satellite image and mask chips (256 pixels in width and height) where 64 3 × 3 filters are applied to each chip. At each subsequent level of the encoding side of the U-Net, the number of filters is doubled, doubling the number of bands of the images and the resolution halved until the bottom level where 1024 filters are applied to images 16 × 16 pixels in size.

The decoding stage also uses two 3 × 3 convolution operations but upsamples the data and concatenates the corresponding information in the encoding stage to double the resolution of the images, eventually restoring the original resolution of the input images in the final level. The final step is to conduct a 1 × 1 convolution using a sigmoid activation to produce a single band output probability classification with values ranging from 0 to 1. Values closer to 1 are more likely to be banana plantations. Using this configuration of the U-Net allows for the training of 31.4 million parameters overall.

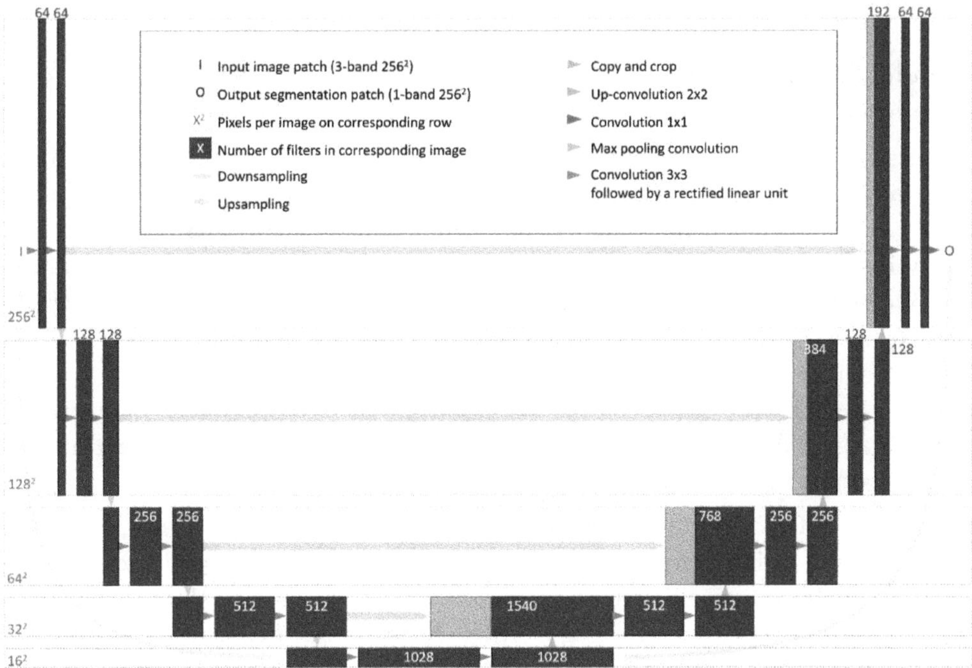

Figure 4. The U-Net architecture, starting with an input image patch (top-left) and ending with an output segmentation patch (top-right) [33].

2.7. U-Net Training

The purpose of the training stage is to allow the model to learn how to identify banana plantations. This is achieved by iterating over the training image and mask chips to determine their relevant color, texture, and context attributes [50]. As the images were captured over a range of image dates, subjected to color balancing and not corrected to surface reflectance, the training patches were randomly augmented by flipping, rotating, and changing the brightness of the image. This creates a more robust model for these image types [51,52].

A loss function of binary cross entropy and the Jaccard Index was used to judge the performance of the model while training. The Nesterov Adam optimizer [47] with an initial learning rate of 1×10^{-5} was used and reduced to 1×10^{-6} at epoch 47 as the model accuracy was no longer improving. The model was trained from scratch for a total of 50 epochs. One epoch represents one complete iteration over all training images and masks. The model took approximately 30 h to train on one Nvidia Tesla V100 GPU.

2.8. U-Net Inference

Sun et al. [37] found in previous studies that the edges of each image chip have a lower accuracy than the center region. To overcome this, a two-pass ensemble inference strategy was adopted. This was done by breaking the whole image into 256 × 256 image chips, iteratively applying the model to the original patch and three rotated versions of the patch and averaging the results. The second pass of the image was conducted, offset by 128 pixels. The result from the two passes are combined using a weighted average with pixels toward the center of the patch given a higher weight than the pixels toward the edge.

A prediction probability threshold of 90% was used to classify areas of banana plantations. Small features and gaps of 0.01 hectares or less were removed or filled and the data were converted into a polygon feature class with the edges smoothed to remove the square edges of individual pixels.

To increase the performance of the classification, the original aerial images were split up into overlapping tiles, allowing the inference to be conducted on all eight Nvidia Tesla V100 GPUs. It took approximately 12 h to run the model on the 2015 imagery.

2.9. Accuracy Assessment

Two independent assessment measures were conducted to assess the accuracy of the U-Net and QLUMP classifications of banana plantations in the Tully River catchment.

The first assessment was based on a stratified random sample of 9805 points using the ALUM Classification tertiary classes as the strata, following the method described in [53]. As the scale of the QLUMP data was different to the high-resolution imagery used to map land use classes, every point was visually inspected to ensure they were correctly classified as a banana plantation or other land use. If there were inconsistencies between the classification of a point and the imagery, the point was reclassified to the correct land use class. For example, if a banana plantation point fell on an area of fallow or a narrow road between the banana plantations, these points were reclassified as land-in-transition or road. The points were used to calculate the user's, producer's, and total accuracies for the U-Net and QLUMP classifications of banana plantations. In total, 701 out of 9805 points were located on banana plantations (7.15%).

The second measure of accuracy for both classifications was conducted using a similarity coefficient, the Jaccard Index defined in Equation (1). The benefit of using the Jaccard index for measuring the accuracy of the banana plantation classification is that it accounts for all the validation data excluded from the training of the U-Net model for the Tully River catchment. It is a similarity index and compares how well the validation and classification banana plantation locations match.

$$J(C,G) = \frac{C \cap G}{C \cup G} \quad (1)$$

where C is the classification (U-Net or QLUMP) and G is the validation data.

3. Results

3.1. U-Net Training

After 50 epochs, the model achieved a Jaccard Index of 0.961 and a loss of 0.01 (Figure 5). Because of the high-quality training data used, the model achieved a Jaccard Index of 0.8 after 4 epochs and 0.9 after 11 epochs.

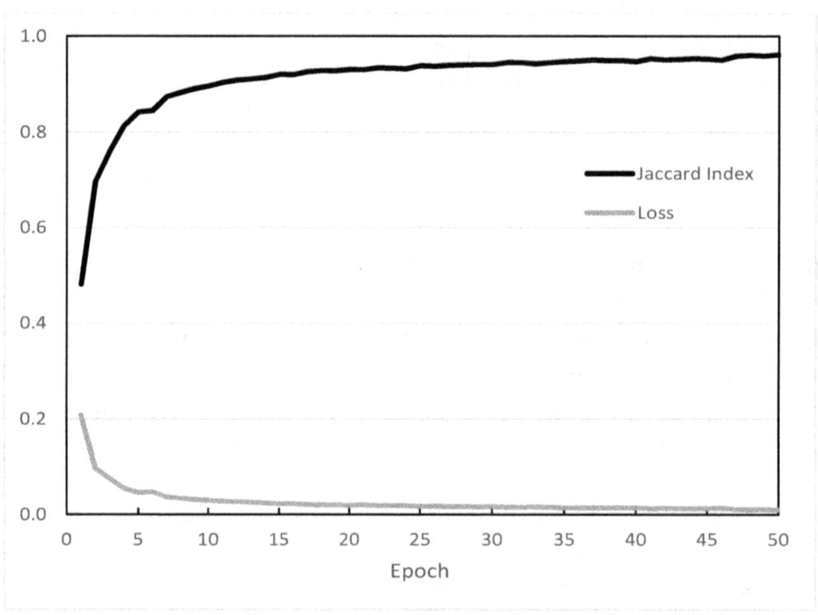

Figure 5. The Jaccard Index and loss function for each epoch. The number of epochs was restricted to 50. A marginal improvement may have been recorded if allowed to continue.

3.2. U-Net Classification

The model was applied to the Tully River catchment section of the aerial imagery that was excluded from the training. It took approximately two hours to create the output rasters and an additional 2.5 h to threshold, filter, and merge the classification tiles to a single polygon feature class. The result was the model-inferred extent of banana plantations, with a probability of 90% or greater.

Figure 6 shows examples of the U-Net classification of banana plantations in the Tully River catchment. In all examples the U-Net classification matches the validation data except in Figure 6(4c) where the U-Net classification was identified a new banana plantation that was missed in the validation dataset because of the young age of the plants and low canopy cover resulting in this area not being identified as a banana plantation by the human operator.

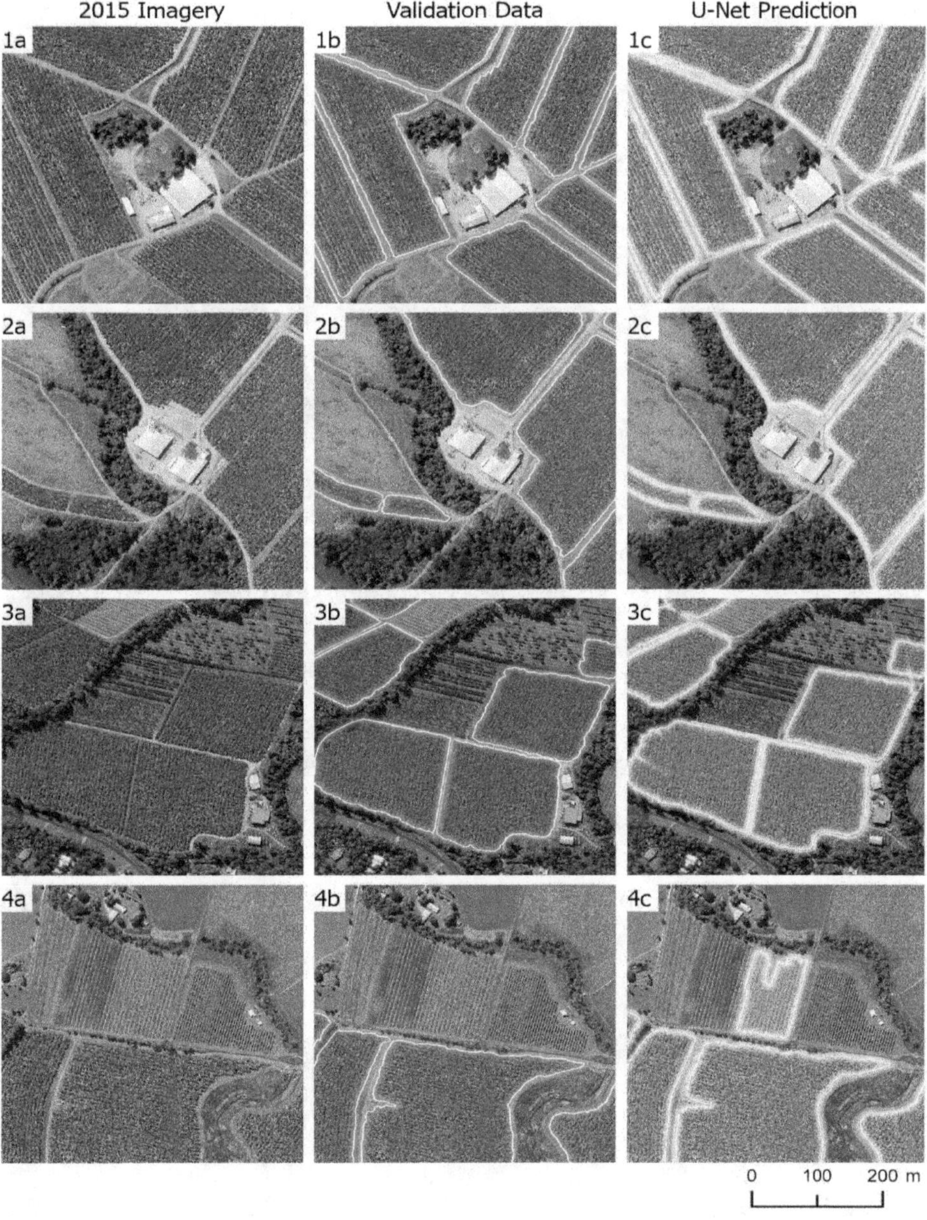

Figure 6. Examples (1–4) show correctly classified areas of banana plantations for the U-Net model. The left column (**a**) shows the 3-band true-color aerial photography. The middle column (**b**) shows the validation data and the right column (**c**) shows the output U-Net model classification. Note in Figure 6(4c), the U-Net classification identified a new plantation that was missed in the validation dataset.

Figure 7 shows areas where the U-Net classification did not perform well. Figure 7(1,2) show a papaya plantation and sugarcane crop respectively which the U-Net has incorrectly classified both as banana plantations. Figure 7(3) shows an area of misclassification possibly related to the effects of shadow on the banana plantation, and Figure 7(4) shows the U-Net classification not correctly classifying blurred areas of the aerial photography.

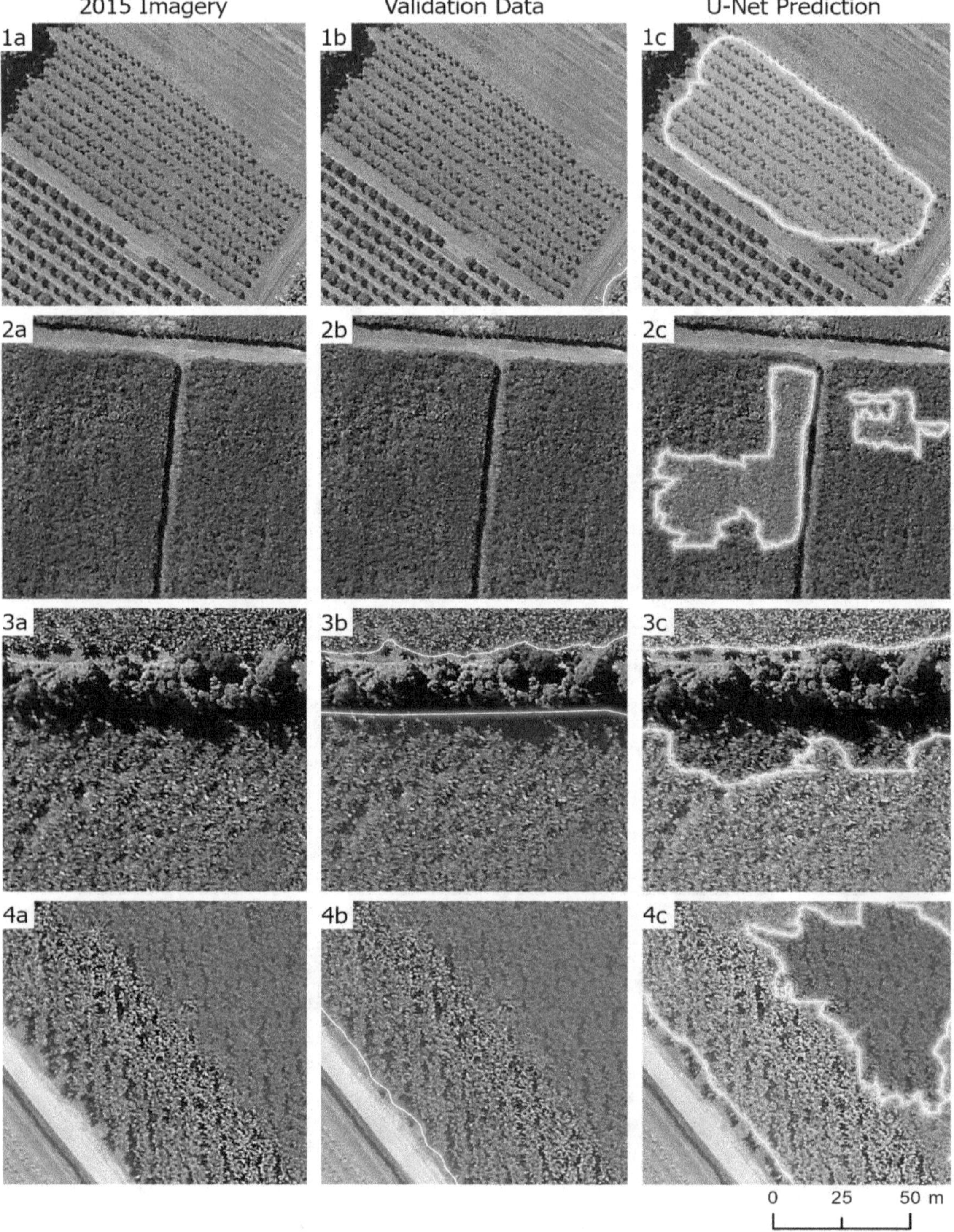

Figure 7. Examples (1–4) show incorrectly classified areas of banana plantations for the U-Net classification. The left column (**a**) shows the three-band true-color aerial photography. The middle column (**b**) shows the validation data and the right column (**c**) shows the output U-Net classification.

3.3. Accuracy Results

The results from the accuracy assessments can be found in Table 1. Overall, both classifications had a high total accuracy of >0.995. As the focus of this work is specifically mapping banana plantations which only represent 2.3% of the total area of the Tully River catchment, Table 1 shows the user's and producer's accuracy for this class. We found the QLUMP banana plantation classification had a user's and producer's accuracy of 0.862 and 0.921 respectively and the U-Net banana plantation classification had a user's and producer's accuracy of 0.983 and 0.959 respectively (Table 1). The Jaccard Index for the QLUMP classification was 0.341 and U-Net classification was 0.943.

Table 1. Accuracy assessment results showing the Jaccard Index, User's, Producer's, and total accuracies, assessed on independent validation data. The Total, User's, and Producer's accuracies were based on a stratified random sample of 9805 points, 701 of which were banana plantation points. The Jaccard Index used all the validation data for the Tully River catchment.

Classification	Total Accuracy	User's Accuracy	Producer's Accuracy	Jaccard Index
QLUMP Banana Plantations [1]	0.996	0.862	0.921	0.341
U-Net Banana Plantations	0.999	0.983	0.959	0.943

[1] The cartographic scale of the QLUMP mapping is smaller than the scale of the U-Net classification.

As Table 1 indicates that both classifications are more likely to miss areas of banana plantations (false negatives) than classify other land use features as banana plantations (false positives). It must be reiterated that the QLUMP data were compiled using lower resolution imagery and mapped land use features at a scale of 1:50,000. As a result, small-area and narrow land uses (e.g., farm sheds and roads) were aggregated into surrounding land uses. Also, banana plantations were mapped as bananas, regardless of them being active or fallow in the imagery. The scale of QLUMP data and fallow plantations affected the accuracy of the mapping, when compared with the U-Net classification. When specifically analyzing the location and extent of banana plantations, the accuracy assessment results suggest the U-Net classification is more accurate than the QLUMP data.

The quality of the imagery for U-Net is restricted to data processed by a commercial third-party. Inconsistencies in the geometric and radiometric corrections and post-processing operations (such as tile mosaicing and color balancing) have affected the output classification.

4. Discussion

In this paper we have presented an automated approach to mapping banana plantations using the U-Net CNN architecture [33] to assist the biosecurity response to Foc TR4. The U-Net has been successfully applied to other land uses around the world, but not at an operational level [40,54,55]. Until this study, there were no existing automated classifications to detect banana plantations using high-resolution aerial photography in Queensland, Australia or globally.

One perceived benefit of manually mapping is that humans can draw on undefined experiences or obscure learnings, such as past, present, or regional knowledge about land uses [15]. Despite this, when comparing the CNN approach to existing methods, we found the new classification technique is more accurate (>0.94). This is consistent with other similar studies mapping land uses using deep learning [34,40,54] except in this study we have applied our model over a broad geographical area.

We have also found the CNN method to be more rapid, allowing the automatic classification of banana plantations within hours when compared with existing methods requiring weeks of manually digitizing features. There were some misclassifications associated with blurred sections of aerial imagery due to third party post-processing, which caused some sections of banana plantations to be missed. Additionally, some papaya plantations and small areas of sugarcane crops were misclassified as banana plantations. To address these issues and to create a more robust model, we suggest additional training data be generated in these problem areas to allow the CNN to better learn these features. It is important for a model classifying land use features within high resolution aerial photography to have the ability to account for these types of artefacts as past and future imagery captures are likely to contain similar issues. Stratifying the random generation of training data by land uses or targeting similar land uses (such as papaya plantations) would also ensure the CNN has enough examples of non-banana plantations, and will improve the classification result.

While the U-Net has enabled greater accuracy and a more rapid land use product, it is limited by the availability of extensive training data which is a prerequisite. Initially this work was made possible using the data produced by the QLUMP methodology which was used to map the initial extent of banana plantations in 2015. This will also be the case for any future land use classes mapped using the U-Net. Therefore, there are implications for both the QLUMP and U-Net methodologies going forward. A logical option would be to iteratively update and improve the original QLUMP data, using the proposed U-Net methodology.

Future work will focus on updating the location and extent of all banana plantations within the Wet Tropics to 2018 which was only partially available for this work (Figure 1). We will also be expanding this method to other land uses and commodities, developing methods for monitoring land use change and investigating if these methods can be used to detect damaged plantations either from wilt associated with Foc TR4 or as a result of wind damage from natural disaster such as Tropical Cyclones. Developing a framework to automatically map land use features would benefit mapping programs in Australia and globally. The automated and efficient classification of land use features from high-resolution imagery will be extremely valuable in responding to current and future biosecurity incidents as well as other events requiring a rapid response such as natural disasters. This will have applications in agricultural productivity and sustainability, land use planning, natural resource condition monitoring and investment, biodiversity conservation, and improving water availability and quality [14].

5. Conclusions

Current methods to map land use in Queensland are based on manual image interpretation, which are time- and resource-intensive. Land use information is fundamental for informing the response to biosecurity incidents, such as the detection of Panama TR4 in the Tully River catchment. Advances in big data and imagery availability have created an opportunity to develop methods to automatically and efficiently classify land use features over large geographical areas. This allows for higher spatial and temporal resolutions, and a more detailed classification for commodity-level observations.

In this paper we have presented an automated and efficient classification technique for detecting the location and extent of banana plantations in the Wet Tropics, which we have shown is an improvement on the existing mapping methodology. The new classification approach used a refined version of the existing QLUMP mapping to train a CNN using the U-Net architecture [33].

Author Contributions: Conceptualization, A.C.; methodology, A.C.; software, A.C.; validation, A.C.; formal analysis, A.C.; investigation, A.C. and J.M.; data curation, A.C.; writing—original draft preparation, A.C. and J.M.; writing—review and editing, A.C. and J.M.; visualization, A.C. and J.M.; supervision, A.C.; project administration, A.C. and J.M. All authors have read and agreed to the published version of the manuscript.

Acknowledgments: The authors would like to thank Grant Moule for introducing us to the U-Net architecture. The authors acknowledge Robert Denham (Joint Remote Sensing Research Program) who contributed to this research by providing advice on the accuracy assessment. The authors also appreciate the reviewers' constructive feedback and suggestions, namely Matthew Pringle (Queensland Department of Environment and Science), Stuart Phinn (The University of Queensland), and Peter Scarth (Joint Remote Sensing Research Program). The authors would also like to acknowledge the Queensland Government Department of Environment and Science for providing access to the HPC infrastructure and supporting this research.

References

1. Zheng, S.-J.; García-Bastidas, F.A.; Li, X.; Zeng, L.; Bai, T.; Xu, S.; Yin, K.; Li, H.; Fu, G.; Yu, Y.; et al. New geographical insights of the latest expansion of *Fusarium oxysporum* f.sp. cubense tropical race 4 into the greater mekong subregion. *Front. Plant Sci.* **2018**, *9*, 457. [CrossRef] [PubMed]
2. Ploetz, R.C. Management of Fusarium wilt of banana: A review with special reference to tropical race 4. *Crop Prot.* **2015**, *73*, 7–15. [CrossRef]
3. Bentley, S.; Pegg, K.G.; Moore, N.Y.; Davis, R.D.; Buddenhagen, I.W. Genetic variation among vegetative compatibility groups of *Fusarium oxysporum* f. sp. *cubense* analyzed by dna fingerprinting. *Phytopathology* **1998**, *88*, 1283–1293. [CrossRef] [PubMed]
4. Carvalhais, L.C.; Henderson, J.; Rincon-Florez, V.A.; O'Dwyer, C.; Czislowski, E.; Aitken, E.A.B.; Drenth, A. Molecular Diagnostics of Banana Fusarium Wilt Targeting Secreted-in-Xylem Genes. *Front. Plant Sci.* **2019**, *10*, 547. [CrossRef] [PubMed]
5. Daniells, J.W.; Lindsay, S.J. TR4 as a driver of agroecological approaches in banana production. *Acta Hortic.* **2018**, *1196*, 203–210. [CrossRef]
6. Ordonez, N.; Seidl, M.F.; Waalwijk, C.; Drenth, A.; Kilian, A.; Thomma, B.P.H.J.; Ploetz, R.C.; Kema, G.H.J. Worse comes to worst: Bananas and panama disease—When plant and pathogen clones meet. *PLoS Pathog.* **2015**, *11*, e1005197. [CrossRef] [PubMed]
7. Food and Agriculture Organisation of the United Nations Banana Facts and Figures. Available online: http://www.fao.org/economic/est/est-commodities/bananas/bananafacts/en/#.XXh7-Sgzb-h (accessed on 11 September 2019).
8. Ploetz, R.C. Diseases and pests: A review of their importance and management. *InfoMusa* **2004**, *13*, 11–16.
9. Acil Allen Consulting. *Panama TR4 Program Review—Final Report*; Acil Allen Consulting: Brisbane, Australia, 2018.
10. Department of Agriculture and Fisheries Current Panama TR4 Situation. Available online: https://www.daf.qld.gov.au/business-priorities/biosecurity/plant/eradication-surveillance-control/panama-disease/current-situation (accessed on 2 March 2020).
11. Department of Science, Information Technology, Innovation and the Arts. *Land Use Summary 1999–2009 for the Great Barrier Reef Catchments*; Queensland Government: Queensland, Australia, 2012.
12. Horticulture Innovation Australia. *Ah15001-Australian-Horticulture-Statistics-Handbook-Fruit-.pdf*; Horticulture Innovation Australia Limited: Sydney, Australia, 2019.
13. Australian Bureau of Agricultural and Resource Economics and Sciences. *The Australian Land Use and Management Classification Version 8*; Australian Bureau of Agricultural and Resource Economics and Sciences: Canberra, Australia, 2016.
14. Thackway, R. *Land Use in Australia*; ANU Press: Canberra, Australia, 2018; ISBN 978-1-921934-42-1.
15. Blaschke, T.; Kelly, M.; Merschdorf, H. Object-Based Image Analysis: Evolution, History, State of the Art, and Future Vision. In *Remotely Sensed Data Characterization, Classification, and Accuracies*; UC Berkeley: Berkeley, CA, USA, 2015; ISBN 978-1-4822-1787-2.
16. Myint, S.W.; Gober, P.; Brazel, A.; Grossman-Clarke, S.; Weng, Q. Per-pixel vs. object-based classification of urban land cover extraction using high spatial resolution imagery. *Remote Sens. Environ.* **2011**, *115*, 1145–1161. [CrossRef]
17. Pandey, P.C.; Koutsias, N.; Petropoulos, G.P.; Srivastava, P.K.; Dor, E.B. Land use/land cover in view of earth observation: Data sources, input dimensions, and classifiers—A review of the state of the art. *Geocarto Int.* **2019**. [CrossRef]
18. Hossain, M.D.; Chen, D. Segmentation for Object-Based Image Analysis (OBIA): A review of algorithms and challenges from remote sensing perspective. *ISPRS J. Photogramm. Remote Sens.* **2019**, *150*, 115–134. [CrossRef]
19. Grounds, S. Pineapples, Pixels and Objects: A Comparison of Remote Sensing Classification Techniques for Land Use Mapping. Masters' Thesis, University of New England, Armidale, Australia, 2007.
20. Johansen, K.; Sohlbach, M.; Sullivan, B.; Stringer, S.; Peasley, D.; Phinn, S. Mapping Banana Plants from High Spatial Resolution Orthophotos to Facilitate Plant Health Assessment. *Remote Sens.* **2014**, *6*, 8261–8286. [CrossRef]
21. Johansen, K.; Phinn, S.; Witte, C.; Philip, S.; Newton, L. Mapping Banana Plantations from Object-oriented Classification of SPOT-5 Imagery. *Photogramm. Eng. Remote Sens.* **2009** *75*, 1069–1081. [CrossRef]

22. Schmidhuber, J. Deep learning in neural networks: An overview. *Neural Netw.* **2015**, *61*, 85–117. [CrossRef] [PubMed]
23. Ma, L.; Liu, Y.; Zhang, X.; Ye, Y.; Yin, G.; Johnson, B.A. Deep learning in remote sensing applications: A meta-analysis and review. *ISPRS J. Photogramm. Remote Sens.* **2019**, *152*, 166–177. [CrossRef]
24. Jensen, J.R. *Introductory Digital Image Processing: A Remote Sensing Perspective*; Prentice Hall: Upper Saddle River, NJ, USA, 2005; ISBN 978-0-13-145361-6.
25. Rao, V.; Rao, H. *C++ Neural Networks and Fuzzy Logic*; BPB Publications: New Delhi, India, 1996; ISBN 978-81-7029-694-2.
26. Zhu, X.X.; Tuia, D.; Mou, L.; Xia, G.-S.; Zhang, L.; Xu, F.; Fraundorfer, F. Deep learning in remote sensing: A review. *IEEE Geosci. Remote Sens. Mag.* **2017**, *5*, 8–36. [CrossRef]
27. Litjens, G.; Kooi, T.; Bejnordi, B.E.; Setio, A.A.A.; Ciompi, F.; Ghafoorian, M.; van der Laak, J.A.W.M.; van Ginneken, B.; Sánchez, C.I. A survey on deep learning in medical image analysis. *Med. Image Anal.* **2017**, *42*, 60–88. [CrossRef]
28. Khalel, A.; El-Saban, M. Automatic Pixelwise Object Labeling for Aerial Imagery Using Stacked U-Nets. *arXiv* **2018**, arXiv180304953.
29. Othman, E.; Bazi, Y.; Alajlan, N.; Alhichri, H.; Melgani, F. Using convolutional features and a sparse autoencoder for land-use scene classification. *Int. J. Remote Sens.* **2016**, *37*, 2149–2167. [CrossRef]
30. Krizhevsky, A.; Sutskever, I.; Hinton, G.E. ImageNet classification with deep convolutional neural networks. *Commun. ACM* **2012**, *2*, 1097–1105. [CrossRef]
31. Mnih, V. *Machine Learning for Aerial Image Labeling*; University of Toronto: Toronto, ON, USA, 2013.
32. Romero, A.; Gatta, C.; Camps-Valls, G. Unsupervised Deep Feature Extraction for Remote Sensing Image Classification. *IEEE Trans. Geosci. Remote Sens.* **2016**, *54*, 1349–1362. [CrossRef]
33. Ronneberger, O.; Fischer, P.; Brox, T. U-Net: Convolutional networks for biomedical image segmentation. *arXiv* **2015**, arXiv150504597.
34. Kestur, R.; Meduri, A.; Narasipura, O. MangoNet: A deep semantic segmentation architecture for a method to detect and count mangoes in an open orchard. *Eng. Appl. Artif. Intell.* **2019**, *77*, 59–69. [CrossRef]
35. Stoian, A.; Poulain, V.; Inglada, J.; Poughon, V.; Derksen, D. Land Cover Maps Production with High Resolution Satellite Image Time Series and Convolutional Neural Networks: Adaptations and Limits for Operational Systems. *Remote Sens.* **2019**, *11*, 1986. [CrossRef]
36. Zhang, P.; Ke, Y.; Zhang, Z.; Wang, M.; Li, P.; Zhang, S. Urban Land Use and Land Cover Classification Using Novel Deep Learning Models Based on High Spatial Resolution Satellite Imagery. *Sensors* **2018**, *18*, 3717. [CrossRef] [PubMed]
37. Sun, Y.; Tian, Y.; Xu, Y. Problems of encoder-decoder frameworks for high-resolution remote sensing image segmentation: Structural stereotype and insufficient learning. *Neurocomputing* **2019**, *330*, 297–304. [CrossRef]
38. Yang, Y.; Newsam, S. Bag-of-visual-words and spatial extensions for land-use classification. In Proceedings of the 18th SIGSPATIAL International Conference on Advances in Geographic Information Systems—GIS '10, San Jose, CA, USA, 3–5 November 2010; p. 270.
39. Risojević, V.; Momić, S.; Babić, Z. Gabor Descriptors for Aerial Image Classification. In *Adaptive and Natural Computing Algorithms*; Dobnikar, A., Lotrič, U., Šter, B., Eds.; Springer: Berlin/Heidelberg, Germany, 2011; Volume 6594, pp. 51–60. ISBN 978-3-642-20266-7.
40. Du, Z.; Yang, J.; Ou, C.; Zhang, T. Smallholder Crop Area Mapped with a Semantic Segmentation Deep Learning Method. *Remote Sens.* **2019**, *11*, 888. [CrossRef]
41. Pringle, M.J.; Schmidt, M.; Tindall, D.R. Multi-decade, multi-sensor time-series modelling—Based on geostatistical concepts—To predict broad groups of crops. *Remote Sens. Environ.* **2018**, *216*, 183–200. [CrossRef]
42. Flood, N.; Watson, F.; Collett, L. Using a U-net convolutional neural network to map woody vegetation extent from high resolution satellite imagery across Queensland, Australia. *Int. J. Appl. Earth Obs. Geoinf.* **2019**, *82*, 101897. [CrossRef]
43. Wagner, F.H.; Sanchez, A.; Tarabalka, Y.; Lotte, R.G.; Ferreira, M.P.; Aidar, M.P.M.; Gloor, E.; Phillips, O.L.; Aragão, L.E.O.C. Using the U-net convolutional network to map forest types and disturbance in the Atlantic rainforest with very high resolution images. *Remote Sens. Ecol. Conserv.* **2019**, *5*, 360–375. [CrossRef]
44. Liu, S.; Qi, Z.; Li, X.; Yeh, A. Integration of Convolutional Neural Networks and Object-Based Post-Classification Refinement for Land Use and Land Cover Mapping with Optical and SAR Data. *Remote Sens.* **2019**, *11*, 690. [CrossRef]

45. NVIDIA. *NVIDIA CUDA*. Available online: https://developer.nvidia.com/cuda-zone (accessed on 16 March 2020).
46. NVIDIA. *NVIDIA CUDA Deep Neural Network Library*. Available online: https://developer.nvidia.com/cudnn (accessed on 16 March 2020).
47. Chollet, F. *Keras: The Python Deep Learning Library*. Available online: https://keras.io (accessed on 16 March 2020).
48. Abadi, M.; Agarwal, A.; Barham, P.; Brevdo, E.; Chen, Z.; Citro, C.; Corrado, G.S.; Davis, A.; Dean, J.; Devin, M.; et al. TensorFlow: Large-Scale Machine Learning on Heterogeneous Distributed Systems. *arXiv* **2016**, arXiv160304467.
49. The Department of Environment and Science. *Land Use Summary for the Burnett Mary NRM Region*; The Queensland Government: Brisbane, Australia, 2019.
50. Zhang, L.; Zhang, L.; Du, B. Deep Learning for Remote Sensing Data: A Technical Tutorial on the State of the Art. *IEEE Geosci. Remote Sens. Mag.* **2016**, *4*, 22–40. [CrossRef]
51. Wieland, M.; Li, Y.; Martinis, S. Multi-sensor cloud and cloud shadow segmentation with a convolutional neural network. *Remote Sens. Environ.* **2019**, *230*, 111203. [CrossRef]
52. Dosovitskiy, A.; Springenberg, J.T.; Brox, T. Unsupervised feature learning by augmenting single images. *arXiv* **2013**, arXiv13125242.
53. Stehman, S.V. Estimating area and map accuracy for stratified random sampling when the strata are different from the map classes. *Int. J. Remote Sens.* **2014**, *35*, 4923–4939. [CrossRef]
54. Freudenberg, M.; Nölke, N.; Agostini, A.; Urban, K.; Wörgötter, F.; Kleinn, C. Large Scale Palm Tree Detection In High Resolution Satellite Images Using U-Net. *Remote Sens.* **2019**, *11*, 312. [CrossRef]
55. Wurm, M.; Stark, T.; Zhu, X.X.; Weigand, M.; Taubenböck, H. Semantic segmentation of slums in satellite images using transfer learning on fully convolutional neural networks. *ISPRS J. Photogramm. Remote Sens.* **2019**, *150*, 59–69. [CrossRef]

8

Geometric and Radiometric Consistency of Parrot Sequoia Multispectral Imagery for Precision Agriculture Applications

Marica Franzini [1,*], Giulia Ronchetti [2], Giovanna Sona [2] and Vittorio Casella [1]

1 Department of Civil Engineering and Architecture, University of Pavia, Via Ferrata, 3, 27100 Pavia, Italy; vittorio.casella@unipv.it
2 Department of Civil and Environmental Engineering, Polytechnic University of Milan, Piazza Leonardo da Vinci 32, 20133 Milan, Italy; giulia.ronchetti@polimi.it (G.R.); giovanna.sona@polimi.it (G.S.)
* Correspondence: marica.franzini@unipv.it

Abstract: This paper is about the geometric and radiometric consistency of diverse and overlapping datasets acquired with the Parrot Sequoia camera. The multispectral imagery datasets were acquired above agricultural fields in Northern Italy and radiometric calibration images were taken before each flight. Processing was performed with the Pix4Dmapper suite following a single-block approach: images acquired in different flight missions were processed in as many projects, where different block orientation strategies were adopted and compared. Results were assessed in terms of geometric and radiometric consistency in the overlapping areas. The geometric consistency was evaluated in terms of point cloud distance using iterative closest point (ICP), while the radiometric consistency was analyzed by computing the differences between the reflectance maps and vegetation indices produced according to adopted processing strategies. For normalized difference vegetation index (NDVI), a comparison with Sentinel-2 was also made. This paper will present results obtained for two (out of several) overlapped blocks. The geometric consistency is good (root mean square error (RMSE) in the order of 0.1 m), except for when direct georeferencing is considered. Radiometric consistency instead presents larger problems, especially in some bands and in vegetation indices that have differences above 20%. The comparison with Sentinel-2 products shows a general overestimation of Sequoia data but with similar spatial variations (Pearson's correlation coefficient of about 0.7, p-value $< 2.2 \times 10^{-16}$).

Keywords: geometric consistency; radiometric consistency; point clouds; ICP; reflectance maps; vegetation indices; Parrot Sequoia

---

1. Introduction

1.1. Key Topics

Precision agriculture (PA) [1] is a very significant societal challenge and promises to enable several significant improvements: increase of productivity; optimal, and thus reduced, use of pesticides and fertilizers; and decreased use of water. These will translate into substantial benefits, including making more food available for mankind, increasing environmental sustainability, and contributing to the mitigation of climate change effects [2]. One key component of precision agriculture is crop health diagnostic capability. Within this context, in the last 5 years the use of lightweight unmanned aerial vehicles (UAVs) equipped with multispectral sensors has become quite popular. UAV-based surveys offer unprecedented ground resolution and operational capability. The second feature is particularly significant when periodic monitoring has to be performed, as the operator is free to choose the optimal time to fly.

1.2. Background

The processing of large datasets that cover wide areas and which need to be acquired by several UAV missions is still a challenging task. As for photogrammetric projects, these types of datasets, composed of various sub-blocks, require a careful assessment of the accuracy of the final products, from both geometric and radiometric points of view. This is even more true when time-series are analyzed; the consistency between data is mandatory in these cases.

Regarding geometric issues, the recent evolution of UAVs has provided low-cost systems with direct georeferencing (DG) capability. DG has several advantages: it allows flights in remote areas where access could be difficult or impossible [3], and reduces mission time and costs, since no ground control points (GCPs) need to be installed and measured. Unfortunately, navigation-grade GPS/GNSS receivers (Global Positioning System/Global Navigation Satellite System), such as the one integrated with the Parrot Sequoia sensor [4], are not of sufficient quality in the solution position for georeferencing of images. Some authors investigated this topic [3,5,6], obtaining metric errors. GCPs are traditionally suggested for georeferencing purposes. The number and distribution of GCPs have been explored by several authors [7–10]. Independently from the extent of the surveyed area, they state that a small number of GCPs is useful when they are only needed to perform datum transformation, while a larger number is necessary when camera self-calibration must be performed. For this second aim, their spatial distribution is important too, as ground points must cover the whole area of interest. However, GCPs cannot always be guaranteed in some applications, such as in precision agriculture, where inaccessibility is a frequent condition due to crops' stages of growth. In this case, other ground information can be useful, such as pre-existing orthophotos [11].

Independently from the strategy used for images orientation, the geometric quality of the results must be assessed both in terms of accuracy and consistency. Such analysis can be focused on exterior orientation parameters (EOPs) or on the photogrammetric products, such as dense point clouds or orthophotos. EOPs are traditionally evaluated by using a set of check points (CPs), which are considered during the bundle block adjustment as simple tie points; residuals between the photogrammetrically obtained object coordinates of markers and those preliminarily determined by surveying are then evaluated [12–14]. Photogrammetric products are assessed using additional information acquired by alternative systems, such as GNSS receivers and total stations [15,16] or light detection and ranging (LIDAR) [17–19].

Data quality can also be assessed in terms of consistency. This term means the agreement between different (partially overlapping) datasets acquired at the same time or at different times. Consistency can be assessed on various photogrammetric products, such as point clouds or orthophotos. The mentioned criterion is particularly significant when time-series are processed or when different processing strategies are tested, as in our case. Within this framework, [20] evaluated volumetric changes of a landslide areas using point clouds over a time-series, while [21] assessed the consistency of UAV-derived point clouds in relation to the focal length and target set.

While geometry is almost always considered when quality assessment is performed, radiometry is less often investigated but plays a key role in several applications, such as precision agriculture and environmental pollution detection. Regarding radiometry, some critical issues remain unsolved, such as which corrections must be considered and modelled, and several authors recently started to investigate these aspects. Honkavaara et al. [22] studied and assessed a processing methodology for biomass estimation in agriculture with a lightweight UAV spectral camera under varying illumination conditions. In [23], authors captured images from an UAV with Parrot Sequoia and assessed canopy reflectance consistency in avocado and banana orchards in Australia, while in [24] reflectance anisotropy of potato canopies in the Netherlands was mapped with a frame camera mounted on an UAV.

Although in recent years sensor manufacturers have improved in describing sensor performance and providing tools for performing radiometric corrections [25], the radiometric quality of data is still uncertain. The reliability of spectral information acquired by multispectral sensors mounted on UAVs is not completely clear [26]. Absolute accuracy might be insufficient for some applications, so that calibration procedures will be required [27].

The proposed radiometric calibrations are based on the availability of spectral targets, whose reflectance response is measured in situ with a spectrometer [28–33]. After evaluating costs, surveying and processing times, and required instrumentations and expertise, this calibration methodology cannot be adopted as a routine method in precision agriculture. Indeed, farmers must take into consideration data quality as well as economic impact [26]. To ensure the dissemination and the use of calibration procedures in the agriculture sector, it is advisable to optimize the exploitation of consumer-friendly tools, and best practices must be simplified.

1.3. Motivation

Nowadays equipment vendors are making an effort to supply easy-to-use HW (Hardware) and SW (Software) so that crop monitoring can be performed by individual farmers. The bundle of Parrot Sequoia© (Parrot S.A., Paris, France) and Pix4D© (Pix4D S.A., Prilly, Switzerland) is a clear and popular example of this approach.

The present paper arises from one simple yet crucial question: what is the reliability of the radiometric information and of the related vegetation indices acquired by the Sequoia camera and processed with the bundled Pix4DMapper software? Considering that UAV surveys for precision agriculture typically are multitemporal, the original question can be rephrased: what is the consistency between repeated surveys? In other words, when two datasets highlight differences for a certain part of a field, to what extent is this due to acquisition and processing errors, and to what extent does this point out a variation in the status of the crop? The importance of such questions is confirmed by the fact that only a few papers in the literature have explored them to date.

The present work studies geometric and radiometric consistency of two overlapping datasets, acquired with a Sequoia camera and processed with the bundled software. We focus on geometry to avoid the influence of its inconsistencies on the quality of the radiometry. A distinctive feature of the paper is that the geometric consistency is not assessed by means of a (generally limited) number of check points (CPs), as is usually done. Instead, we assess it by exhaustively evaluating the distance between the whole generated point clouds. We investigate radiometry as well, because it is the main source for agronomic studies. Moreover, we compare datasets acquired almost at the same time. This is a strength, as the difference assessed in vegetation indices can only be attributable to sensor noise, and possibly to issues in the radiometric calibration procedure.

2. Methods

2.1. The Equipment

The dataset was acquired with the HEXA-PRO™ UAV, which is operated by the Laboratory of Geomatics of the University of Pavia and is shown in Figure 1a. The vehicle was made by a small Italian company named Restart® and has the following main characteristics: 6 engines (290 W each one), Arducopter-compliant flight controller, maximum payload of 1.5 kg (partly used by the gimbal, weighting 0.3 kg), flight autonomy of approximately 15 min. The UAV was equipped with a Parrot Sequoia camera (see Figure 1c). Sequoia has a high-resolution RGB camera with a 4608 × 3456 pixel sensor, a pixel size of 1.34 μm, and a focal length of 4.88 mm; the ground sampling distance (GSD) is 1.9 cm at 70 m height above ground level (AGL). Sequoia also has four monochrome cameras that are sensitive to the following spectral bands: green (G, 530–570 nm), red (R, 640–680 nm), red-edge (RE, 730–740 nm), and near-infrared (NIR, 77–810 nm). Their resolution is 1280 × 960, with a pixel size of 3.75 μm and a focal length equal to 3.98 mm; the GSD is 6.8 cm at the 70 m flying height (AGL), which was adopted for the described survey.

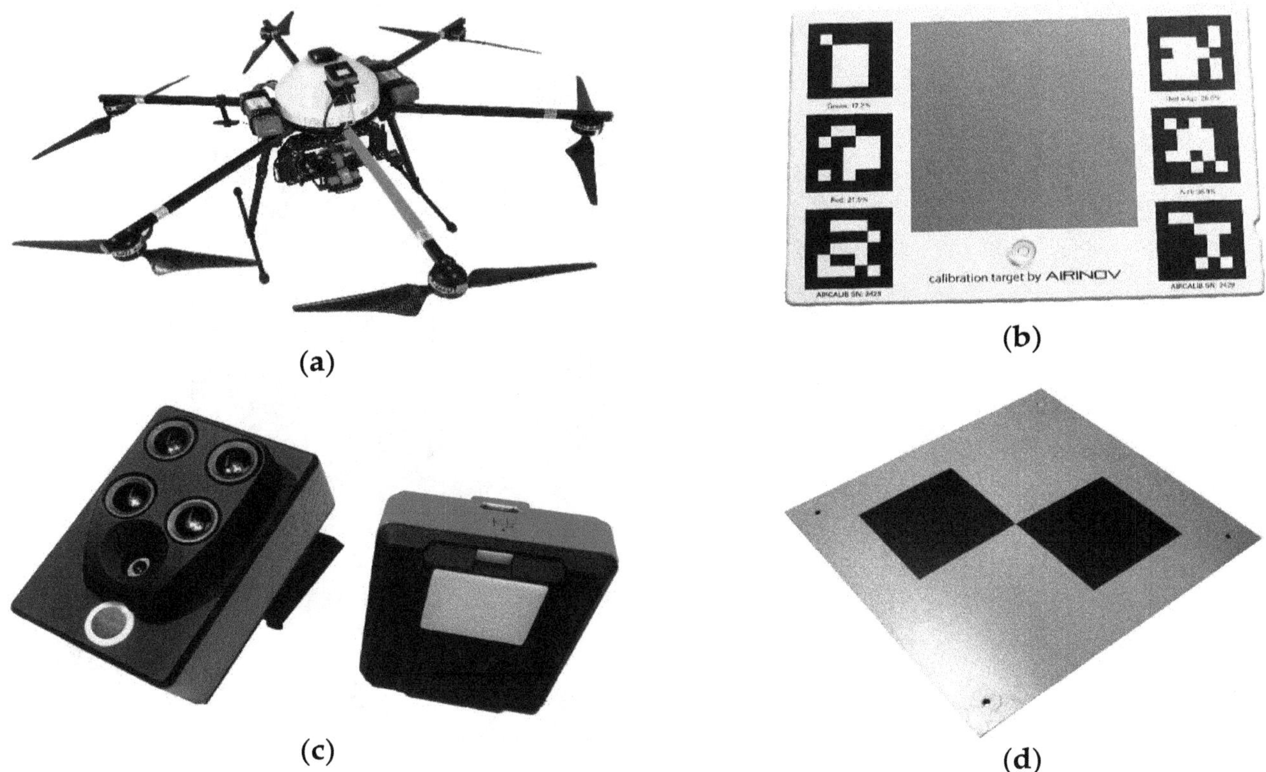

Figure 1. The equipment operated by the Laboratory of Geomatics of the University of Pavia: (**a**) the HEXA-PRO™ unmanned aerial vehicle (UAV) used for the survey; (**b**) the Airinov calibration target supplied with the camera; (**c**) the Parrot Sequoia camera (the imaging and irradiance sensors are shown); (**d**) an example of the artificial markers used.

2.2. The Block Structure

On September 13, 2017, a photogrammetric survey was performed on the Santa Sofia farmstead, near Pavia, Northern Italy (Figure 2a). The test-site is a flat area totaling about 36 ha, used exclusively to cultivate rice. The whole acquisition was obtained by five flight missions, the planning for which is shown in Figure 2b, where the optical orthomosaic, which was used as background, was derived from a previous survey. In total, the project constituted about 1300 multispectral images, each composed of four bands. The AGL height was 70 m and image overlapping was 80% and 60% along- and across-track, respectively. The Sequoia camera was adopted, as previously mentioned. Twelve markers were placed on the ground and surveyed with the network real-time kinematic (NRTK) GPS mode. Virtual reference station (VRS) differential corrections were applied via connecting networked transport of RTCM (Radio Technical Commission for Maritime) via internet protocol (NTRIP - Networked Transport of RTCM via Internet Protocol) to the GNSS positioning service of "Regione Piemonte and Regione Lombardia" [34]. GCP coordinates have a planimetric and altimetric accuracy of 2–3 cm and 4–5 cm, respectively. GCPs were constituted by artificial markers with black and gray diamond shapes (Figure 1d); marker positions are illustrated in Figure 2b. At the beginning of each flight, the recommended radiometric calibration procedure was performed by acquiring the calibration target (Figure 1b).

The present paper will only focus on flights 3 and 4, as these had a methodological purpose. The overlapping area allowed us to deeply analyze geometric and radiometric congruency under several processing scenarios (as described in Section 2.3), because it is quite wide (23 ha) and encompasses 4 GCPs.

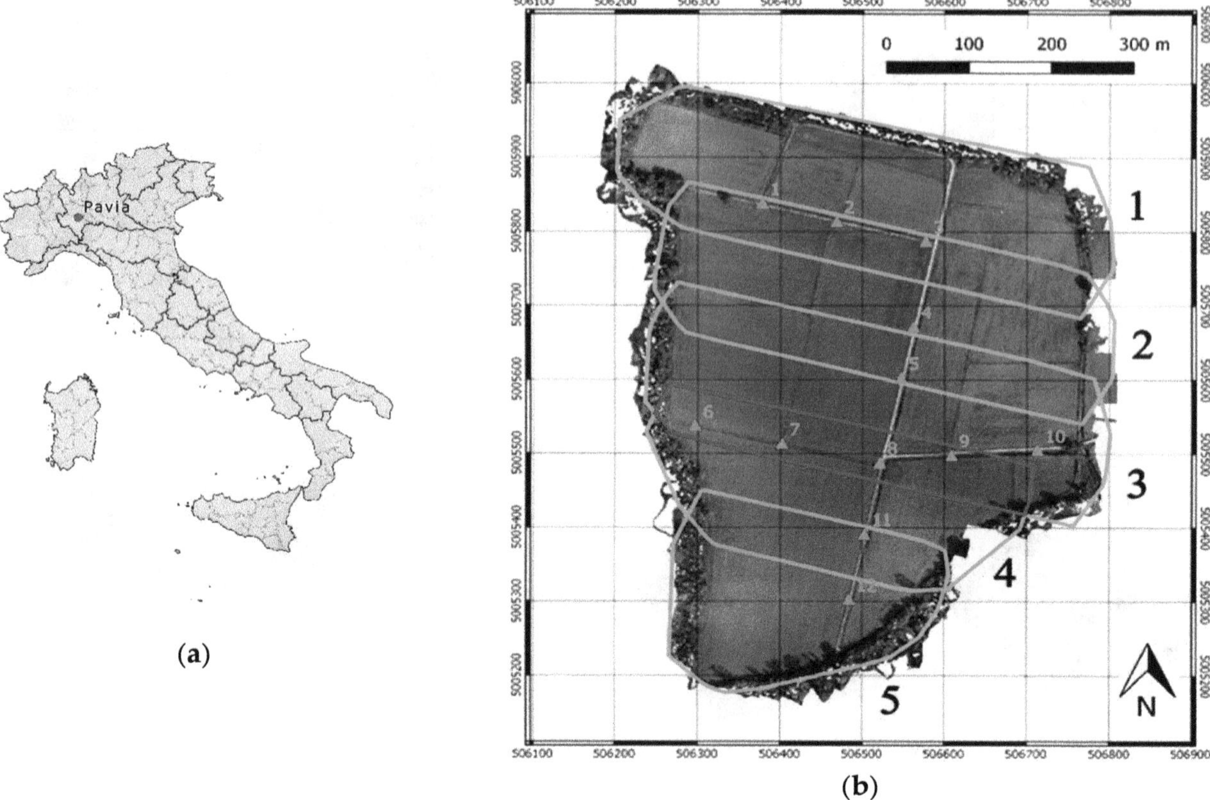

Figure 2. Unmanned aerial vehicle (UAV) survey framework: (**a**) site location; (**b**) the sub-block compositions. Note: light blue lines represent the flight outlines where the overlapping areas are clearly visible. The one considered in the paper is highlighted in red and includes four ground control points (GCPs), named 6, 7, 8, and 9. GCP locations are reported with green triangles. Coordinate reference system (CRS): WGS84/UTM 32N. Central coordinates (E, N): 506500, 5005600.

2.3. The Photogrammetric Processing

The photogrammetric project was carried out with Pix4Dmapper Pro, version 4.4.9. Only the four multispectral channels were considered, having 6.8 cm GSD; higher resolution RGB imagery was disregarded, as it is recorded in the JPEG format with a high compression factor, and has low quality compared to photogrammetry requirements. The processing followed the usual pipeline [35,36]: image alignment, tie point extraction, manual measurement of GCPs and CPs, bundle block adjustment (BBA), generation of dense point clouds, digital surface modeling (DSM), and creation of orthomosaic and reflectance maps. The software allows only one set of calibration target images to be used per project, so the photogrammetric processing followed a single-block approach. Four scenarios were depicted based on georeferentiation methodology and radiometric processing:

1. Direct georeferencing (DG) scenario: no GCPs were inserted in the BBA and each sub-block was processed by direct photogrammetry using positions from the Sequoia integrated GPS receiver. This scenario was used only in geometric assessment.
2. Independent georeferentiation/independent radiometric processing (Ig/Ir) scenario: the two blocks were independently processed in terms of geometry and radiometry. This scenario was used both in geometric and radiometric assessment.
3. Independent georeferentiation/joint radiometric processing (Ig/Jr) scenario: this scenario is a variation of the previous one, in which orientation parameters were computed for each block independently, as in the second scenario, but the two flights were then merged for dense point cloud and reflectance maps generation. This scenario coincides with the so-called "merge option" in Pix4Dmapper software, and it is the recommended procedure for processing photogrammetric

blocks with a large number of images and an overlapping area. It should ensure that radiometric differences caused by a misalignment in the dense point clouds are avoided. Scenario Ig/Jr was used only in radiometric assessment.

4. Joint georeferentiation/independent radiometric processing (Jg/Ir) scenario: the two blocks were jointly orientated, and the obtained exterior orientation parameters were then transferred to a single-block project for generation of dense point clouds and reflectance maps. In this scenario, possible radiometric inconsistencies due to separate computation of interior and exterior orientation parameters are eliminated. This scenario was used in both geometric and radiometric assessment.

The bundle block adjustment parameters were set according to the described scenario, since they differ in terms of calibration method and camera optimization. In DG scenario, the calibration method was set to the "alternative" option. This choice is recommended when the surveyed area is flat (as in this case) and there is availability of good image geolocation; for the Sequoia sensor, the used geolocation comes from the on-board GPS receiver, even if its quality is low, as discussed before. For camera optimization, external parameters were all re-estimated, while for the internal ones they were adopted from the camera model that is delivered by Sequoia directly into the EXIF (Exchangeable image file) section of each image. As we knew from the Pix4D technical support, the parameters delivered into the EXIF are individually determined for each item at the factory. Their reliability is good, as reported in [11], in which the changes between nominal and optimized camera parameters were as low as 0.01%. In Ig/Ir and Ig/Jr scenarios, the calibration method was again set to "alternative". For camera optimization, since the GCPs were imported and measured on each of the two blocks, both external and internal parameters were optimized. Finally, Jg/Ir is a two-step scenario in which the two blocks were jointly processed, and so the obtained internal and external parameters were used to separately generate the dense point clouds for each block. For the first step (image orientation), the parameters were set as equal to Ig/Ir; for the second step (single-block dense point cloud generation), the calibration method was set to geolocation-based, since accurate positioning and orientation are available from the first step. Besides, in this case, neither interior nor external parameters were optimized because they were directly imported in the first step of the project.

All dense point cloud generation was performed by adopting the default options: half image size resolution images, point density was set to optimal, and a cloud point was accepted only if it was positively matched in at least three images. The average density was between 11 to 14 points per m^3. In a preliminary test, the original image size resolution was also evaluated, but higher point density did not significantly improve the generation of orthophotos and reflectance maps; the requirements for precise agriculture are lower in comparison to other applications, such as 3D mapping, and the obtained resolution was considered satisfactory for the research aims.

Pix4Dmapper allows generation of orthophotos and reflectance maps during step 3 of the processing, together with the computation of the DSM. In this study, products were generated with GSD equal to 0.10 m and project settings were the same for all considered scenarios. Reflectance maps were generated by setting camera and sun irradiance correction in the radiometric processing and calibration panel. This allows one to apply corrections to the camera parameters stored in the image metadata (i.e., vignetting, dark current, ISO), as well as for the sun irradiance information acquired with the proper sensor (see Figure 1c). Images of the calibration target are required to perform corrections. Hence, during the survey, the prescribed radiometric calibration procedure of the Parrot Sequoia camera was performed and the suitable calibration target (see Figure 1b) was imaged several times, with different exposure times. Acquisitions were taken at the beginning of each flight, so that different calibration data were stored for each flight, ensuring similar sky and illumination conditions between calibration images and flight images.

For the radiometric processing and calibration, calibration images with the highest value of exposure time were retained and the software automatically detected target on them, defining the proper reflectance values for each spectral band as equal to 0.172, 0.215, 0.266 and 0.369 for green, red, red-edge, and NIR, respectively.

2.4. Geometric Consistency Assessment

The iterative closest point (ICP) methodology was adopted to register overlapping point clouds, evaluate their distance, and estimate their geometric consistency. As is well-known in literature [37–39], ICP is a procedure aiming to align point clouds without requiring the identification of homologous points. It starts by associating each point of cloud A to its closest point belonging to cloud B, then a coordinate transformation (typically a roto-translation, having six parameters, also known as a rigid body transformation in literature [40]) is estimated, based on the obtained coupled points, and applied to one point set. The procedure is iterated until the latest estimated transformation is negligible. A dedicated Matlab procedure was specifically developed at the University of Pavia, implementing ICP and including some unique features. Procedure flowcharts are reported in Figure 3 and Figure 5.

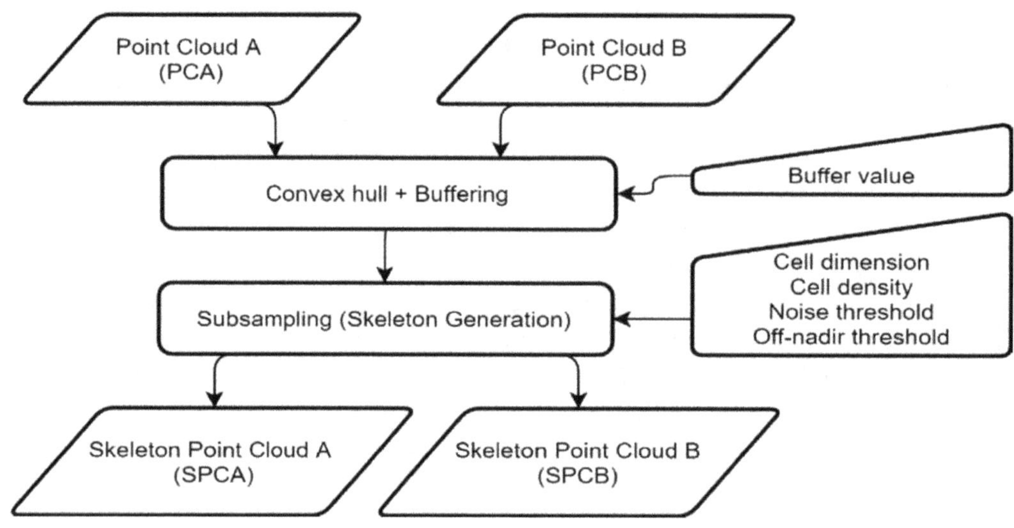

Figure 3. Flowchart for data preparation.

Preliminarily (Figure 3), the common enveloped area for the two point clouds is determined. The clouds are then trimmed according to this area, adding a further precautionary buffer to avoid edge effects; the buffer value is set manually. Therefore, a subset of points is extracted (the so-called skeleton) from each of the point clouds that have a variable density. The skeleton is constituted by squared meshes with sides measuring 2 m and belonging to two classes. There are skeleton points inside meshes. Those lying on flat terrain contain 1 pt/m^2, and therefore the spacing is 1 m. The others, which lie where there are ditches and escarpments, contain 64 pt/m^2, with a spacing of 0.125 m. The skeleton was adopted to reduce the complexity of the calculation and to avoid flat terrain parts, where most of the original points are de-facto overweighed.

The classification of each mesh of the skeleton was performed by selecting all the original cloud points lying in the mesh and estimating the interpolating plane. By imposing suitable criteria concerning the residuals (low residuals mean flat areas) and the deviation from the vertical of the plane normal, the two classes (flat and variable terrain) were decided quite effectively. In the current version, the thresholds used for skeleton classification must be tuned by the operator and inserted manually; an improved one is under development, based on machine-learning. An example of the skeleton structure is reported in Figure 4 for sub-block 3, showing the skeleton points for flat and steep parts of the terrain.

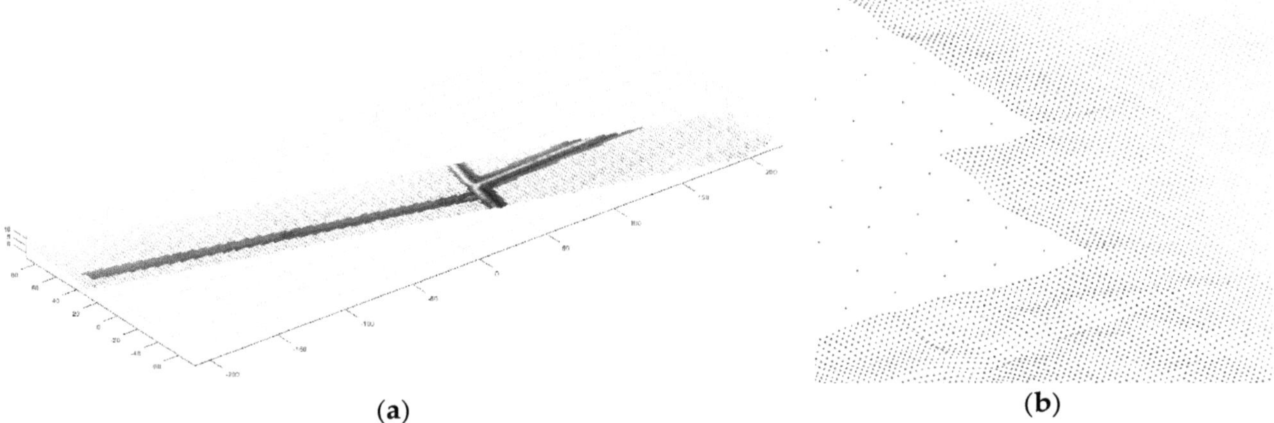

Figure 4. The skeleton structure for sub-block 3. (**a**) Overview: the parts shown in light grey have a lower resolution of 1 pt/m^2; the darker parts have a higher resolution of 64 pt/m^2. (**b**) Detailed view: individual skeleton points are visible, and their different densities can be easily detected.

After defining the skeleton as described above, we used it to define a new point cloud. By this, we mean a set of points for which we know the 3D coordinates, the normal vector of the surface at their position, and the color, totaling 9 descriptors. Original point clouds A and B will be referred with the acronyms PCA and PCB, respectively, while point clouds obtained from skeletons will be named skeleton point clouds A and B, shortened to SPCA and SPCB, respectively. Each element of the skeleton point clouds has 9 descriptors, as stated before; some were known from the definition and some were calculated. For each skeleton mesh, the fitting plane was used to estimate the height of the interior skeleton points. The plane's cartesian equation was also used to obtain its normal vector. Finally, the color was determined for each skeleton point by picking that of the closest original point.

To perform quality assessment, data filtering, and further analysis, data used for each plane estimation was stored in a complex data structure. It is named cell array in the Matlab environment and can be thought of as a matrix where each element can store any kind of data structure. We created a cell array containing as many rows as the meshes constituting the skeleton and with 9 columns (the same number as the descriptors associated with each point cloud, but purely by chance—there is no relationship). For each mesh, the associated 9 cells contain the planimetric coordinates of the central point; mesh size; mesh point density, where we counted the number of the cloud points lying in the mesh and divide this by the area; mesh planimetric bounding box; mesh edge, intended as the points defining the corresponding polygon; coordinates of the original cloud points lying in the mesh; interpolating plane's normal vector; parameters of the plane and its fitting goodness (the so-called Matlab gof (goodness of fit) data structure); and the descriptors of the skeleton points located inside.

Once SPCA and SPCB were created, they were used to fit the ICP transformation. The procedure (Figure 5) is iterative; it starts with the original SPCA and SPCB and finally produces the parameters of a six-parameter 3D rigid transformation aligning SPCB to SPCA. Each iteration:

- has in its input the running SPCB (produced by the previous iteration; in the first iteration, this is SPCB itself), the SPCA (remaining unaltered for all the process), and the running 3D rigid transformation determined so far;
- couples each point of the running SPCB with the closest point belonging to SPCA;
- performs outlier rejection based on the points' distances, the angle between the surface normals at the two points considered, and the norm of the difference between their RGB vectors;
- determines the parameters of a refinement of the 3D rigid transformation aligning running SPCB and SPCA, by solving a non-linear least squares (NLLS) problem defined by Equation (1); in plain words, the NLLS solver evaluates the distance between each B-point (belonging to running SPCB) and the plane passing through the paired A-point (belonging to SPCA) and the normal to SPCA; it determines the unknowns in order to minimize the sum of all the distances;

- applies the determined 3D transformation to the running SPCB;
- composes the newly determined transformation with that received in the input;
- returns the updated running SPCB and coordinate transformation.

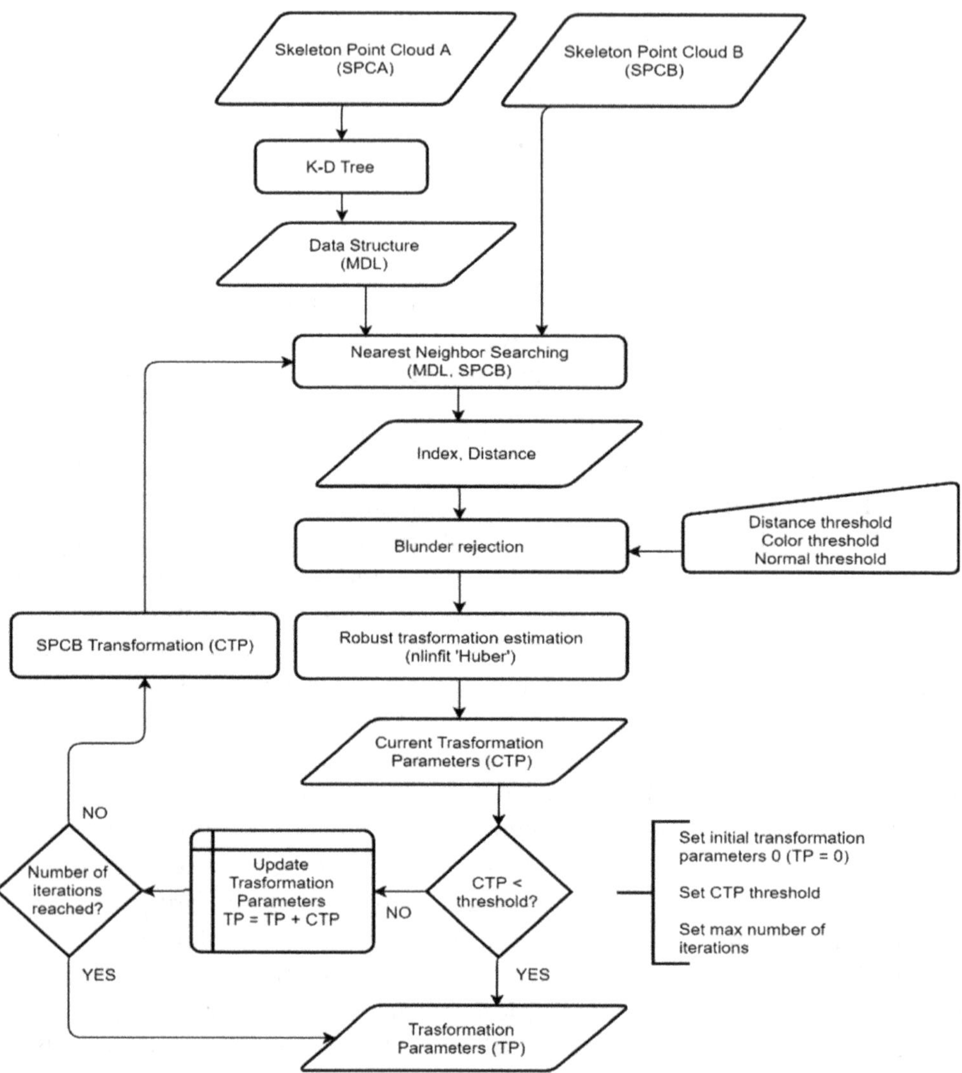

Figure 5. Flowchart for ICP procedure. Note: K-D Tree, K-dimensional tree algorithm; MDL, KD tree model object; CTP, current transformation Parameters; TP, transformation parameters.

The process is stopped when the latest estimated transformation is negligible. The above-described process was coded in a Matlab toolbox coded at the University of Pavia. Among other features, the NLLS problem is solved with a robust approach, based on the Huber method. Moreover, the procedure takes advantage of the k-d tree functionalities (k-dimensional tree) to speed up point coupling, which are available in the used Matlab environment [40]. The k-d tree engine is trained for SPCA, which was kept fixed throughout the procedure.

The mathematical formulation [41,42] of the estimation for the coordinate transformation is

$$CPTopt = arg\min_{T} \sum ((CPT \cdot SPCB_i - SPCA_i) \cdot n_i)^2 \tag{1}$$

where $SPCB_i$ is the generic point of the skeleton B; $SPCA_i$ is the correspondent point of the skeleton A, derived by nearest neighbor searching; n_i is the normal vector at point $SPCAi$; CPT is the 4×4

3D rigid-body transformation matrix estimated from previous iterations; and *CPTopt* is the 4×4 3D rigid-body transformation matrix estimated during the current iteration.

2.5. Radiometric Consistency Assessment

Radiometric consistency was assessed by computing, pixel by pixel, differences for the co-registered reflectance maps in the overlapping area of photogrammetric blocks 3 and 4. Respective statistics were also analyzed. Considering that Sequoia is a sensor mainly dedicated to agricultural applications, assessment was conducted also for some vegetation index (VI) maps, since they commonly represent a proxy of vegetation parameters to be used for agronomy purposes. VI maps were computed in Matlab, by applying an index formula to proper reflectance maps (Table 1).

Table 1. Vegetation indices (VIs) used in this study.

Index	Name	Formula	References
NDVI	Normalized Difference Vegetation Index	$\frac{Nir-Red}{Nir+Red}$	[43]
GNDVI	Green Normalized Difference Vegetation Index	$\frac{Nir-Green}{Nir+Green}$	[44]
NDRE	Normalized Difference Red-Edge Index	$\frac{Nir-RedEdge}{Nir+RedEdge}$	[45]
NDVIre	Red-Edge Normalized Difference Vegetation Index	$\frac{RedEdge-Red}{RedEdge+Red}$	[46]
NGRDI	Normalized Green Red Difference Index	$\frac{Green-Red}{Green+Red}$	[47]

For Ig/Ir and Jg/Ir scenarios, maps derived from blocks 3 and 4 were directly compared, while Ig/Jr scenario was checked with respect to the single blocks of Ig/Ir scenario (see Section 2.3 for more details about scenario characteristics). From here on, maps are identified with the names "3 Ig/Ir", "4 Ig/Ir", "3 Jg/Ir", "4 Jg/Ir", "Ig/Jr", where "3" stands for block 3, and "4" for block 4. DG scenario was not considered for radiometric assessment.

Moreover, since no ground truth was available, the reliability of reflectance and VI maps was evaluated by comparing maps with the one obtained from Sentinel-2 (S2) imagery. Indeed, a Sentinel-2 acquisition two days after the survey (September 19, 2017) was available. Maps derived from the photogrammetric blocks (having a GSD equal to 0.10 m) were upscaled with a nearest-neighbor resampling to 10 m spatial resolution, to match Sentinel-2 imagery resolution. Correlation analysis was applied and statistics were performed on differences in terms of single bands and radiometric indices.

Although a comparison with ground truths calculated with a spectroradiometer would have been more effective, a test on compatibility between Sequoia and S2 data is also of scientific relevance, given the growing interest in the integration of data acquired from satellite and UAV platforms [48] for environmental applications [49,50], including PA [51–53], both from research and applied points of view.

3. Results

3.1. Geometric Consistency

3.1.1. Reliability of the ICP-Derived Transformations

Since geometric consistency is based on ICP, it is mandatory to find a way to quantify the quality of this procedure, because if ICP fails, the estimated distance will not correspond to the actual one. It is known that the ICP estimation is not always reliable, especially when it is used to register almost flat clouds, as in our case. Registration is performed between two point clouds, namely PCA and PCB; it is possible to estimate the transformation PCA-to-PCB, which when applied to PCA, aligns it to PCB. This transformation is constituted by a roto-translation (i.e., the composition of a 3D shift and a 3D rotation). Of course, it is possible to estimate the PCB-to-PCA transformation, which should coincide, aside from uncertainties, with the inverse of the first one. We used the comparison between the estimated and calculated inverse transformation to infer the precision of our estimations.

Results are shown in Table 2, where columns 3–5 report the components of the estimated shift (delta E, delta N, delta H) in meters and columns 6–8 show the rotation angles (ω, φ, κ) in degrees. Rows are grouped in fours, with each chunk being associated with one processing configuration. Row 1 reports the transformation (3-to-4) aligning point cloud 3 to cloud number 4. Row 2 shows the parameters of the calculated inverse transformation. Row 3 displays the parameters of the 4-to-3 estimated transformation and row 4 shows the differences between rows 2 and 3.

Table 2. Reliability of the iterative closest point (ICP)-estimated transformations to be used for cloud registration.

		Translation Components [m]			Rotation Angles [deg]		
Scenario	**Direction**	**delta E**	**delta N**	**delta H**	**delta ω**	**delta φ**	**delta κ**
DG	3-to-4	−1.526	0.406	−2.417	0.3322	0.3721	−0.1110
	calc-4-to-3	1.511	−0.389	2.429	−0.3314	−0.3727	0.1088
	4-to-3	1.626	−0.352	2.326	−0.4372	−0.3867	0.0840
	differences	−0.115	−0.037	0.103	0.1057	0.0134	0.0248
Ig/Ir	3-to-4	0.119	0.055	0.139	0.2388	−0.0648	−0.0082
	calc-4-to-3	−0.120	−0.055	−0.139	−0.2388	0.0648	0.0085
	4-to-3	−0.072	−0.044	−0.181	−0.2810	0.0662	0.0139
	differences	−0.048	−0.011	0.042	0.0422	−0.0016	−0.0054
Jg/Ir	3-to-4	0.081	0.069	0.045	0.0704	−0.0057	−0.0197
	calc-4-to-3	−0.081	−0.069	−0.045	−0.0704	0.0057	0.0197
	4-to-3	−0.022	−0.033	0.010	−0.0110	0.0064	0.0194
	differences	0.059	0.036	0.055	0.0594	0.0007	−0.0003

Scenario abbreviations: direct georeferencing (DG); independent georeferentiation/independent radiometric processing (Ig/Ir); joint georeferentiation/independent radiometric processing (Jg/Ir).

Excluding the DG scenario, in which direct georeferencing is adopted and point clouds are slightly deformed, the worst residual is 6 cm for the shift components and 0.06 deg for rotations. A distance-equivalent error (e) can be computed for the resulting angular residual (α) by assuming a distance (d) of 100 m, corresponding to the half-width of the considered test area. By applying the simple formula $e = d\alpha$, where the angle α is expressed in radiants, it can be found that $e = 12$ cm.

Now, we must consider the granularity of the datasets (i.e., the points' linear spacing). For the skeleton, this is considered for ICP estimation; as already explained, the spacing is 12.5 cm for dense parts and 100 cm elsewhere. As residuals of the transformation equal the discretization size of the considered datasets, we consider the estimated transformations reliable and precise.

3.1.2. Assessment of the Distance between Overlapping Blocks

There are three processing scenarios, and for each of them the distance between the two overlapping clouds (blocks 3 and 4) was assessed. Given two clouds, the ICP procedure was used to estimate the rigid transformation to register point cloud-A to point cloud-B. For the ICP procedure, the skeleton structures were used, as explained in Section 2.4, while for distance evaluation the original point clouds were considered. Once the transformation was applied, each point of B was coupled with the closest point from A; the components and the norm of the connecting vector were stored, together with the two indices addressing the selected point in the lists representing the two point-clouds. To work out the distance before ICP, the original point clouds were used with the same point couplings mentioned before.

Limited data cleaning was performed. First, a buffer (0.9 scale factor) was created along the border of the analyzed regions and points inside were disregarded. The goal was to ignore border effects, where the geometry of photogrammetric measurements is weak, and consequently model deformations might occur. Furthermore, a limited blunder rejection was performed. The empirical cumulative distribution function (CDF) of the 3D distances between coupled points was calculated and the pairs corresponding to values exceeding the 99th percentile were discarded.

Point cloud distances were assessed for the three scenarios, and descriptive statistics were applied to 4-tuples constituted by x, y, and z components of the displacement vector plus its norm. All the results are shown in Table 3. The first column reports the identifier of the scenario and how many point couples were used to evaluate the surface distance; columns 3 to 5 focus on the three components of the original clouds, while column 6 focuses on the 3D distance.

Table 3. Summary statistics of the 3D distance between overlapping point clouds.

		delta E [m]	**delta N [m]**	**delta H [m]**	**delta 3D [m]**
DG# 367982	Min.	−2.210	−0.164	−4.874	1.625
	Max.	−1.352	1.050	−0.368	5.254
	Mean	−1.736	1.214	−2.609	3.236
	STD	0.098	0.193	1.092	0.877
	RMSE	1.739	0.464	2.828	3.353
Ig/Ir# 366061	Min.	−0.241	−0.294	−0.480	0.003
	Max.	0.361	0.377	0.362	0.493
	Mean	0.061	0.051	−0.043	0.167
	STD	0.080	0.089	0.098	0.066
	RMSE	0.100	0.103	0.107	0.179
Jg/Ir# 378018	Min.	−0.253	−0.265	−0.291	0.001
	Max.	0.272	0.350	0.250	0.369
	Mean	0.006	0.038	−0.022	0.132
	STD	0.079	0.093	0.058	0.053
	RMSE	0.079	0.101	0.062	0.143

The DG scenario shows large values, as expected, as the overlapping point clouds have an average 3D distance of 3.24 m. The considered scenario is based on direct georeferencing and the reported results confirm that the Sequoia's on-board GPS receiver is unfit for georeferencing photogrammetric products. This is not a surprise for the authors, nor should it be for any aware user. However, in times of widespread use of photogrammetry [54], we thought it was worth noting. Ig/Ir and Jg/Ir scenarios both adopt GCPs within different adjustment strategies. The RMSE values of the residuals for the single components x, y, and z are range between 6 and 11 cm. Considering the already mentioned granularity of the analyzed datasets, the reported figures highlight that the overlapping point clouds are optimally aligned and consistent.

Maps of the 3D distances between the overlapping point clouds are meaningful. Figure 6 shows them for all the three scenarios assessed. Remarkably, the three sub-figures shown adopt the same color map, even if Figure 6b,c only shows a small part of it. We also remark that values shown in Table 3 and plots reported in Figure 6 are related to the original point clouds, as they were generated by the photogrammetric procedure. ICP was only used to properly couple points belonging to different clouds in order to conveniently evaluate their distance.

Figure 6a highlights that in scenario DG, the clouds are quite far. Moreover, the map of the distances is a sort of a ramp, meaning that the two clouds are not simply displaced but are also affected by a significant rotation. The other two sub-figures are related to Ig/Ir and Jg/Ir scenarios and confirm that the distances are limited in size and are substantially constant. Moreover, Jg/Ir scenario shows lower values above the fields, where the terrain is flat, while distances are slightly greater beside dirt roads and ditches, where low vegetation is present.

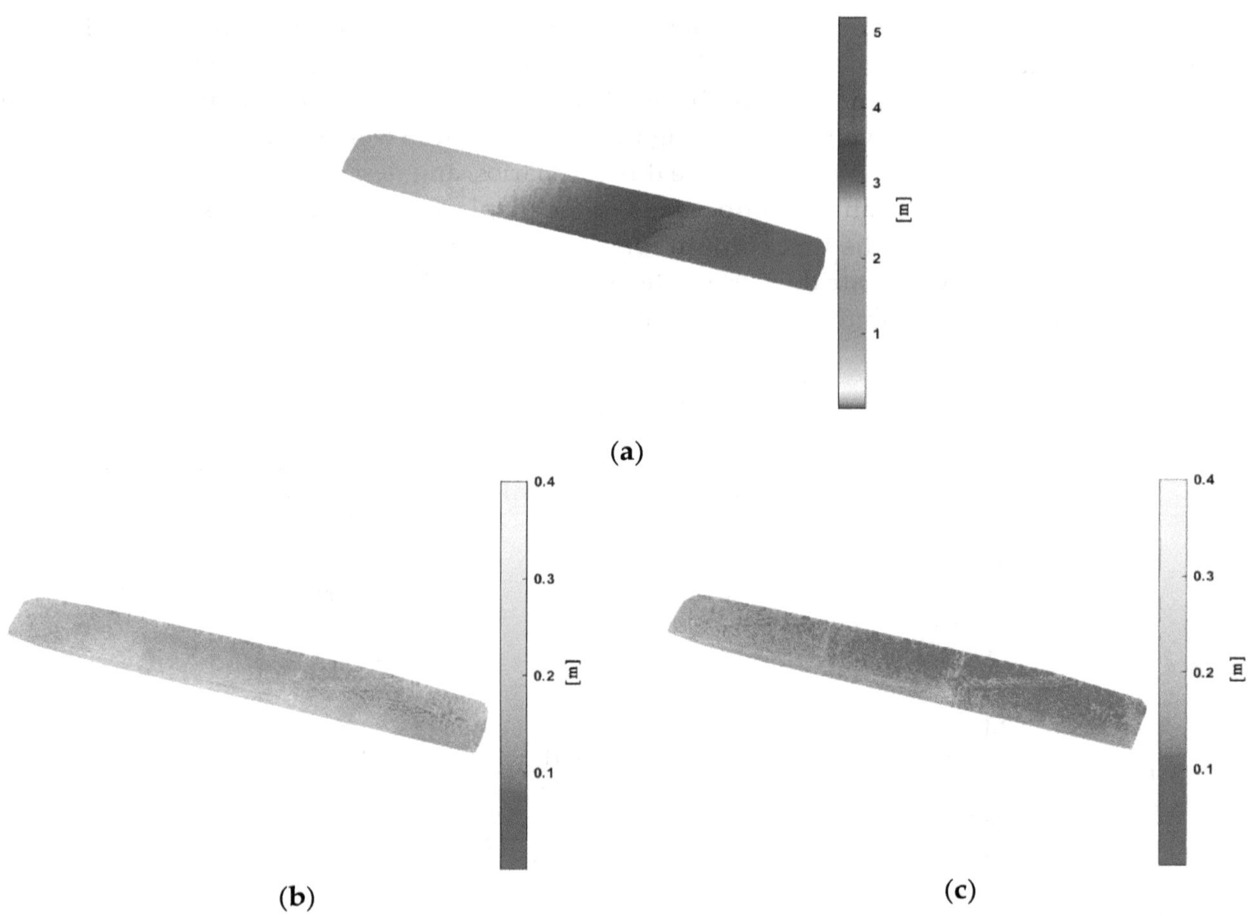

Figure 6. The cloud 3D distance maps between areas 3 and 4, expressed in meters: (**a**) direct georeferencing (DG); (**b**) independent georeferentiation/independent radiometric processing (Ig/Ir); (**c**) joint georeferentiation/independent radiometric processing (Jg/Ir).

3.2. *Radiometric Consistency*

3.2.1. Assessment of the Radiometric Differences between Overlapping Blocks

For the three processing scenarios, differences were calculated pixel by pixel among corresponding reflectance and VI maps in the overlapping area. While Ig/Ir and Jg/Ir scenarios were independently evaluated, Ig/Jr scenario was compared to the single blocks of Ig/Ir scenario (see Section 2.5). Descriptive statistics for differences calculated on reflectance maps are shown in Table 4, and results of VIs maps are reported in Table 5. Although differences have similar ranges, it is important to remember that reflectance maps have values in the range [0, 1], while values for VIs maps are in the range [−1, 1].

The computed RMSE values are quite close to zero for all cases, but significant differences among single reflectance maps and VI maps can be stressed, considering minimum and maximum absolute values. In particular, differences with maximum and minimum values above 0.4 are calculated for the NIR maps, differences reach values close to 0.3 for the red-edge map, and lower values are registered for the green and red maps, with minimum and maximum absolute values below 0.2 for the red maps in some cases. A similar behavior is also evident for the VI maps, where the differences calculated on NDVI maps assume lower RMSE values, while maximum and minimum values even greater than 0.5 are calculated for many VIs. The comparison between the statistics computed for Ig/Ir and Jg/Ir scenarios shows that both reflectance and VI map differences reach very similar values.

Table 4. Summary statistics of the differences between reflectance maps in the overlapping area.

		Green	Red	Red-Edge	NIR
3 Ig/Ir–4 Ig/Ir	Min.	−0.1923	−0.1640	−0.2957	−0.5217
	Max.	0.2822	0.2194	0.4168	0.5931
	Mean	0.0088	−0.0013	0.0268	0.0368
	STD	0.0572	0.0272	0.0367	0.1103
	RMSE	0.0579	0.0272	0.0454	0.1163
3 Ig/Ir–Ig/Jr	Min.	−0.0947	−0.1042	−0.2336	−0.3432
	Max.	0.2538	0.1763	0.3557	0.6642
	Mean	0.0166	0.0061	0.0182	0.0482
	STD	0.0305	0.0131	0.0208	0.0692
	RMSE	0.0348	0.0144	0.0276	0.0843
4 Ig/Ir–Ig/Jr	Min.	−0.1791	−0.1964	−0.3547	−0.3863
	Max.	0.1915	0.1696	0.2769	0.5047
	Mean	0.0079	0.0075	−0.0086	0.0115
	STD	0.0338	0.0195	0.0213	0.0613
	RMSE	0.0347	0.0208	0.0230	0.0624
3 Jg/Ir–4 Jg/Ir	Min.	−0.1826	−0.1761	−0.3217	−0.5519
	Max.	0.2670	0.1874	0.4511	0.5831
	Mean	0.0087	−0.0014	0.0270	0.0365
	STD	0.0572	0.0276	0.0446	0.1104
	RMSE	0.0579	0.0277	0.0522	0.1163

Table 5. Summary statistics of the differences between VI maps in the overlapping area.

		NDVI	GNDVI	NDRE	NDVIre	NGRDI
3 Ig/Ir–4 Ig/Ir	Min.	−0.4431	−0.3859	−0.2651	−0.5629	−0.4726
	Max.	0.4749	0.5066	0.3567	0.4930	0.6011
	Mean	0.0293	0.0064	−0.0003	0.0361	0.0392
	STD	0.0242	0.0678	0.0767	0.0616	0.0910
	RMSE	0.0380	0.0681	0.0767	0.0714	0.0991
3 Ig/Ir–Ig/Jr	Min.	−0.3822	−0.2891	−0.2033	−0.4939	−0.3508
	Max.	0.5734	0.5924	0.4426	0.4208	0.4884
	Mean	0.0142	−0.0012	0.0248	−0.0008	0.0245
	STD	0.0183	0.0310	0.0513	0.0311	0.0538
	RMSE	0.0232	0.0310	0.0570	0.0311	0.0591
4 Ig/Ir–Ig/Jr	Min.	−0.2675	−0.2200	−0.2502	−0.400	−0.5166
	Max.	0.3856	0.3130	0.3924	0.4575	0.3907
	Mean	−0.0151	−0.0077	0.0252	−0.0370	−0.0146
	STD	0.0208	0.0450	0.0418	0.0467	0.0481
	RMSE	0.0257	0.0457	0.0488	0.0596	0.0503
3 Jg/Ir–4 Jg/Ir	Min.	−0.4354	−0.3559	−0.4354	−0.6301	−0.4934
	Max.	0.5024	0.5452	0.7340	0.6121	0.6811
	Mean	0.0293	0.0062	0.0003	0.0359	0.0394
	STD	0.0271	0.0689	0.0861	0.0688	0.0942
	RMSE	0.0400	0.0691	0.0861	0.0775	0.1021

To assess the significance of the calculated values, the differences are presented in the form of box and whisker plots. Figure 7 reports box and whisker plots for differences computed on reflectance maps, while in Figure 8 VI results are shown. The plots do not refer to the Jg/Ir scenario, as similar results are obtained with respect to Ig/Ir scenario.

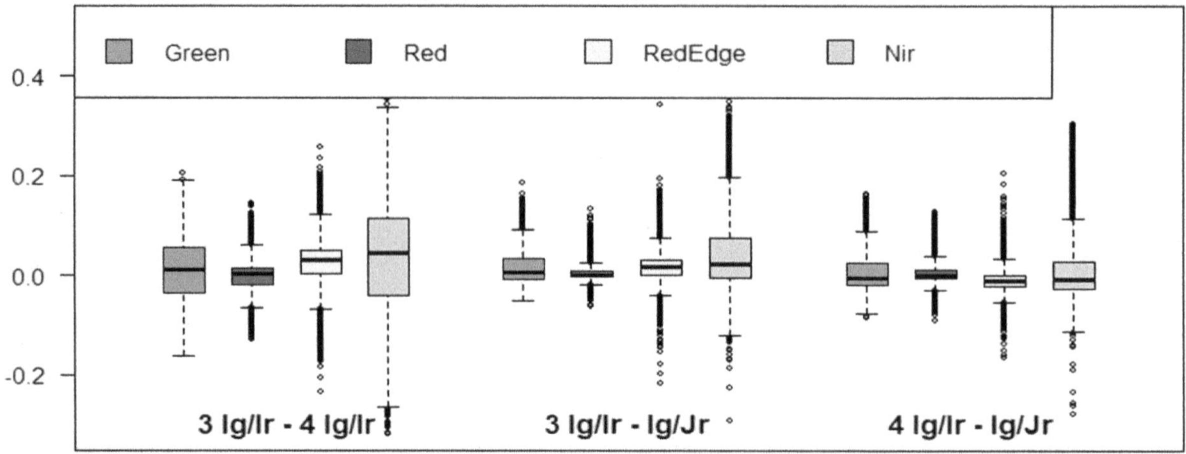

Figure 7. Box and whisker plots of differences computed on different reflectance maps in the overlapping area.

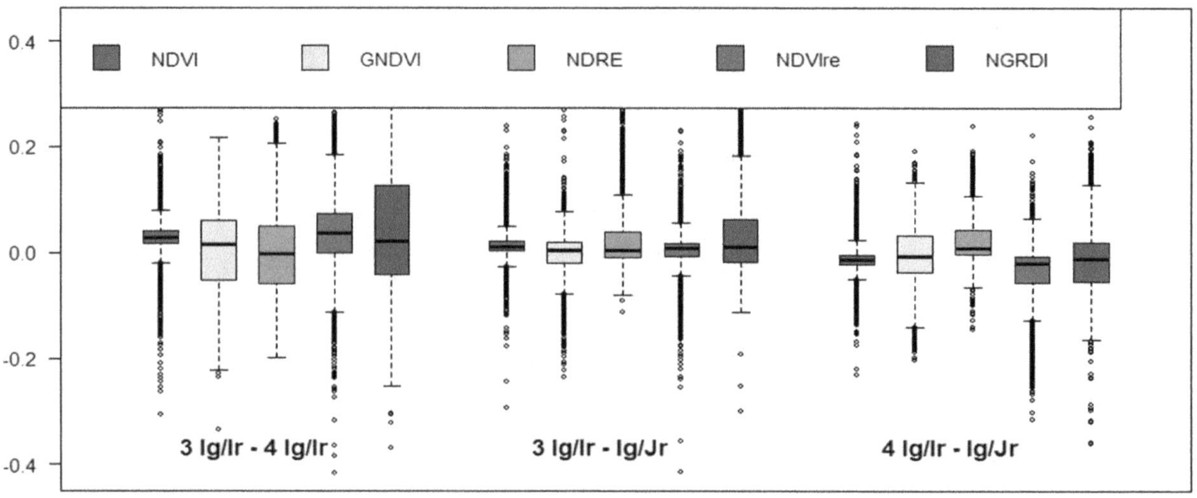

Figure 8. Box and whisker plots of differences computed on different VI maps in the overlapping area.

From the plots it is evident that results vary from map to map, but few general considerations can be drawn. Median values are overall around 0, while maximum and minimum values are outside of the confidence intervals and can be considered as outliers. For most cases, the variability of the differences is contained in the range [−0.2, 0.2]; thus, this interval of values is retained as significant for further analysis. The VIs can mitigate the effects of single reflectance maps, specifically the high differences registered for NIR maps are rather compensated in the NDVI maps. Moreover, with respect to the differences computed between single blocks (i.e., 3 Ig/Ir–4 Ig/Ir), results obtained considering Ig/Jr scenario (i.e., 3 Ig/Ir–Ig/Jr and 4 Ig/Ir–Ig/Jr) have narrower confidence intervals.

Spatial distribution of the differences in the overlapping area is shown in Figures 9–11. For the sake of brevity, only the most significative results are presented. As a matter of fact, similar results were registered for Ig/Ir and Jg/Ir scenarios. As regarding Ig/Jr scenario, green, NIR, and NDVI maps are shown, since the other maps have a similar spatial behavior. The remaining results are reported in Supplementary Materials (Figures S1–S3).

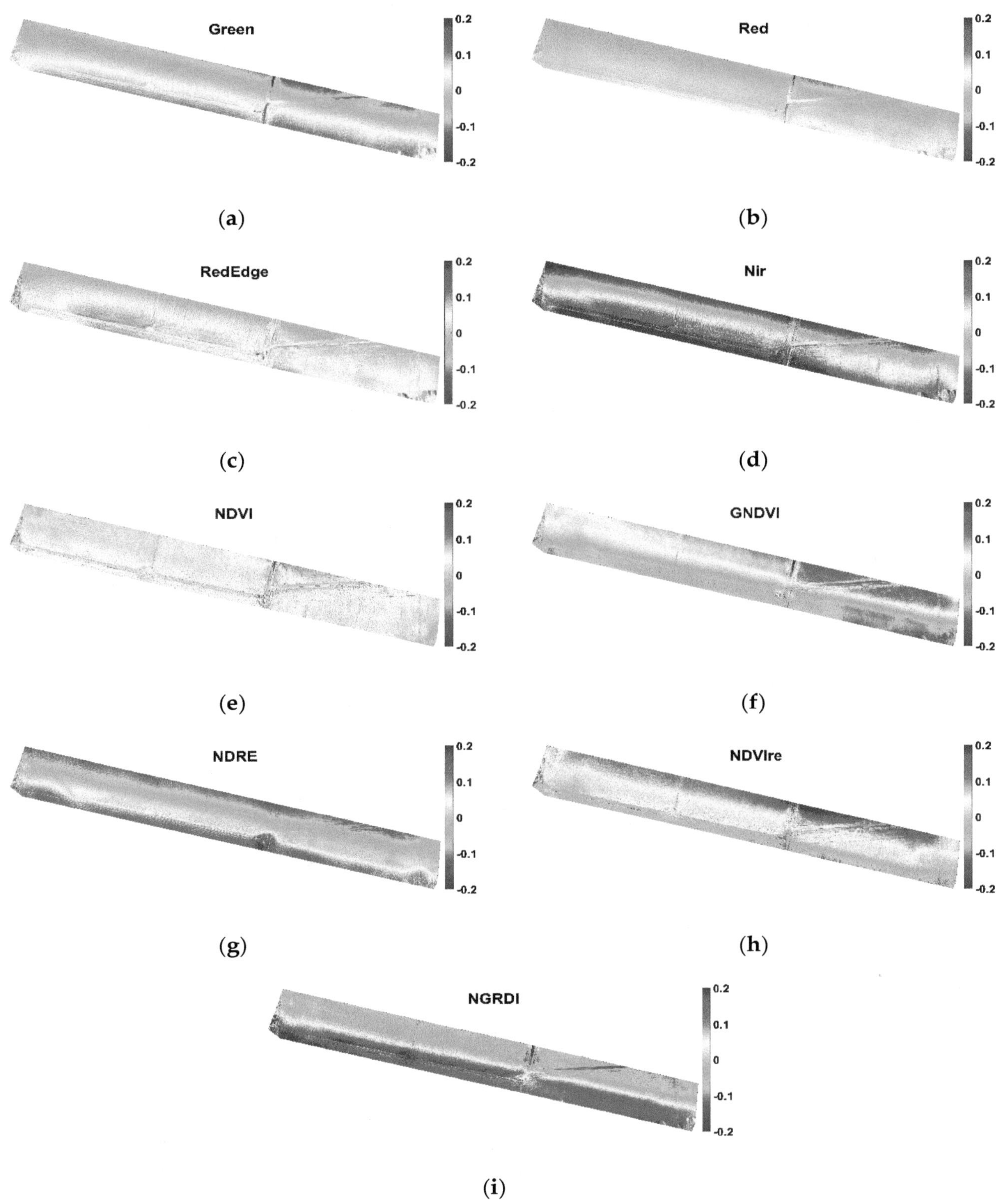

Figure 9. Spatial distribution of differences in the overlapping area. Ig/Ir scenario: green (**a**), red (**b**), red-edge (**c**), NIR (**d**), NDVI (**e**), GNDVI (**f**), NDRE (**g**), NDVIre (**h**), and NGRDI (**i**).

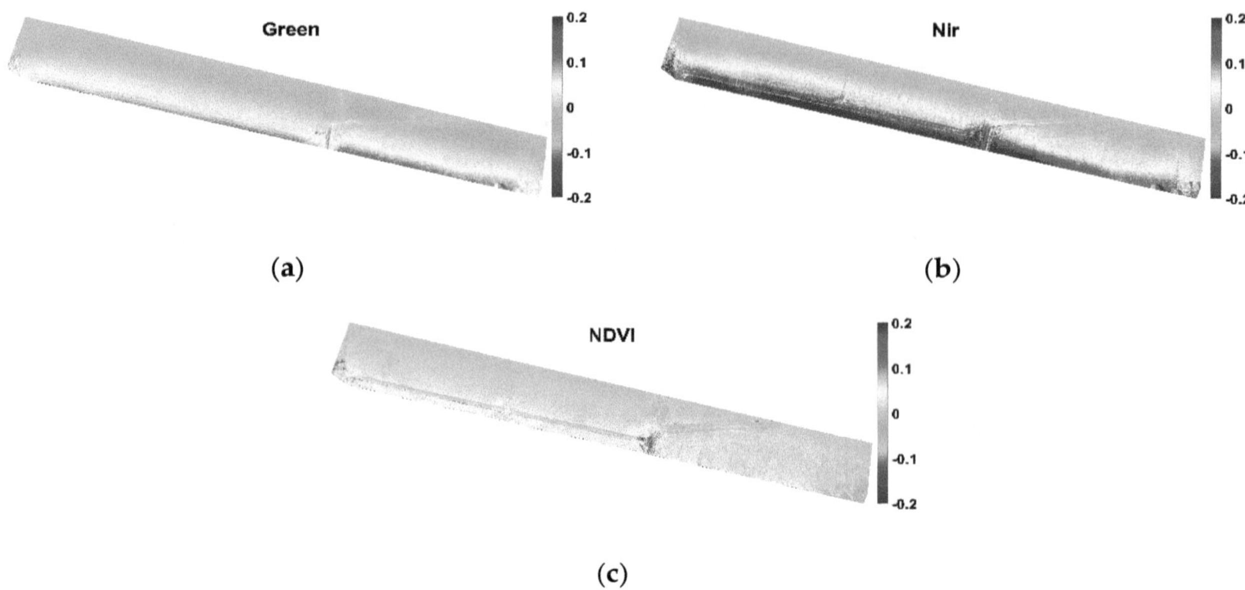

Figure 10. Spatial distribution of differences in the overlapping area for Ig/Jr scenario, with respect to block 3: green (**a**), NIR (**b**), NDVI (**c**).

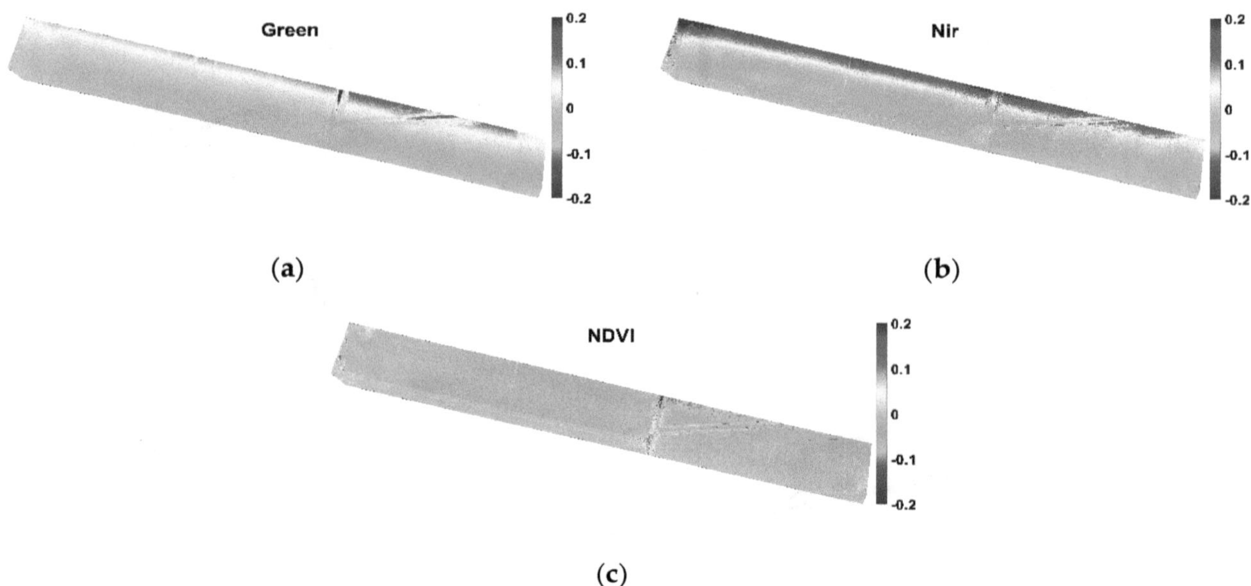

Figure 11. Spatial distribution of differences in the overlapping area for Ig/Jr scenario, with respect to block 4: green (**a**), NIR (**b**), NDVI (**c**).

A clear spatial pattern can be noted from the plots—the reflectance values tend to be overestimated as moving away from the center of the block (i.e., approaching the borders of the block); an analogous effect is visible in VI difference maps. This effect is more evident considering the differences calculated between the single blocks of Ig/Ir scenario (Figure 9). Tt is less evident when introducing also Ig/Jr scenario (Figures 10 and 11). No difference or very small differences are found in NDVI maps for all considered cases, which are uniformly distributed with no specific spatial profile in the overlapping area of blocks 3 and 4.

3.2.2. Comparison with Sentinel-2 Imagery

As described in Section 2.5, the reliability of Sequoia maps was assessed with respect to Sentinel-2 data to evaluate the feasibility of data integration. First, an upscaling of maps derived from Sequoia imagery was required, then correlation analysis was computed (N = 265 samples). Results for the correlation analysis are reported in Figure 12 and map statistics are summarized in Table 6. For the sake of brevity, only results for NDVI are shown. As a matter of fact, other studies are present in the literature focusing on the comparison of NDVI only [51,52]. RMSE values reported in Table 6 were calculated by considering the NDVI map from S2 imagery as a reference.

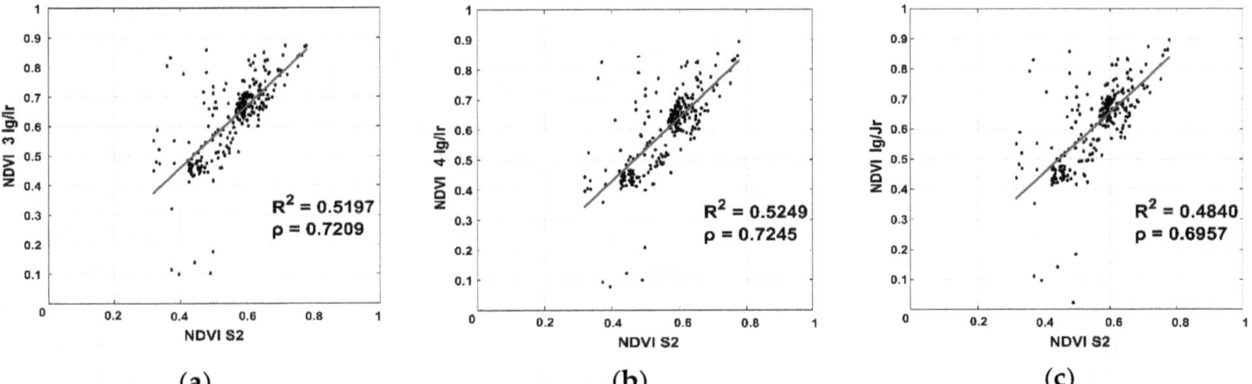

(a) (b) (c)

Figure 12. Scatter plot and regression line for NDVI maps computed on S2 imagery with respect to Sequoia imagery: 3 Ig/Ir (**a**), 4 Ig/Ir (**b**), Ig/Jr (**c**). For each graph, the coefficient of determination (R^2) and the Pearson's correlation coefficients (ϱ) are reported (p-value $< 2.2 \times 10^{-16}$).

Table 6. Summary statistics of the NDVI maps computed from S2 imagery and Sequoia imagery, in the overlapping area.

		S2	3 Ig/Ir	4 Ig/Ir	Ig/Jr
NDVI	Min.	0.3197	0.0998	0.0764	0.0221
	Max.	0.7787	0.8731	0.8926	0.8946
	Mean	0.5583	0.6262	0.5945	0.6119
	STD	0.0906	0.1323	0.1331	0.1324
	RMSE	-	0.1140	0.0984	0.1090

The correlation with NDVI map from S2 imagery shows a good correspondence: coefficients of determination are 0.5197 for 3 Ig/Ir, 0.5249 for 4 Ig/Ir, and 0.4840 for Ig/Jr. The NDVI map with the highest correspondence against S2 imagery is the one derived from 4 Ig/Ir data, with Pearson's correlation coefficient ϱ and RMSE equal to 0.7245 and 0.0984, respectively. Nevertheless, the regression lines show a slight overestimate of Sequoia data compared to S2; NDVI maps from Sequoia imagery report higher values with respect to the S2 map (as also summarized by higher values for max. and mean in Table 6) and cover wider ranges (lower values for min. and higher values for STD in Table 6).

4. Discussion

4.1. Geometric Consistency

Geometric consistency was evaluated by computing the distance between the whole dense point clouds for the DG, Ig/Ir, and Jg/Ir scenarios. The Ig/Jr scenario was disregarded because although it is interesting for the study of radiometry, it is based on the extraction of a unique, common point cloud, and therefore there is nothing to check. Thanks to the high numerosity of the samples (the used clouds have between 10^5 and 10^6 points), we obtained very significant statistical figures. In addition, we could perform a sort of continuous evaluation of the distance between the overlapping surfaces.

The datasets considered are quite challenging for ICP, as the terrain is quite flat and the only height variations are due to medium-sized streams. Nevertheless, our developed algorithm proved to be reliable; indeed, we performed the closure check by composing direct and inverse transformations and obtained residual parameters (shift components and rotation angles), which are limited in size and equivalent to the granularity of the considered point clouds, having an average distance of around 12 cm (see Section 3.1.1).

Cross checks of overlapping point clouds can be usefully applied when no or few CPs are available. Moreover, this allows detection of fine-grained deformations. Scenario DG is related to direct georeferencing, performed through the measurements of the Sequoia's on-board GPS receiver. Results are poor, not surprisingly, as the average distance between the clouds is above 3 m. The point sets are significantly shifted and rotated. This result was largely expected, but we think it is worth mentioning to warn newer photogrammetrists. Such georeferencing precision is not acceptable, even for precision agriculture applications. In Ig/Ir and Jg/Ir scenarios, bundle block adjustment was performed by means of 4 GCPs. Similar results are shown for both scenarios. Indeed, RMSE values of the components of the point-wise shift vectors are within 10 cm. It is the order of magnitude of the granularity of point clouds and of the pixel size of the generated orthomosaics. Such figures assure us that no difference or negligible radiometric differences are induced by geometric consistency issues.

In summary, two main findings come from geometry assessment: we have been able to reliably estimate a rigid 3D transformation by robust ICPs between almost flat point clouds; and we have demonstrated that the geometric consistency is good, so that the inconsistencies shown by radiometry have a different origin.

4.2. *Radiometric Consistency*

As already stressed by many authors in the literature [28–30,55], radiometric corrections are necessary when using sensors mounted on UAV for PA, but the ease of use and diffusion is limited. The radiometric processing available in Pix4Dmapper software for the Sequoia camera provides most of the corrections, including vignetting, dark current, exposure time, and sunlight irradiance, but omits other possible causes of radiometric variations [23]. First, this research points out that radiometric inconsistencies due to differences in the acquisition geometry remain unsolved. Reflectance values of pixels at the borders of the blocks tend to be overestimated, as a consequence of the inclination of the point of view during the photogrammetric survey. From Figure 9, it is evident that differences are not uniformly distributed, but present a clear spatial pattern. Higher difference values (absolute values) are measured at the borders of the overlapping area, while the lowest values approach the center of the area. This demonstrates the presence of a high edge effect on the reflectance maps, which must be considered during flights planning. In practical use, it is advisable to plan UAV surveys covering an area wider than the one of interest. Enlarging the survey area should guarantee uniformity in the acquisition geometry even in the edges, otherwise characterized by non-negligible radiometric distortions.

Radiometric differences are not affected by different geometric processing of the blocks, as confirmed by the similar values of the differences computed for Ig/Ir and Jg/Ir scenarios for both reflectance and VI maps (Tables 4 and 5). As can be noted from RMSE values reported in the tables, differences calculated for some VI maps are lower than values obtained for reflectance maps, meaning that some indices can decrease inconsistencies of single reflectance bands [56]. Considering the Ig/Jr scenario, the differences are moderate with respect to Ig/Ir scenario for both reflectance and VI maps, with mean values overall close to zero. The edge effect is also still evident from the spatial distributions shown in Figures 10 and 11, however with lower values, as is evident in the box plots in Figures 7 and 8. As a matter of fact, it should be recalled that the Ig/Jr scenario corresponds to the procedure recommended by Pix4Dmapper software to process large photogrammetric blocks. For adjacent blocks acquired with separate but temporally close flights, the recommended merging option can partially correct the effect of illumination geometry and mitigate radiometric inconsistencies in the overlapping areas between blocks.

There are still uncertainties regarding the obtained absolute values of reflectance and for the derived indices [56], and consequently in the quantitative use of the Sequoia data for the possible calculation of biophysical parameters. From the results reported in this study, it should be noted that in some areas, differences have values close or even larger than 0.2 (absolute value). Therefore, the different processing scenarios have an impact on the results in terms of radiometry. A difference of this magnitude cannot be neglected in the operational phase for precision agriculture applications; even more so if used for multitemporal surveys. As a matter of fact, the map that shows the most homogeneous values in all cases is NDVI, which is widely used in most agriculture applications [57–60].

Regarding the comparison with S2, which is limited in this paper to NDVI, it should be mentioned that despite the analysis being affected by the different geometric resolutions of sensors and acquisition platforms, a significative correlation is found between Sequoia and S2 maps. Following the approach of [52], the Pearson's correlation coefficient can be adopted as a map similarity measure. The obtained coefficients, which are close to 0.7, prove a coherence between the data collected from the different platforms and show similar spatial variability values of NDVI maps, which are to be interpreted as the same behavior in terms of crop vigor [61]. Therefore, the compatibility and integration of NDVI maps obtained by Sequoia and Pix4D systems should be feasible along with the Sentinel-2 products.

5. Conclusions

Even though producers and developers have made great efforts to enhance them, radiometric corrections leave significant radiometric distortions in orthomosaics obtained by Sequoia and Pix4D systems, which can result in biased absolute values. This study shows that relevant differences are found depending on flight geometry and block processing choices, with differences that can reach 20% of pixel values for single reflectance bands or VIs, thus reducing the effective use in PA. Moreover, available radiometric corrections do not guarantee uniform accuracy and consistency of results, and this can cause difficulties in comparing surveys carried on out different lightening conditions. Careful planning of the survey, together with proper choices of image processing, can enhance the results. Very high image overlap yields uniformity over a single block, and edge distortions can be reduced by surveying a wider area that includes the study area.

Nevertheless, for large surveys that imply the acquisition and processing of separated sub-blocks, the merge option suggested by Pix4D is effective in reducing radiometric inconsistencies in adjacent areas. This fact, together with the high correlation found with S2 products, proves that Sequoia is suitable for agronomic purposes, but great attention must be paid to the planning of the survey and to the data processing.

Therefore, it is necessary to increase the awareness in the use of sensors and semi-automatic data processing to deeply understand the strengths and weaknesses of UAV usage for PA. In this study, the choice to process the dataset following the proposed scenarios (Section 2.3) instead of a standard workflow was driven by the apparently impossibility of attributing the corresponding calibration set of images to each block. The new Sequoia+ sensor should bypass this issue because no calibration target is needed; imagery processing exploits a new fully automatic calibration pipeline in Pix4D. The authors do not have experience with this new camera release, however, they consider the proposed processing method an interesting and simple way to assess the performance of new sensors.

Finally, the applicability of the proposed method can be extended. In this paper, geometric and radiometric consistency was evaluated comparing results obtained from two almost contemporaneous flights, processed following a single-block approach. The same method can be used to evaluate consistency between two or more blocks acquired days or month apart; in other words, the method can be used to assess time-series.

Supplementary Materials
Figure S1: Spatial distribution of differences in the overlapping area for scenario Ig/Jr with respect to block 3: red (**a**), red-edge (**b**), GNDVI (**c**), NDRE (**d**), NDVIre (**e**), NGRDI (**f**). Figure S2: Spatial distribution of differences in the overlapping area for scenario Ig/Jr with respect to block 4: red (**a**), red-edge (**b**), GNDVI (**c**), NDRE (**d**), NDVIre (**e**), NGRDI (**f**). Figure S3: Spatial distribution of differences in the overlapping area. Scenario Jg/Ir: green (**a**), red (**b**), red-edge (**c**), NIR (**d**), NDVI (**e**), GNDVI (**f**), NDRE (**g**), NDVIre (**h**), NGRDI (**i**).

Author Contributions: Conceptualization, M.F. and V.C.; formal analysis, M.F., G.R., G.S., and V.C.; methodology, G.S. and V.C.; project administration, M.F.; software, M.F. and G.R.; supervision, G.S. and V.C.; visualization, M.F. and G.R.; writing—original draft, M.F., G.R., G.S., and V.C.; writing—review and editing, M.F., G.R., G.S., and V.C.

Acknowledgments: Paolo Marchese is a technician of the Laboratory of Geomatics of the University of Pavia. He is the pilot of the UAV, plans the flight missions, carries out maintenance, and performs sensor integration; he is gratefully acknowledged here. We want to thank the farmhouse owner too, Roberto Manfredi, for always kindly allowing us to enter his property and carry out our activities.

References

1. Candiago, S.; Remondino, F.; De Giglio, M.; Dubbini, M.; Gattelli, M. Evaluating multispectral images and vegetation indices for precision farming applications from UAV images. *Remote Sens.* **2015**, *7*, 4026–4047. [CrossRef]
2. Schieffer, J.; Dillon, C. The economic and environmental impacts of precision agriculture and interactions with agro-environmental policy. *Precis. Agric.* **2015**, *16*, 46–61. [CrossRef]
3. Turner, D.; Lucieer, A.; Watson, C. An automated technique for generating georectified mosaics from ultra-high resolution unmanned aerial vehicle (UAV) imagery, based on structure from motion (SfM) point clouds. *Remote Sens.* **2012**, *4*, 1392–1410. [CrossRef]
4. Parrot. Available online: https://community.parrot.com/t5/Sequoia/bd-p/Sequoia (accessed on 13 June 2019).
5. Chiang, K.W.; Tsai, M.L.; Chu, C.H. The development of an UAV borne direct georeferenced photogrammetric platform for ground control point free applications. *Sensors* **2012**, *12*, 9161–9180. [CrossRef]
6. Lussem, U.; Bareth, G.; Bolten, A.; Schellberg, J. Feasibility study of directly georeferenced images from low-cost unmanned aerial vehicles for monitoring sward height in a long-term experiment on grassland. *Grassl. Sci. Eur.* **2017**, *22*, 354–356.
7. Tahar, K.N. An evaluation on different number of ground control points in unmanned aerial vehicle photogrammetric block. *Int. Arch. Photogramm. Remote Sens. Spat. Inf. Sci.* **2013**, *40*, 93–98. [CrossRef]
8. James, M.R.; Robson, S.; d'Oleire-Oltmanns, S.; Niethammer, U. Optimising UAV topographic surveys processed with structure-from-motion: Ground control quality, quantity and bundle adjustment. *Geomorphology* **2017**, *280*, 51–66. [CrossRef]
9. Mesas-Carrascosa, F.J.; Torres-Sánchez, J.; Clavero-Rumbao, I.; García-Ferrer, A.; Peña, J.M.; Borra-Serrano, I.; López-Granados, F. Assessing optimal flight parameters for generating accurate multispectral orthomosaicks by UAV to support site-specific crop management. *Remote Sens.* **2015**, *7*, 12793–12814. [CrossRef]
10. Assmann, J.J.; Kerby, J.T.; Cunliffe, A.M.; Myers-Smith, I.H. Vegetation monitoring using multispectral sensors—best practices and lessons learned from high latitudes. *J. Unmanned Veh. Syst.* **2018**, *7*, 54–75. [CrossRef]
11. Fernández-Guisuraga, J.; Sanz-Ablanedo, E.; Suárez-Seoane, S.; Calvo, L. Using unmanned aerial vehicles in postfire vegetation survey campaigns through large and heterogeneous areas: Opportunities and challenges. *Sensors* **2018**, *18*, 586. [CrossRef]
12. Caroti, G.; Martínez-Espejo Zaragoza, I.; Piemonte, A. Accuracy assessment in structure from motion 3D reconstruction from UAV-born images: The influence of the data processing methods. *Int. Arch. Photogramm. Remote Sens. Spat. Inf. Sci.* **2015**, *40*. [CrossRef]
13. Benassi, F.; Dall'Asta, E.; Diotri, F.; Forlani, G.; Morra di Cella, U.; Roncella, R.; Santise, M. Testing accuracy and repeatability of UAV blocks oriented with GNSS-supported aerial triangulation. *Remote Sens.* **2017**, *9*, 172. [CrossRef]
14. Martínez-Carricondo, P.; Agüera-Vega, F.; Carvajal-Ramírez, F.; Mesas-Carrascosa, F.J.; García-Ferrer, A.; Pérez-Porras, F.J. Assessment of UAV-photogrammetric mapping accuracy based on variation of ground control points. *Int. J. Appl. Earth Obs.* **2018**, *72*, 1–10. [CrossRef]
15. Harwin, S.; Lucieer, A. Assessing the accuracy of georeferenced point clouds produced via multi-view stereopsis from unmanned aerial vehicle (UAV) imagery. *Remote Sens.* **2012**, *4*, 1573–1599. [CrossRef]
16. Casella, V.; Franzini, M. Modelling steep surface by various configurations of nadir and oblique photogrammetry. *ISPRS Ann. Photogramm. Remote Sens. Spat. Inf. Sci.* **2016**, *3*. [CrossRef]
17. Jaud, M.; Passot, S.; Le Bivic, R.; Delacourt, C.; Grandjean, P.; Le Dantec, N. Assessing the accuracy of high-resolution digital surface models computed by PhotoScan®and MicMac®in sub-optimal survey conditions. *Remote Sens.* **2016**, *8*, 465. [CrossRef]
18. Fritz, A.; Kattenborn, T.; Koch, B. UAV-based photogrammetric point clouds—Tree stem mapping in open stands in comparison to terrestrial laser scanner point clouds. *Int. Arch. Photogramm. Remote Sens. Spat. Inf. Sci.* **2013**, *40*, 141–146. [CrossRef]

19. Wallace, L.; Lucieer, A.; Malenovský, Z.; Turner, D.; Vopěnka, P. Assessment of forest structure using two UAV techniques: A comparison of airborne laser scanning and structure from motion (SfM) point clouds. *Forests* **2016**, *7*, 62. [CrossRef]
20. Turner, D.; Lucieer, A.; de Jong, S. Time series analysis of landslide dynamics using an unmanned aerial vehicle (UAV). *Remote Sens.* **2015**, *7*, 1736–1757. [CrossRef]
21. Clapuyt, F.; Vanacker, V.; Van Oost, K. Reproducibility of UAV-based earth topography reconstructions based on Structure-from-Motion algorithms. *Geomorphology* **2016**, *260*, 4–15. [CrossRef]
22. Honkavaara, E.; Saari, H.; Kaivosoja, J.; Pölönen, I.; Hakala, T.; Litkey, P.; Mäkynen, J.; Pesonen, L. Processing and Assessment of Spectrometric, Stereoscopic Imagery Collected Using a Lightweight UAV Spectral Camera for Precision Agriculture. *Remote Sens.* **2013**, *5*, 5006–5039. [CrossRef]
23. Tu, Y.-H.; Phinn, S.; Johansen, K.; Robson, A. Assessing Radiometric Correction Approaches for Multi-Spectral UAS Imagery for Horticultural Applications. *Remote Sens.* **2018**, *10*, 1684. [CrossRef]
24. Roosjen, P.P.J.; Suomalainen, J.M.; Bartholomeus, H.M.; Kooistra, L.; Clevers, J.G.P.W. Mapping Reflectance Anisotropy of a Potato Canopy Using Aerial Images Acquired with an Unmanned Aerial Vehicle. *Remote Sens.* **2017**, *9*, 417. [CrossRef]
25. Parrot for Developers. Available online: https://forum.developer.parrot.com/t/parrot-announcement-release-of-application-notes/5455?source_topic_id=6558 (accessed on 18 June 2019).
26. Borgogno Mondino, E.; Gajetti, M. Preliminary considerations about costs and potential market of remote sensing from UAV in the Italian viticulture context. *Eur. J. Remote Sens.* **2017**, *50*, 310–319. [CrossRef]
27. Aasen, H.; Honkavaara, E.; Lucieer, A.; Zarco-Tejada, P.J. Quantitative Remote Sensing at Ultra-High Resolution with UAV Spectroscopy: A Review of Sensor Technology, Measurement Procedures, and Data Correction Workflows. *Remote Sens.* **2018**, *10*, 1091. [CrossRef]
28. Guo, Y.; Senthilnath, J.; Wu, W.; Zhang, X.; Zeng, Z.; Huang, H. Radiometric Calibration for Multispectral Camera of Different Imaging Conditions Mounted on a UAV Platform. *Sustainability* **2019**, *11*, 978. [CrossRef]
29. Mafanya, M.; Tsele, P.; Botai, J.O.; Manyama, P.; Chirima, G.J.; Monate, T. Radiometric calibration framework for ultra-high-resolution UAV-derived orthomosaics for large-scale mapping of invasive alien plants in semi-arid woodlands: Harrisia pomanensis as a case study. *Int. J. Remote Sens.* **2018**, *39*, 5119–5140. [CrossRef]
30. Honkavaara, E.; Khoramshahi, E. Radiometric Correction of Close-Range Spectral Image Blocks Captured Using an Unmanned Aerial Vehicle with a Radiometric Block Adjustment. *Remote Sens.* **2018**, *10*, 256. [CrossRef]
31. Miyoshi, G.T.; Imai, N.N.; Tommaselli, A.M.G.; Honkavaara, E.; Näsi, R.; Moriya, E.A.S. Radiometric block adjustment of hyperspectral image blocks in the Brazilian environment. *Int. J. Remote Sens.* **2018**, *39*, 4910–4930. [CrossRef]
32. Johansen, K.; Raharjo, T. Multi-temporal assessment of lychee tree crop structure using multi-spectral RPAS imagery. *Int. Arch. Photogramm. Remote Sens. Spatial Inf. Sci.* **2017**, *42*, 165–170. [CrossRef]
33. Ahmed, O.S.; Shemrock, A.; Chabot, D.; Dillon, C.; Williams, G.; Wasson, R.; Franklin, S.E. Hierarchical land cover and vegetation classification using multispectral data acquired from an unmanned aerial vehicle. *Int. J. Remote Sens.* **2017**, *38*, 2037–2052. [CrossRef]
34. GNSS Positioning Service Portal of Regione Piemonte and Regione Lombardia. Available online: https://www.spingnss.it/spiderweb/frmindex.aspx (accessed on 16 August 2019).
35. Casella, V.; Chiabrando, F.; Franzini, M.; Manzino, A.M. Accuracy assessment of a photogrammetric UAV block by using different software and adopting diverse processing strategies. In Proceedings of the 5th International Conference on Geographical Information Systems Theory, Applications and Management, Heraklion, Crete, Greece, 3–5 May 2019.
36. Ronchetti, G.; Pagliari, D.; Sona, G. DTM generation through UAV survey with a fisheye camera on a vineyard. *Int. Arch. Photogramm. Remote Sens. Spatial Inf. Sci.* **2018**, *42*, 983–989. [CrossRef]
37. Rusinkiewicz, S.; Levoy, M. Efficient variants of the ICP algorithm. *3DIM* **2001**, *1*, 145–152.
38. Gelfand, N.; Ikemoto, L.; Rusinkiewicz, S.; Levoy, M. Geometrically stable sampling for the ICP algorithm. *3DIM* **2003**, *1*, 260–267. [CrossRef]
39. Glira, P.; Pfeifer, N.; Briese, C.; Ressl, C. Rigorous strip adjustment of airborne laser scanning data based on the ICP algorithm. *Int. Ann. Photogramm. Remote Sens. Spat. Inf. Sci.* **2015**, *2*, 73–80. [CrossRef]
40. M ATLAB. *Statistics and Machine Learning Toolbox*; The MathWorks, Inc.: Natick, MA, USA, 2019.

41. Low, K.L. *Linear Least-Squares Optimization for Point-to-Plane ICP Surface Registration*; Technical report, 4; University of North Carolina: Chapel Hill, NC, USA, 2004.
42. Chen, Y.; Medioni, G. Object modelling by registration of multiple range images. *Image Vis. Comput.* **1992**, *10*, 145–155. [CrossRef]
43. Rouse, J.W., Jr.; Haas, R.H.; Schell, J.A.; Deering, D.W. Monitoring Vegetation System in the Great Plains with ERTS. In Proceedings of the Third Earth Resources Technology Satellite-1 Symposium, Greenbelt, MD, USA, 10–14 December 1974.
44. Gitelson, A.A.; Kaufman, Y.J.; Merzlyak, M.N. Use of a green channel in remote sensing of global vegetation from EOS-MODIS. *Remote Sens. Environ.* **1996**, *58*, 289–298. [CrossRef]
45. Barnes, E.M.; Clarke, T.R.; Richards, S.E.; Colaizzi, P.D.; Haberland, J.; Kostrzewski, M.; Waller, P.; Choi, C.; Riley, E.; Thompson, T.; et al. Coincident detection of crop water stress, nitrogen status and canopy density using ground based multispectral data. In Proceedings of the Fifth International Conference on Precision Agriculture, Bloomington, MN, USA, 16–19 July 2000.
46. Gitelson, A.A.; Merzlyak, M.N. Spectral reflectance changes associated with autumn senescence of Aesculus hippocastanum L. and Acer platanoides L. leaves. Spectral features and relation to chlorophyll estimation. *J. Plant Physiol.* **1994**, *143*, 286–292. [CrossRef]
47. Gitelson, A.A.; Kaufman, Y.J.; Stark, R.; Rundquist, D. Novel algorithms for remote estimation of vegetation fraction. *Remote Sens. Environ.* **2002**, *80*, 76–87. [CrossRef]
48. Dash, J.P.; Pearse, G.D.; Watt, M.S. UAV Multispectral Imagery Can Complement Satellite Data for Monitoring Forest Health. *Remote Sens.* **2018**, *10*, 1216. [CrossRef]
49. Puliti, S.; Saarela, S.; Gobakken, T.; Ståhl, G.; Næsset, E. Combining UAV and Sentinel-2 auxiliary data for forest growing stock volume estimation through hierarchical model-based inference. *Remote Sens. Environ.* **2018**, *204*, 485–497. [CrossRef]
50. Stark, D.J.; Vaughan, I.P.; Evans, L.J.; Kler, H.; Goossens, B. Combining drones and satellite tracking as an effective tool for informing policy change in riparian habitats: A proboscis monkey case study. *Remote Sens. Ecol. Conserv.* **2018**, *4*, 44–52. [CrossRef]
51. Matese, A.; Toscano, P.; Di Gennaro, S.F.; Genesio, L.; Vaccari, F.P.; Primicerio, J.; Belli, C.; Zaldei, A.; Bianconi, R.; Gioli, B. Intercomparison of UAV, Aircraft and Satellite Remote Sensing Platforms for Precision Viticulture. *Remote Sens.* **2015**, *7*, 2971–2990. [CrossRef]
52. Khaliq, A.; Comba, L.; Biglia, A.; Ricauda Aimonino, D.; Chiaberge, M.; Gay, P. Comparison of Satellite and UAV-Based Multispectral Imagery for Vineyard Variability Assessment. *Remote Sens.* **2019**, *11*, 436. [CrossRef]
53. Zhang, S.; Zhao, G.; Lang, K.; Su, B.; Chen, X.; Xi, X.; Zhang, H. Integrated Satellite, Unmanned Aerial Vehicle (UAV) and Ground Inversion of the SPAD of Winter Wheat in the Reviving Stage. *Sensors* **2019**, *19*, 1485. [CrossRef]
54. Strecha, C. From expert to everyone: Democratizing photogrammetry. In Proceedings of the ISPRS Technical Commission II Symposium "Toward Photogrammetry 2020", Riva del Garda, Italy, 3–7 June 2018.
55. Iqbal, F.; Lucieer, A.; Barry, K. Simplified radiometric calibration for UAS-mounted multispectral sensor. *Eur. J. Remote Sens.* **2018**, *51*, 301–313. [CrossRef]
56. Stow, D.; Nichol, C.J.; Wade, T.; Assmann, J.J.; Simpson, G.; Helfter, C. Illumination Geometry and Flying Height Influence Surface Reflectance and NDVI Derived from Multispectral UAS Imagery. *Drones* **2019**, *3*, 55. [CrossRef]
57. González-Piqueras, J. Radiometric Performance of Multispectral Camera Applied to Operational Precision Agriculture. In Proceedings of the IGARSS 2018—2018 IEEE International Geoscience and Remote Sensing Symposium, Valencia, Spain, 22–27 July 2018; pp. 3393–3396. [CrossRef]
58. Teal, R.K.; Tubana, B.; Girma, K.; Freeman, K.W.; Arnall, D.B.; Walsh, O.; Raun, W.R. In-season prediction of corn grain yield potential using normalized difference vegetation index. *Agron. J.* **2006**, *98*, 1488–1494. [CrossRef]
59. Katsigiannis, P.; Misopolinos, L.; Liakopoulos, V.; Alexandridis, T.K.; Zalidis, G. An autonomous multi-sensor UAV system for reduced-input precision agriculture applications. In Proceedings of the 24th Mediterranean Conference on Control and Automation (MED), Athens, Greece, 21–24 June 2016; pp. 60–64. [CrossRef]
60. Freidenreich, A.; Barraza, G.; Jayachandran, K.; Khoddamzadeh, A.A. Precision Agriculture Application for Sustainable Nitrogen Management of Justicia brandegeana Using Optical Sensor Technology. *Agriculture* **2019** *9*, 98. [CrossRef]

9

An Augmented Reality Tool for Teaching Application in the Agronomy Domain

Dolores Parras-Burgos [1,*], Daniel G. Fernández-Pacheco [1], Thomas Polhmann Barbosa [1], Manuel Soler-Méndez [2] and José Miguel Molina-Martínez [2]

[1] Departamento de Estructuras, Construcción y Expresión Gráfica, Universidad Politécnica de Cartagena, C/Doctor Fleming s/n, 30202 Cartagena, Murcia, Spain; daniel.garcia@upct.es (D.G.F.-P.); thomaspolhmannbarbosa@gmail.com (T.P.B.)

[2] Grupo de I+D+i Ingeniería Agromótica y del Mar, Universidad Politécnica de Cartagena, C/Ángel s/n, Ed. ELDI E1.06, 30203 Cartagena, Murcia, Spain; josem.molina@upct.es (M.S.-M.); manuel.ia@agrosolmen.es (J.M.M.-M.)

* Correspondence: dolores.parras@upct.es

Abstract: Nowadays, the combination of new technologies and the use of mobile devices opens up a new range of teaching–learning strategies in different agricultural engineering degrees. This article presents an augmented reality tool that allows for improved spatial viewing for students who have certain difficulties with viewing graphic representations of agronomic systems and devices. This tool is known as ARTID (Augmented Reality for Teaching, Innovation and Design) and consists in a free-access mobile application for devices using the Android operating system. The proposed method provides each exploded drawing or overall drawing with a QR code that can be used by students to view their 3D models by augmented reality in their own mobile devices. An evaluation experience was carried out to assess the validity of the tool on different devices and the acceptance and satisfaction level of this kind of resources in subjects of graphic expression in engineering. Finally, an example of application in the agronomic domain is provided by the 3D virtual model of portable ferticontrol equipment that comprises the different structures and tanks, which, if viewed by conventional graphical representations, may entail a certain level of difficulty. Thanks to this tool, reality can be merged with the virtual world to help favour the understanding of certain concepts and to increase student motivation in agronomy studies.

Keywords: 3D model; spatial vision; fertirrigation; teaching–learning

---

1. Introduction

Nowadays, augmented reality (AR) is being introduced into many areas of society, especially since advertising has mostly driven its development given the effects that it has on potential customers [1–3]. Indeed, more and more AR applications are being used, especially as smartphones and Internet access have boomed because, thanks to location, users can be shown data about their position [4], seek new experiences and sensations [5] or encourage different types of family activities [6].

In the teaching area, this technology is becoming particularly relevant and is applied to different levels of education and subjects [7–11]. Viewing 3D objects in space is one of the main challenges that engineering students face during their academic training. Rapid technological development, along with growing popularity and easy access to smart mobile devices, has enabled the development of new opportunities to improve teaching quality [12,13], which, in turn, leads to more motivation and higher learning levels for the students who resort to them [14]. It has been demonstrated that using such technologies inside and outside classrooms can increase the understanding and motivation of the students registered for technical drawing subjects in engineering degrees [15]. Specifically, a study that uses virtual reality in engineering students to assess spatial capacity can be found in literature, showing that the use of this type of technology significantly improves these capacities in the students who use them. The only drawback of this technology is the need for special devices, such as virtual reality glasses with a smartphone [16].

In the literature, we can find various resources where AR focused on education is used with very interesting results, such as a mathematics textbook for secondary education, which could be validated and considered as an appropriate tool for the teaching–learning process [17]. In other fields, the younger generation tends to see history as an uninteresting and boring subject, and, for this reason, the development of books with AR can favour interest and the acquisition of knowledge in a more satisfactory way [18,19]. Libraries have also investigated the use of web-based AR by means of ISBN barcodes on the books themselves [20] or generating the editorial design of an educational book about environmental issues, including AR for children from 6 to 8 years [21]. This type of technology has also been incorporated into language learning, where the experimental results revealed that AR-based learning activities can better help students in understanding and expressing themselves about learning contents instead of passively listening to the teacher or reading books [22]. In subjects related to graphic expression in engineering, several didactic materials that use AR with results that demonstrate a positive impact on students' ability in their spatial skills can be found [23].

The advantage of AR is that it allows students to zoom in/out and rotate 3D models, and to perceive contents in a more appealing manner, which encourage them to learn. Configuration and viewing are not easy processes because different elements need to be taken into account. Of all the analysed resources, it is worth mentioning those that use AR by scanning QR codes with smartphones [24]. This is the case of a colouring book application that was developed for an aquarium musem. The mobile application, along with the Android operating system and AR, allowed users to express, create and interact with their creativity through colouring activities [25].

In the graphic expression domain, the spatial viewing skill needs to be trained in with most cases. Former studies have verified that using different technologies, including AR, can favour this skill in STEM (Science, Technology, Engineering and Mathematics) studies [26–30].

For this reason, this paper describes the creation of a mobile application to visualize, using AR and QR codes, 3D virtual models of figures used in graphic expression subjects. A test was carried out in the following subjects taught at the Technical University of Cartagena during the course 2018–19:

- Graphic Expression: Degree in Mechanical Engineering, Degree in Electrical Engineering, Degree in Automation and Industrial Electronic Engineering, Degree in Chemical and Industrial Engineering and Degree in Industrial Technologies Engineering.
- Industrial Design: Degree in Mechanical Engineering.

This test validated the suitability of using the application in different devices, as well as the level of student satisfaction after using the tool. In order to help gain an understanding of graphic representations with exploded drawings and overall drawings in agricultural engineering degrees, the mobile application ARTID (Augmented Reality for Teaching, Innovation and Design) was developed for viewing virtual 3D models by AR. The objective was to use such tools inside and outside classrooms to favour students' understanding and motivation in different university degree subjects.

2. ARTID Tool

The main objectives to create and develop the AR ARTID tool for its application in graphic expression subjects in the agronomic domain were to:

- Develop a support tool for those teachers teaching the subject inside and outside classrooms;
- Turn classes into more interactive and dynamic activities to motivate students;
- Help students to develop their spatial viewing and to, thus, understand the exercises given in class;

- Create and make available an open access tool for all those students interested in the topic.

To develop the ARTID application, we resorted to using different software, which allowed us to obtain an easy-to-use and open access tool. Its development process can be divided into three main phases (see Figure 1): (i) creating QR stamps to scan and reproduce parts; (ii) 3D modelling of parts; (iii) joining stamps and parts in the Unity programme.

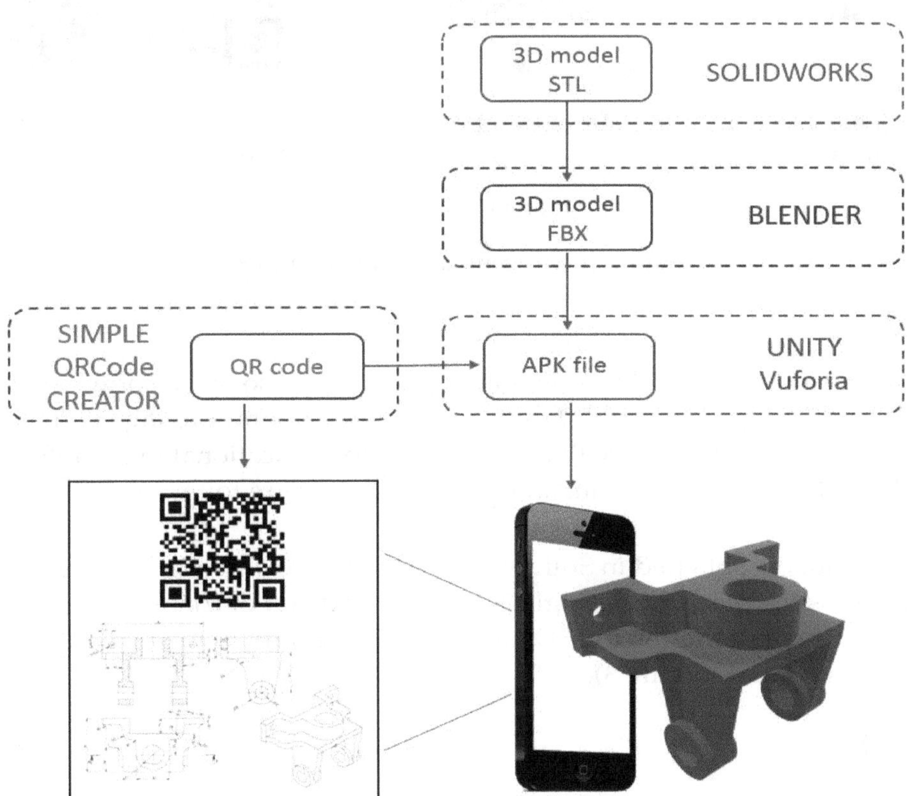

Figure 1. Scheme of the proposed methodology.

2.1. Stamp Creation

The stamp corresponds to the image that must be scanned via the mobile phone to project a part in AR. To create it, the following indications need to be considered:

(a) The stamp must be unique. It is important to avoid the 3D model taking a long time to appear or another object appearing in its place. So, an abstract, but well-defined, image must be selected. Using QR images in ARTID was chosen because it makes two equal images look somewhat difficult.

(b) Stamp size must be optimum to project the model. If the stamp is small, it can be projected more quickly, but lacks details. If it is very big, the part might not be generated or take too long to appear (Figure 2). The defined optimum size for this tool would be a stamp whose approximate dimensions would be 11 × 11 cm so it fits in an A4 format.

Different options are available to create the QR code as both websites and applications can be used. For this tool, the *Simple QRCode Creator* application was employed, which is free and easy to use, and is obtained from the Microsoft Store. To use it, the only thing that users had to do was to introduce a text (the more characters in the QR definition name, the more unique it will be) and the encryption level, which was set at *high* (selection of the ECC (Error Correction Feature) level, the higher the level, the lower the probability of reading failure).

Figure 2. Different levels and qualities.

2.2. 3D Modelling Parts

The second stage involves the 3D modelling of parts. To do so, we resorted to the design and 3D modelling SolidWorks 2018 programme (Dassault Systèmes). In order to import the 3D parts in Unity (Unity Technologies, a licensed software that can be used for educational and not-for-profit purposes), they must take the FBX (FilmBoX) format, and the steps below are followed:

(a) Export the 3D models obtained in Solidworks to the STL (STereoLithography) format. The STL format transforms the piece into a single 3D model made up of triangles; the greater the number of triangles, the more definition the 3D model will have, but, in turn, it will be more difficult to project it onto the stamp (Figure 3).

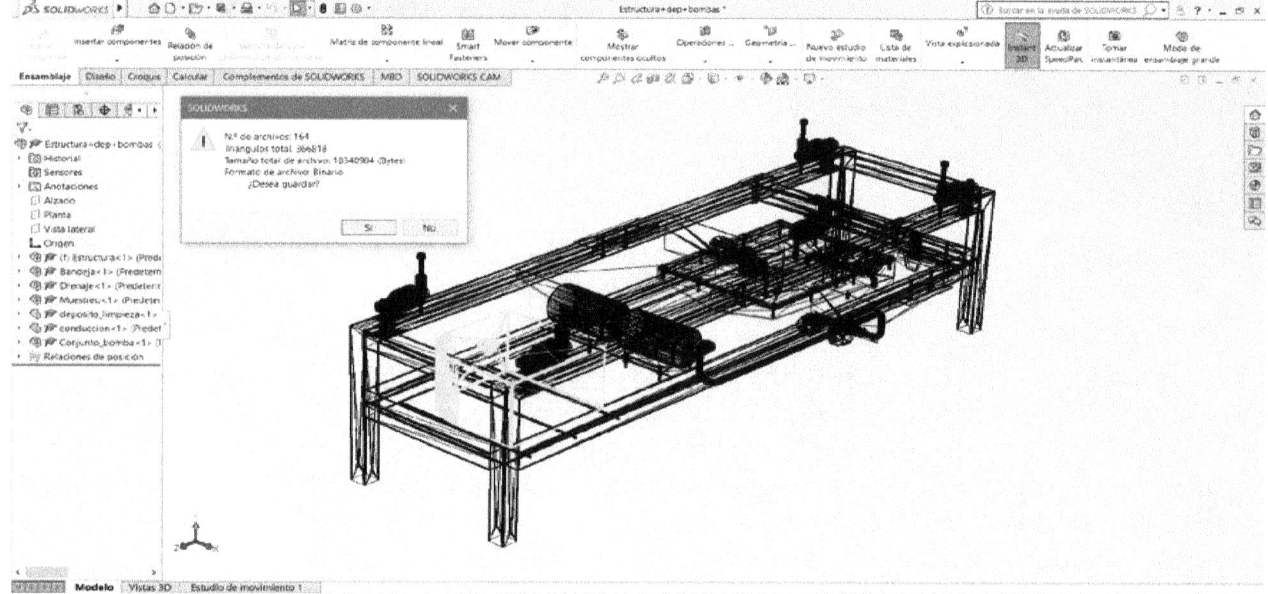

Figure 3. Screenshot of the SolidWorks software.

(b) Import the STL file in the Blender software (Fundación Blender, free multiplatform software) to add colour or textures. Blender allows you to apply textures and colors on the piece as well as allowing you to go from STL format to FBX (Figure 4).

(c) Export the STL file in Blender to the FBX format. If models are directly drawn in Blender, they do not need the first conversion.

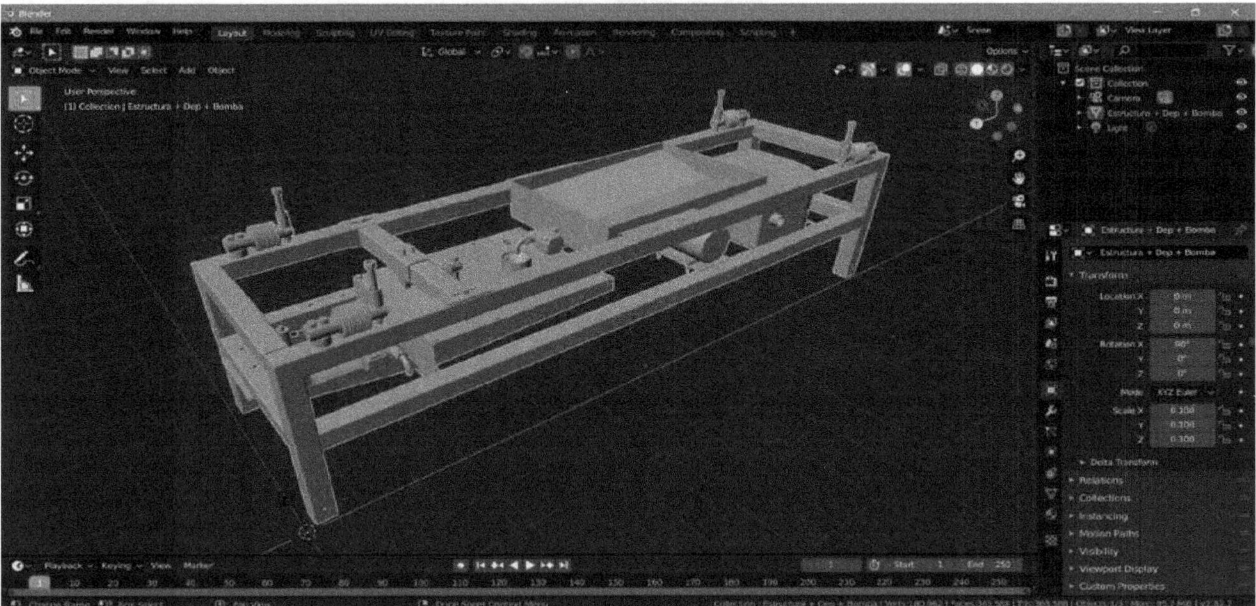

Figure 4. Screenshot of the Blender software.

2.3. *Creating the Augmented Reality Tool*

Vuforia is a software development kit (SDK) that allows work to be done with AR. Both Vuforia and Unity are licensed software that can be used for education and not-for-profit purposes, whose authorisation was requested to do this work. This stage involves two work phases:

(a) Firstly, the stamps created in the previous stage must be uploaded. To do so, the Vuforia website (https://developer.vuforia.com) is employed. A file is created that has to be added to Unity (Figure 5).

(b) Secondly, the work area in the Vuforia tool is found in the *GameObjects* tab in Unity. The file generated in Vuforia must be executed when Unity is open, and the chosen stamps must be added. Vuforia is accessed by the *GameObjects* tab and *ARCamera* is selected. Stamps are added in *ARCamera* (ImageTarget), followed by the 3D models (Figure 6).

(c) Once the 3D models and QR codes are associated, the model must be customized in terms of size, brightness, shadows and position on the QR. In case any model has a texture or image applied to its surface, it must be implemented in Unity creating a "Material" and applying it to the model surface.

In Unity, it is necessary to follow a series of configuration procedures to generate the application:

- Select the Android version required for the application to run. In this study, only versions greater than 4.0 were chosen.
- Choose the cameras used on the device (front/rear). In this study, we opt for the rear camera because it is usually the camera with the best quality.
- Configure whether the application can be used vertically or horizontally (Portrait or Landscape).
- Define the amount of graphics memory used on the device. The more memory it has, the faster the model will be displayed.
- And other minor settings, s application logo, encryption keys to protect the application or author assignment, among others.

Figure 5. Screenshot of the Vuforia software.

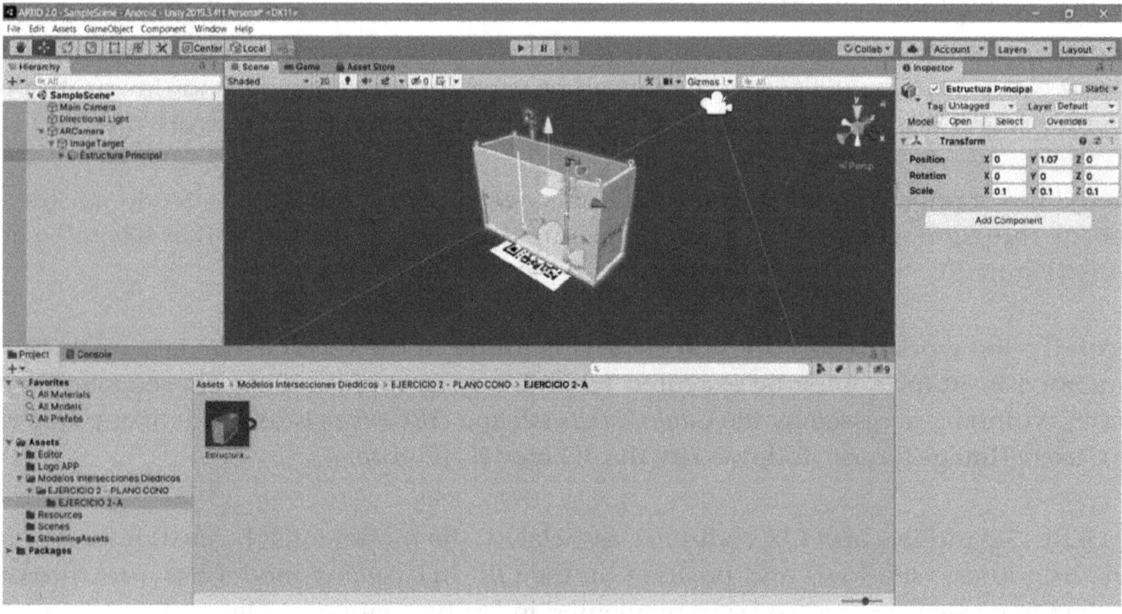

Figure 6. Screenshot of the Unity software.

After following all these steps, the *Build* button is clicked on to create the ".apk" file of the final application to be installed in the mobile device.

Currently, 3D models are embedded within the .apk file. In future studies, the application will be developed to download via streaming the required 3D models when the camera focuses on a QR code.

In addition, a website (www.artid.es) where you can find information related to augmented reality and download this application has been developed. The application is currently passing the Google approval process to be introduced within the Google Play Store, enabling an easier way of downloading it onto mobile devices.

3. Experience with Students

3.1. Participants

In order to verify the suitability of the created application, a test was performed to the students of the following subjects taught at the Technical University of Cartagena during the course 2018–19 (Table 1):

- Graphic Expression (Degree in Mechanical Engineering, Degree in Electrical Engineering, Degree in Automation and Industrial Electronic Engineering, Degree in Chemical and Industrial Engineering and Degree in Industrial Technologies Engineering).
- Industrial Design (Degree in Mechanical Engineering).

Table 1. Participants by subjects and degrees.

Subjects	Degrees	Participants	%
Graphic Expression	Degree in Mechanical Engineering	64	34.04
	Degree in Industrial Technologies Engineering	11	5.85
	Degree in Automation and Industrial Electronic Engineering	36	19.14
	Degree in Chemical and Industrial Engineering	17	9.04
	Degree in Electrical Engineering	11	5.85
Industrial Design	Degree in Mechanical Engineering	49	26.06
	TOTAL	**188**	**100**

3.2. Procedure

The test was carried out by teachers in the area of graphic expression in engineering, conducting a pilot experience in each of the subjects indicated above. The main objectives of this pilot test were the following:

(a) Collect technical information on the use of the ARTID application on different mobile devices with the Android operating system. With this type of information, the aim was to determine: (i) the accessibility and management of the application, (ii) the feasibility of using geometric 3D models and QR codes, and (iii) the effectiveness of the application, analyzing possible failures and errors.

(b) Collect teaching information about the ARTID tool in drawing subjects. This type of information is more subjective and sought to determine: (i) the acceptance of this type of technology by students in graphic expression and industrial design subjects, (ii) the perception of usefulness that students could have with this type of tool, and (iii) the level of satisfaction of these less conventional activities focused on teaching.

During the test, which lasted 20 min, teachers interacted with students to answer questions and exchange opinions. Participation in the test was totally voluntary and consisted of three phases:

i A presentation of the tool, where students were given with the instructions for the download and installation of the application on their mobile devices through the web www.artid.es.

ii The completion of a practical test where a set of four planes with QR codes was provided to each student in order to probe the tool (Figures 7 and 8).

iii After completing the use test, the students accessed through the web and made an anonymous and voluntary online survey to assess technical aspects of the tool and collect their opinion about the use of augmented reality applications in the graphic expression and industrial design subjects. The questions raised were open in order to collect preliminary information and data on this tool that could serve to continue advancing in this research line.

Figure 7. One of the plans with a QR code used in the test.

(**a**) (**b**)

Figure 8. (**a**) Photograph of the class where the test was performed, (**b**) one student capturing the QR code during the test.

3.3. Results

The activity was carried out by a total of 188 students whose mobile device was endowed with Android Operating System v4.2 or higher. The results obtained in the survey (see Table 2) show: (i) a proper operation of the application in the majority of the mobile devices used, (ii) agility in the visualization of the 3D models, and (iii) an adequate interaction of the students with the application. On the other hand, the combination of AR together with the handling of mobile devices is very attractive to students, so they perceive a great utility of the tool in the teaching field. During this activity, many of the students who participated showed their interest and asked for and extended version with more exercises. The collection of this preliminary information favours the development of materials and specific exercises with AR in the different subjects of graphic expression in engineering. Once these materials have been produced, a more objective and analytical study can be performed to determine the cognitive and practical advantages of this tool when used by students in relation to their spatial vision and learning.

Table 2. Results of the survey conducted by the students.

Questions	Yes		No	
	np	%	np	%
Has the application worked properly on your device?	171	90.96	17	9.04
Have all the stamps been read correctly?	170	90.43	18	9.57
Have the 3D models appeared on the stamp quickly?	174	92.55	14	7.45
Do you consider that this tool favors the learning of basic concepts of the subject?	186	98.94	2	1.06
Do you consider that spatial vision is favored with this type of tools?	187	99.47	1	0.53
Does this experience give you interesting results?	187	99.47	1	0.53
Does this tool seem useful to you?	185	98.40	3	1.60
Would you like to repeat this activity?	188	100.00	0	0.00

np = number of participants.

4. ARTID Tool in the Agronomic Domain

In addition to the test performed to the students, another example of using this technology in agronomic studies was developed. In this case, a ferticontrol equipment, based on the weighing lysimeter developed in the Agromatic and Sea Engineering Research Group of the Universidad Politécnica de Cartagena (Figure 9), was employed as a 3D model. This device is formed by several structures and is located on a farm. The crop container, which holds a volume of reconstituted soil taken from the plot, reproduces natural conditions to determine crop evapotranspiration while it grows, as well as water/nutrient uses and losses through leaching. Apart from acting as an infiltrometer and pluviometer, it determines waterlogging and/or surface runoff. This allows optimum fertirrigation management with no losses through surface runoff or drainage.

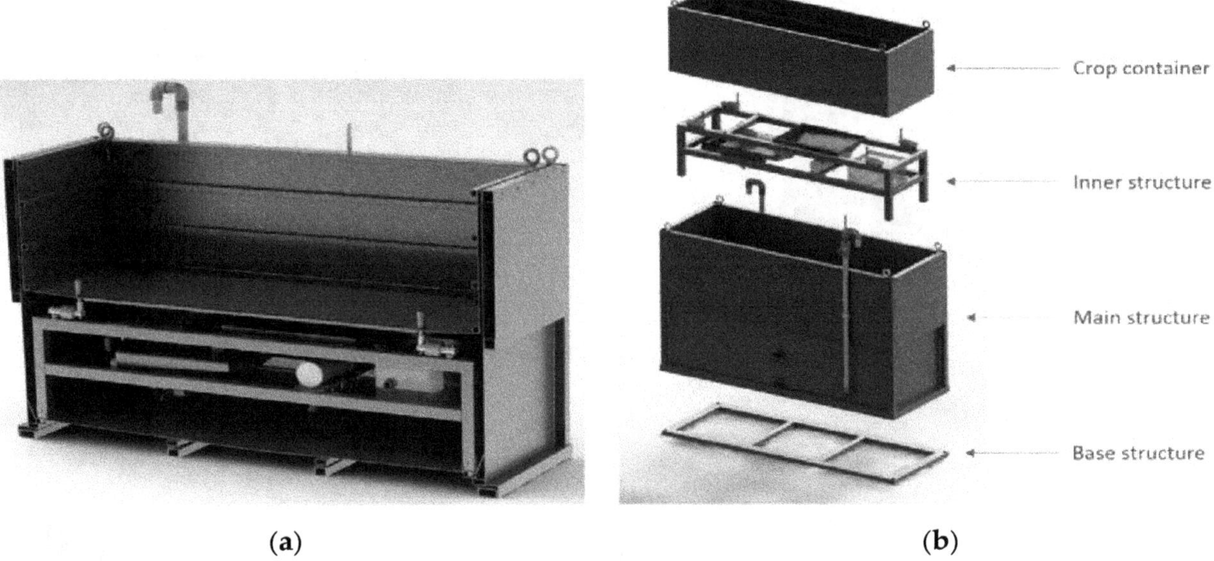

(a) (b)

Figure 9. Three-dimensional model of the ferticontrol equipment based on weighing lysimeter: (**a**) cross-sectional view of the equipment; (**b**) exploded view of the equipment with its main structures.

The 3D models of the proposed ferticontrol equipment and the QR codes were developed following the procedure described in the previous chapter (Figure 10). Next, a series of overall drawings and exploded drawings was produced with 2D and 3D views, to which the corresponding QR codes were added to view them. In this case, the ferticontrol equipment was formed by different structures, which were represented on several drawings, similarly to those offered in Figure 11. The QR code that comes with each drawing can be scanned when students require further graphical information or wish to view the structure from a new perspective, which encourages students' interest and understanding.

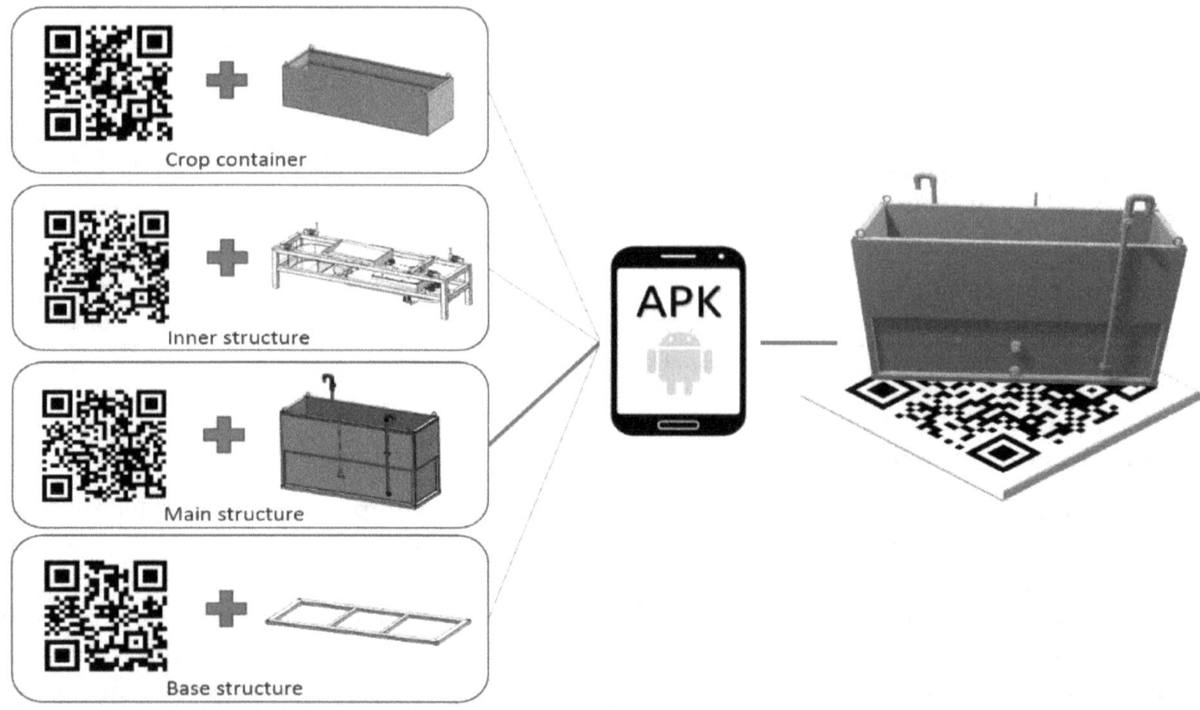

Figure 10. Procedure with proposed ferticontrol equipment.

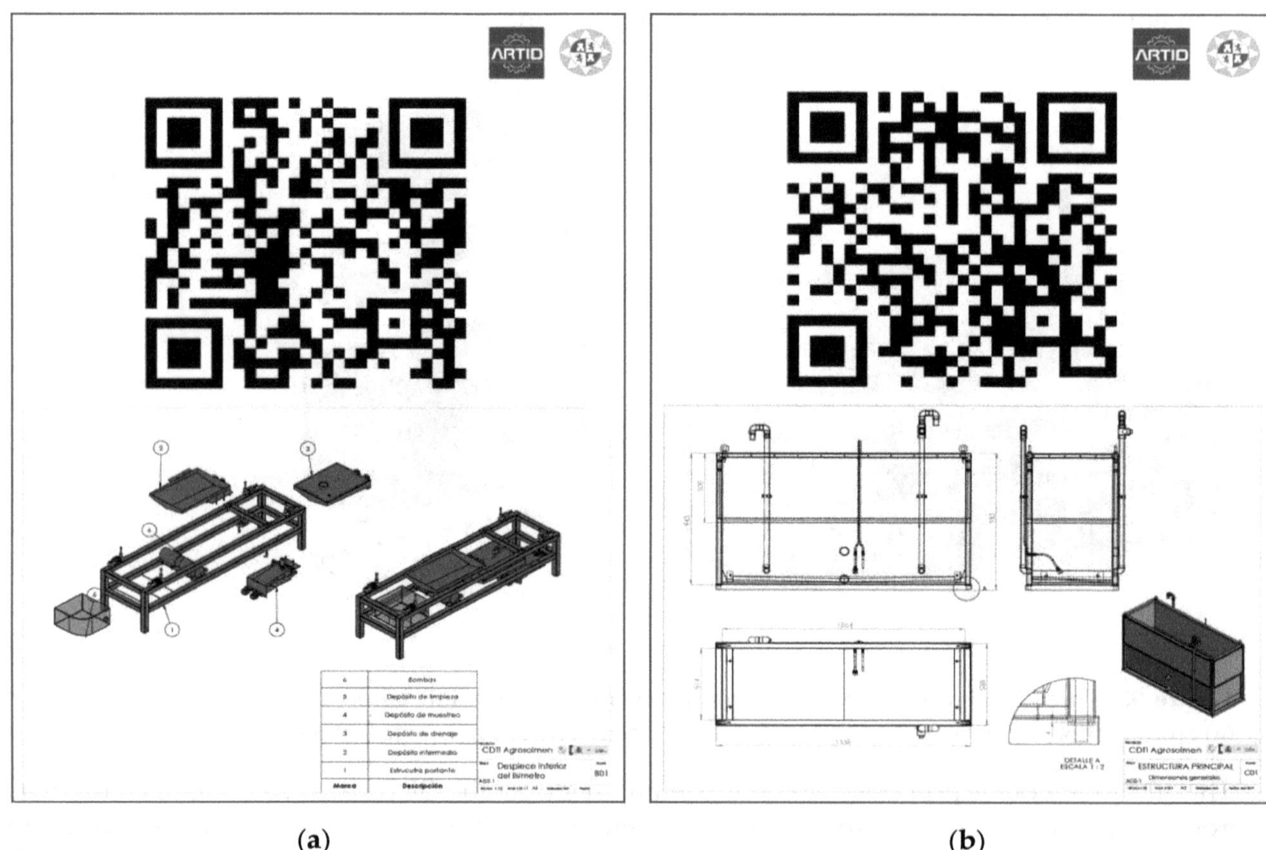

(**a**) (**b**)

Figure 11. Example of the ferticontrol equipment drawings with QR codes: (**a**) inner structure; (**b**) main structure.

Using these codes consists in focusing them by means of the ARTID application in a mobile device where the related 3D model is automatically shown (Figure 12). The position of the mobile device and its angle in relation to the drawing can be changed to view 3D models from several perspectives, but by always focusing these codes.

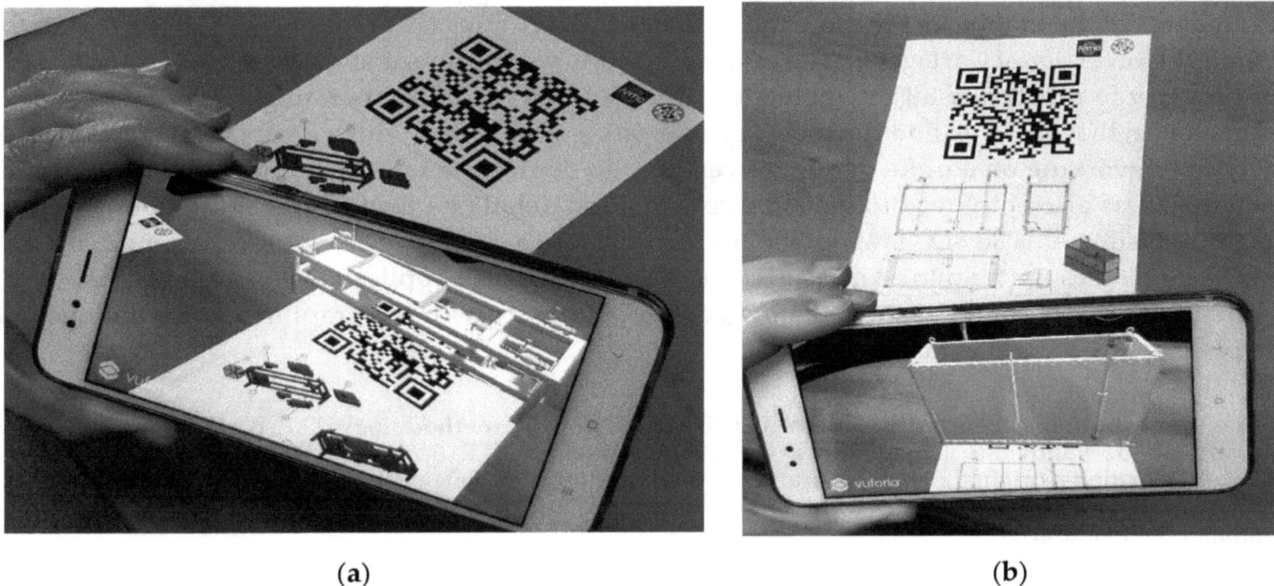

(a) (b)

Figure 12. Viewing two ferticontrol equipment structures with the Augmented Reality for Teaching, Innovation and Design (ARTID) application: (**a**) inner structure; (**b**) main structure.

5. Discussion

Nowadays, augmented reality is a widely used tool in certain technological sectors. On the one hand, it is used in the field of entertainment, such as video games, trying to provide the user with a more interactive experience. On the other hand, it is applied in various industrial sectors to show designs and projects in a more graphical way. All this gives the tool great potential to be developed, so the companies that develop software for programming AR-based applications ask for a fee for its use. When creating an augmented reality application for non-profit purposes, it is difficult to find free software tools.

One of the main problems found in this study was the devices that are normally used for augmented reality, in this case, mobile phones. Projecting 3D elements with great definition on a screen is much easier using a computer that has more processing capacity compared to a mobile phone or tablet. It is a problem that, over time, is solved thanks to the fact that mobile devices have increasingly better characteristics; even so, it is still a barrier. On the other hand, developing an application is not complicated thanks to the wide variety of tutorials and programming courses on Android, but tutorials for something as specific as augmented reality are scarce. The ARTID application stands out for being direct and easy to use, so no menus or interfaces have been included, which also improves its performance and functionality.

The ARTID tool can be extrapolated to any teaching field, technological or not, or to other fields, such as archaeology, medicine or agronomy. In addition, it allows the use of 3D models created in any design software, not only the one proposed in this study.

According to the experience carried out with the industrial engineering students, this interactive way of learning produces greater motivation, since they use tools that are very attractive to them. This can allow students to better assimilate content and visualize more complex graphical representations more effectively. This hypothesis, that has already been validated in some studies referenced in this article, will be a future research line with the aim of validating the ARTID tool.

6. Conclusions

This study describes the process of creating a tool based on augmented reality and called ARTID. The main objective of this new tool is to obtain an open source application that allows to visualize, through augmented reality and QR codes, different types of pieces and sets usually used in technical drawing subjects. To verify the proper functioning of the application, a test was carried out by 188 students of the graphic expression and industrial design subjects taught in several degrees at the Technical University of Cartagena. The obtained results were very satisfactory, since the application ran properly in practically all the mobile devices with the Android operating system. The survey results reveal that students find the tool very interesting, useful and with a great scope that enhances to continue working on a more definitive version. To demonstrate its usefulness in the agronomic domain, it was also employed to view the construction detail of a ferticontrol equipment based on weighing lysimeter. The 3D viewing of objects in space is a challenge for engineering students and these technologies can help to acquire this skill. Developing such applications opens up a wide range of possibilities in teaching as they allow students to interact with the graphic representations that complete degree subject contents.

Author Contributions: Conceptualization, D.P.-B. and D.G.F.-P.; methodology, D.P.-B., D.G.F.-P. and T.P.B.; software, T.P.B.; validation, D.P.-B. and D.G.F.-P.; investigation, D.P.-B. and T.P.B.; resources, M.S.-M. and J.M.M.-M.; writing—original draft preparation, D.P.-B.; writing—review and editing, D.G.F.-P. and J.M.M.-M; visualization, T.P.B.; supervision, M.S.-M. and J.M.M.-M. All authors have read and agreed to the published version of the manuscript.

References

1. Azuma, R.T. A survey of augmented reality. *Presence Teleoperators Virtual Environ.* **1997**, *6*, 355–385. [CrossRef]
2. Fan, X.; Chai, Z.; Deng, N.; Dong, X. Adoption of augmented reality in online retailing and consumers' product attitude: A cognitive perspective. *J. Retail. Consum. Serv.* **2020**, *53*. [CrossRef]
3. Yang, S.; Carlson, J.R.; Chen, S. How augmented reality affects advertising effectiveness: The mediating effects of curiosity and attention toward the ad. *J. Retail. Consum. Serv.* **2020**, *54*. [CrossRef]
4. Uribe, F.C. Realidad aumentada aplicada a la enseñanza de la geometría descriptiva. *AUS* **2017**, 18–22.
5. Park, S.; Stangl, B. Augmented reality experiences and sensation seeking. *Tour. Manag.* **2020**, *77*. [CrossRef]
6. Shin, H.; Gweon, G. Supporting preschoolers' transitions from screen time to screen-free time using augmented reality and encouraging offline leisure activity. *Comput. Hum. Behav.* **2020**, *105*. [CrossRef]
7. Arora, R.; Kazi, R.H.; Grossman, T.; Fitzmaurice, G.; Singh, K. SymbiosisSketch: Combining 2D & 3D sketching for designing detailed 3D objects in situ. In Proceedings of the Conference on Human Factors in Computing Systems—Proceedings, Montreal, QC, Canada, 21–26 April 2018.
8. Atrio Cerezo, S.; Guardado Moreno, E. La realidad aumentada y su presencia en un modelo docente tecnológico para la didáctica de la Química en Bachillerato. *Educación y Tecnología* **2012**, *1*, 9–38.
9. Chang, H.Y.; Hsu, Y.S.; Wu, H.K.; Tsai, C.C. Students' development of socio-scientific reasoning in a mobile augmented reality learning environment. *Int. J. Sci. Educ.* **2018**, 1–22. [CrossRef]
10. Ibáñez, M.B.; Delgado-Kloos, C. Augmented reality for STEM learning: A systematic review. *Comput. Educ.* **2018**, *123*, 109–123. [CrossRef]
11. Suh, A.; Prophet, J. The state of immersive technology research: A literature analysis. *Comput. Hum. Behav.* **2018**, *86*, 77–90. [CrossRef]
12. Ferrer Torregrosa, J.; Jiménez Rodríguez, M.Á.; Barcia González, J.M.; Torralba Estellés, J. La realidad aumentada en la docencia de ciencias de la salud. *Nuevos Caminos para la Comprensión* **2014**.
13. Lineros, M.L.; Jiménez, F.J.S.; Carballo, F.M.; Tatay, C.V. Realidad Aumentada Como Herramienta Docente en la Asignatura de Construcción en las Enseñanzas de Ingeniería. Proceedings of 24 Congreso de Innovación Educativa en las Enseñanzas Técnicas, Cádiz, Spain, 21–23 September 2016; Universidad de Cádiz: Cádiz, Spain; p. 10.
14. Redondo, B.; Cózar-Gutiérrez, R.; González-Calero, J.A.; Sánchez Ruiz, R. Integration of Augmented Reality in the Teaching of English as a Foreign Language in Early Childhood Education. *Early Child. Educ. J.* **2020**, *48*, 147–155. [CrossRef]

15. Juan, C.; YuLin, W.; Tjondronegoro Dian, W.; Wei, S. Construction of interactive teaching system for course of mechanical drawing based on mobile augmented reality technology. *Int. J. Emerg. Technol. Learn.* **2018**, *13*, 126–139. [CrossRef]
16. Molina-Carmona, R.; Pertegal-Felices, M.L.; Jimeno-Morenilla, A.; Mora-Mora, H. Virtual Reality learning activities for multimedia students to enhance spatial ability. *Sustainability* **2018**, *10*, 1074. [CrossRef]
17. Buchori, A.; Prasetyowati, D.; Wijayanto. Design of the magic book math media based on augmented reality. *Test Eng. Manag.* **2020**, *82*, 1480–1489.
18. Aljojo, N.; Munshi, A.; Zainol, A.; Al-Amri, R.; Al-Aqeel, A.; Al-khaldi, M.; Khattabi, M.; Qadah, J. Lens application: Mobile application using augmented reality. *Int. J. Interact. Mob. Technol.* **2020**, *14*, 160–177. [CrossRef]
19. Azhar, N.H.M.; Diah, N.M.; Ahmad, S.; Ismail, M. Development of augmented reality to learn history. *Bull. Electr. Eng. Inform.* **2019**, *8*, 1425–1432. [CrossRef]
20. Lou, D. Two fast prototypes of web-based augmented reality enhancement for books. *Libr. Hi Tech News* **2019**, *36*, 19–24. [CrossRef]
21. Borja-Galeas, C.; Guevara, C.; Varela-Aldás, J.; Castillo-Salazar, D.; Arias-Flores, H.; Fierro-Saltos, W.; Rivera, R.; Hidalgo-Guijarro, J.; Yandún-Velasteguí, M. Editorial Design Based on User Experience Design. In *Advances in Intelligent Systems and Computing*; Springer: Cham, Switzerland, 2020; Volume 1026, pp. 411–416.
22. Yang, Y.; Wu, S.; Wang, D.; Huang, Y.; Cai, S. Effects of learning activities based on augmented reality on students' understanding and expression in an English class. In Proceedings of the ICCE 2019—27th International Conference on Computers in Education, Kenting, Taiwan, 2–6 December 2019; pp. 685–690.
23. Martín-Gutiérrez, J.; Saorín, J.L.; Contero, M.; Alcañiz, M.; Pérez-López, D.C.; Ortega, M.J.C. Design and validation of an augmented book for spatial abilities development in engineering students. *Comput. Graph.* **2010**, *34*, 77–91. [CrossRef]
24. Nguyen, M.; Le, H.; Lai, P.M.; Yan, W.Q. A web-based augmented reality plat-form using pictorial QR code for educational purposes and beyond. In Proceedings of the ACM Symposium on Virtual Reality Software and Technology, VRST, Parramatta, Australia, 12–15 November 2019.
25. Sing, A.L.L.; Ibrahim, A.A.A.; Weng, N.G.; Hamzah, M.; Yung, W.C. Design and Development of Multimedia and Multi-Marker Detection Techniques in Interactive Augmented Reality Colouring Book. In *Lecture Notes in Electrical Engineering*; Springer: Singapore, 2020; Volume 603, pp. 605–616.
26. Bairaktarova, D.; van Den Einde, L.; Bell, J.E. Using digital sketching and augmented reality mobile apps to improve spatial visualization in a freshmen engineering course. In Proceedings of the ASEE Annual Conference and Exposition, Conference Proceedings, Tampa, FL, USA, 16–19 June 2019.
27. Durlach, N.; Allen, G.; Darken, R.; Garnett, R.L.; Loomis, J.; Templeman, J.; Wiegand, T.E.V. Virtual environments and the enhancement of spatial behavior: Towards a comprehensive research agenda. *Presence Teleoperators Virtual Environ.* **2000**, *9*, 593–615. [CrossRef]
28. Hartman, N.W.; Connolly, P.E.; Gilger, J.W.; Bertoline, G.R.; Heisler, J. Virtual reality-based spatial skills assessment and its role in computer graphics education. In *Proceedings of ACM SIGGRAPH 2006 Educators Program*; Association for Computing Machinery: Boston, MA, USA; p. 46-es.
29. Kaufmann, H.; Steinbügl, K.; Dünser, A.; Glück, J. General training of spatial abilities by geometry education in augmented reality. *Annu. Rev. Cyber Ther. Telemed. A Decade VR* **2005**, *3*, 65–76.
30. Kim, J.; Irizarry, J. Evaluating the Use of Augmented Reality Technology to Improve Construction Management Student's Spatial Skills. *Int. J. Constr. Educ. Res.* **2020**. [CrossRef]

10

Prediction of Fracture Damage of Sandstone using Digital Image Correlation

Fanxiu Chen [1,*], Endong Wang [2], Bin Zhang [1], Liming Zhang [1] and Fanzhen Meng [1]

[1] School of Science, Qingdao University of Technology, Qingdao 266033, Shandong, China; zb901052@163.com (B.Z.); dryad_274@163.com (L.Z.); xuelianmfzh@163.com (F.M.)
[2] Sustainable Construction, The State University of New York, Syracuse, NY 13210, USA; ewang01@esf.edu
* Correspondence: mecfx@163.com

Featured Application: A non-contact digital image correlation technique is used to predict sandstone failure. Failure strain for the tested sandstone is estimated to be around 0.004. Finite element analysis verifies the accuracy of prediction results based on experiments.

Abstract: Investigation on the deformation mechanism of sandstone is crucial to understanding the life cycle patterns of pertinent infrastructure systems considering the extensive adoption of sandstone in infrastructure construction of various engineering systems, e.g., agricultural engineering systems. In this study, the state-of-the-art digital image correlation (DIC) method, which uses classical digital photography, is employed to explore the detailed failure course of sandstone with physical uniaxial compression tests. Four typical points are specifically selected to characterize the global strain field by plotting their corresponding strain–time relationship curves. Thus, the targeted failure thresholds are identified. The Hill–Tsai failure criterion and finite element simulation are then used for the cross-check process of DIC predictions. The results show that, though errors exist between the experimental and the theoretical values, overall, they are sufficiently low to be ignored, indicating good agreement. From the results, near-linear relationships between strain and time are detected before failure at the four chosen points and the failure strain thresholds are almost the same; as low as 0.004. Failure thresholds of sandstone are reliably determined according to the strain variation curve, to forecast sandstone damage and failure. Consequently, the proposed technology and associated information generated from this study could be of assistance in the safety and health monitoring processes of relevant infrastructure system applications.

Keywords: failure strain; sandstone; digital image correlation; Hill–Tsai failure criterion; finite element method

1. Introduction

As one typical type of sedimentary rock, often consisting of sand-scale mineral particles (e.g., quartz, feldspar), sandstone has long been used as a functional construction material in various project types of pavements, hydraulic systems, warehouses, and underground structures for structural purposes in diverse agricultural and industrial sectors [1–4]. The failure mechanism and pattern of sandstone imposes an important role on the safe construction and operation of pertinent structures. Moreover, the material properties of sandstone are often observed and characterized to be of discontinuity, nonlinearity, anisotropy, and non-elasticity [5]. This leaves the prediction of the structural behavior and performance of sandstone as a significant challenge. Consequently, mechanical properties and failure patterns of sandstone have been the common research focus of various disciplines integrating mechanics, material science, and engineering [6,7].

Significant research efforts were committed to understanding and profiling mechanical behavior, material properties, and failure modes of sandstone through various methods, techniques, and tools, e.g., multi-scale analysis, digital image correlation technique, strain gage method, three-point bending test, and empirical simulation in [2,4,8–10]. Traditional engineering measurement methods for motions and strains were often criticized for their limitations in real applications due to their underlying deficiencies resulting from contacting and localized operations [11]. In contrast, the digital image correlation (DIC) technique can compensate with its unique advantages of being contactless and able to capture full-field strain and displacement over the traditional measurement approaches. Moreover, DIC technology utilizes classical digital photography and it is relatively cost-effective when compared to other optical methods, such as laser shearography [12]. Therefore, DIC has been extensively employed in various engineering applications since the 1980s [13–16], with no exception to relevant investigations on sandstone and rock materials. Munoz and Taheri [17] studied stress-strain features of sandstone under monotonic uniaxial compression. Strain pattern development of sandstone with varying aspect ratios was inspected by a three-dimensional digital image correlation technique since it can capture field strain during the whole compression process. Their study revealed that different strain development features are associated with pre- and post-peak regimes. They found that, in the pre-peak regime, strains localize gradually and develop at a slower speed, but in the post-peak regime, strains develop at diverse speeds due to varying impacts of local deformations. Yue et al. [8] perceived that little information is available on whether DIC and the strain gage method are accurate for fracture mechanism characterization. Then, they conducted a study to examine the accuracy of DIC and the strain gage method in characterizing crack patterns for white marble specimens. The DIC technique was discovered to be flexible and stable for characterizing rock failure mechanisms. Song et al. [18] investigated the damage evolution and crack growth of rocks using digital image correlation analysis. It was found that the cyclic value has a significant influence on strain localization and damage evolution when the cyclic loading amplitude exceeds a certain value. By performing tests on Springwell sandstone, Stirling et al. [19] utilized DIC technology to qualitatively and numerically examine how loading method and object geometry can affect strain localization processes over sample faces. In their study, the applicability of DIC to substitute for traditional strain measurement methods for Brazilian testing process was demonstrated. To support the applications of rock and sandstone materials in secure underground construction, Wu et al. [4] performed uniaxial compression tests to understand the mechanical behavior of holed sandstone using the DIC method. Compared to the intact specimens, those artificially holed ones show almost half of the expected mechanical performance depending on hole shapes. DIC technology is able to visualize strain and displacement fields to consistently profile failure patterns. The digital image correlation method was used by Lin and Labuz [20] to successfully assess the mode I fracture parameters including process-zone magnitude and critical opening displacement for Berea sandstone with a three-point bending test. The tensile fracture features covering opening displacement and crack size were identified, showing that over all the tests, process-zone retains a stable length and critical opening. Li et al. [21] examined the micro behavior of mode I crack of sandstone due to varied loading speeds using acoustic emission and digital image correlation. All the previous studies are helpful in facilitating the understanding and characterization of mechanical behavior and performance of sandstone components within various application contexts. Nevertheless, in the existing literature several disadvantages are associated with DIC applications. First, based on a literature search, few studies have been carried out yet to particularly concentrate on the dynamic failure process to obtain full-field strain values of sandstone which can provide more direct, specific, and deeper information on a sandstone failure course. Moreover, to the best of our knowledge, no studies have been performed to systematically examine the specific thresholds of failure strain of sandstone materials. Finally, most studies used the DIC technique for sandstone behavior investigation without knowing the reliability and accuracy of the obtained results.

The present research expects to fill out this research gap. Compared to the existing literature, this study provides a more specialized and detailed investigation into the full-field strain of sandstone under different loading scenarios by examining multiple representative field regions using DIC. Different from previous studies, this research especially explores the specific failure thresholds of sandstone materials. Advantageous over the literature, this study combines three different methods including optical DIC, theoretical analysis, and finite element simulation for the cross-check of result accuracy. By conducting uniaxial compression experiments on sandstone, this paper proposes to use the state-of-the-art digital image correlation method to gain the full-field strain of sandstone model subject to the process of uniaxial compression. In particular, deep investigations are carried out on four typical points selected from the four areas and moments of the full-field failure.

These four points are considered to be the typical representatives characterizing the specimen failures course. A total of 12 specimens with identical dimensions are tested. Dynamic strain–time curves are derived, corresponding to the four chosen points in the course of the compression test in order to detect the range values of failure thresholds for sandstone samples. Based on the obtained thresholds, failure patterns can then be forecasted for sandstone, and further, the pertinent structure components and systems. The DIC processing results are cross-checked by the Hill–Tsai failure criterion and finite element method (FEM), which have been widely used for analyzing failure mechanisms from two distinct perspectives.

2. Digital Image Correlation Technique

The DIC technique represents an optical approach for reliably measuring the two-dimensional or three-dimensional image alterations to quantify deformations and strains via advanced image registration and target tracking algorithms [17]. Compared to other approaches, such as mechanical deflectometers and speckle shearing interferometry, for deformation and strain measurement, DIC is an easy to operate and cost-effective technique. For example, in many cases, it does not need any special working environment and can even work for extreme conditions, e.g., at the temperature higher than 1000 °C. Meanwhile, it is more accurate than manual measurements. DIC can be used to estimate strains in a wide range (e.g., from the lower order of 0.001 to the higher order of 1) with its measurement sensitivity on deformations reaching 1/100,000 of the view field in-plane. With clean cameras and patterned speckles, the strain resolution can be around 20 microstrains. Due to these advantages, along with advances in computer technology, DIC has been extensively adopted for monitoring deformation patterns, failure modes of materials, and structural health without damaging the objects under investigation.

2.1. Working Principle of the DIC Technique

For mechanical performance tests, DIC works to obtain micro details of targeted process information including local and average strain and displacement. To achieve this, DIC processes the digital images collected during physical experiments through image processing techniques and statistical cross-correlations.

More specifically, as in Figure 1a, it first defines the specific pixel intensity subsets referring to the small regions of interest on the raw pre-deformation image and then, as in Figure 1b, the cross-correlations between pre-deformation and post-deformation are computed. With the pre-deformation image as the reference, best cross-correlation coefficients between the subsets, which define the fundamental computing units in DIC, on pre-deformation and post-deformation images are calculated to seek the mapping correspondences between subsets.

After the mapping process, displacements are calculated for each subset center point to get the information of full-field deformation. Strains are then calculated based on the obtained information on displacements, as in Figure 1c. Among many options, the standardized covariance correlation function [22,23] defines an alternative approach for the computation of the two-dimensional discrete correlation $C(u,v)$ (Equation (1)):

$$C(u,v) = \frac{\sum_{x=-M}^{M} \sum_{y=-M}^{M} [f(x,y) - f_m][g(x+u, y+v) - g_m]}{\sqrt{\sum_{x=-M}^{M} \sum_{y=-M}^{M} [f(x,y) - f_m]^2} \sqrt{\sum_{x=-M}^{M} \sum_{y=-M}^{M} [g(x+u, y+v) - g_m]^2}} \quad (1)$$

where $f(x,y)$ and $g(x + u, y + v)$ represent the grey-scale function values of all the pixels on the images taken before and after deformation (x, y indicate point locations prior to deformation.); f_m and g_m are the average grey-scale values of image subsets; u, v mean the displacement values in the center of relevant subset in terms of pixels.

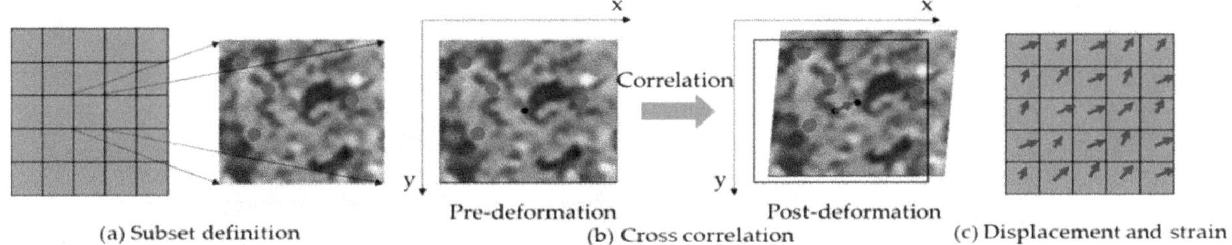

Figure 1. Working principle of digital image correlation.

This assumes that maximum correlations indicate correspondence [24]. That is, for a specific pre-deformation subset, the subset on the post-deformation image having the largest correlation with this designated subset, is regarded as its counterpart after deformation. The corresponding matching accuracy can be in the order of 0.01 pixels. By contrasting the counterpart pairs on pre- and post-deformation images, the desired deformation information at all points can finally be derived. The full-field strain can further be estimated based on the deformations of all these individual points.

In general, the DIC method uses charge-coupled device (CCD) camera systems with specific resolutions to record the images for the irregular distribution profiles of speckles on the surfaces of specimens before and after deformations. These digital images are then processed and analyzed using the chosen numerical methods to acquire such process information as displacement and strain. For image shooting, additional lighting sources may be required where natural lights are not capable of providing enough luminance. The demand of lighting intensity depends on the quality requirements of images taken by the CCD. In addition, for the use of DIC, the cross-correlation calculation requires detailed information of randomly scattered speckles on the surfaces of studied objects. As such, when natural speckles are of insufficient utility, extra speckles can be artificially created by painting the targeted surfaces in either black or white [25].

2.2. *Implementation of DIC for Strain Measurement in This Study*

Based on the principle of DIC described above, this study combines the use of a universal testing machine, automation control system, charge-coupled device camera, and computer system for the characterization and prediction of the failure pattern of sandstone materials. Figure 2 shows the implementation procedure of the study. For each experiment cycle, five steps are essentially executed. Step 1 is to prepare specimens in specific dimensions and set up equipment systems. To ensure test effects, specimens can be artificially treated, e.g., painted to create scattered speckles. The equipment systems used for compression testing, image recording, and storing are set up. Step 2 is loading specimens. Equipment positions are adjusted to enable the studied specimens to be placed at the center of the universal testing machine for accurate experiment processes. Step 3 records pre-deformation images and stores them on the computer as reference images. Step 4 operates the testing machine and obtains post-deformation images at different time points. Step 5 is to conduct DIC analysis to calculate numerical displacement and strain information based on the process in Figure 1. Dynamic strain–time curves are plotted based on the received information. Specific equipment and operation environment information is as follows. A Basler SCA1600 14FM CCD black-white camera was used for image collection. It is equipped with a Sony ICX274 CCD sensor and can supply up to 14 frames per second with a 2 MP resolution and 4.4×4.4 μm pixel size. The adopted universal testing machine has a loading capacity of 0–600 kN. A regular HP-Z600 workstation was used for image storage and processing. VIC-2D software from Correlated Solutions was installed on the workstation for DIC analysis. All the experiments were performed in fall with a dry bulb temperature around 24 °C and a relative humidity about 55% in Qingdao City, China.

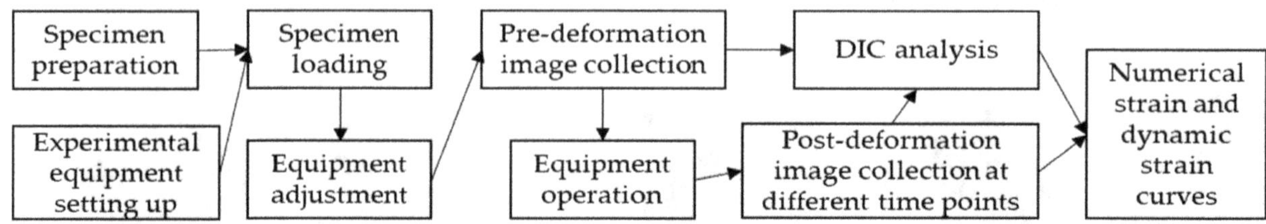

Figure 2. Implementation of digital image correlation (DIC) for strain measurement.

3. Theoretical Modeling of Sandstone Failure with Hill–Tsai Criterion

Sandstone, as one type of natural composite material, often consists of sand-size grains of multiple single materials of varying characteristics with certain combinations. Physical properties and mechanical behavior of sandstone materials vary significantly depending on the random distribution of material constituents. From this perspective, sandstone materials cannot be viewed and analyzed simply through some particular models and theories aiming at single non-composite materials. Models and theories for investigating mechanical performance of composite materials are demanded.

Based on [26,27], in this study, the classical stress-based Hill–Tsai yield criterion (Equation (2)) put forward by Hill in 1948 for characterizing the properties and behavior of anisotropy composite materials [27] was adopted for theoretical analysis of sandstone failure process.

$$\frac{\cos^4\theta}{X^2} + \left(\frac{1}{S^2} - \frac{1}{X^2}\right)\cos^2\theta\sin^2\theta + \frac{\sin^2\theta}{Y^2} = \frac{1}{\sigma^2} \tag{2}$$

where X, Y, S represent axial strength, lateral strength, and shear strength, respectively. For example, the axial strength, lateral strength, and shear strength of sandstone are around 40, 20, and 10 MPa, respectively. The letter θ symbolizes off-axis angle, while σ means failure stress.

Figure 3 shows the relation between stress strength and off-axis angle for sandstone. It can be seen from Figure 3 that the stress strength of sandstone is 40 MPa when $\theta = 0$. As θ increases, stress strength drops at first and then goes up. When $\theta = 60°$, the smallest stress strength reaches 18.3 MPa. When θ is 90°, the stress strength arrives at 20 MPa. That is to say, the minimal stress strength occurs at the position where a 60° angle is formed. This identifies the weakest point where failure easily occurs. The Young's elasticity modulus of this class of sandstone is about $E = 5 \times 10^3$ MPa. Combining the failure stress of 18.3 MPa, as obtained by the Hill–Tsai criterion, through Young's modulus formula (Equation (3)), failure strain ε of the sample sandstone can be calculated as 3.7×10^{-3}.

$$\varepsilon = \frac{\sigma}{E} \tag{3}$$

where ε represents the dimensionless strain and σ means the stress. E is Young's modulus.

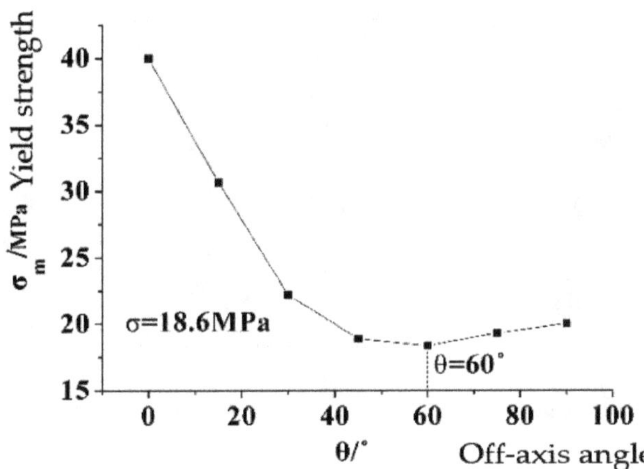

Figure 3. Relationship between yield strength and off-axis angle.

These theoretical results may be utilized as numerical benchmarks against which the testing results obtained from the following physical experiments are to be compared. These mutual comparisons and cross-checks are expected to help to validate the feasibility of the DIC technique for reliable characterization and forecasts on the failure patterns and modes of sandstone materials.

4. Experiment Design

The chosen rock test-piece was a sandstone cylinder 50 mm in diameter and 100 mm in height. As mentioned, a CCD black-white camera Basler SCA1600 14FM was used for shooting images. It is a 1/1.8″ CCD Firewire camera having 1628 × 1236 pixel resolution with a maximum frame rate of 14 fps and a pixel size of 4.4 × 4.4 μm. All the during-process figures were automatically recorded and stored onto an HP-Z600 workstation in bmp format at the frequency of 2 fps. The loading range of the adopted universal testing machine was 0–600 kN. During the compression process, the compressing head of the machine can automatically adjust its pressure direction corresponding to the dynamic specimen movements so as to satisfy the required experimental conditions of uniaxial compression. Our specific goal was to obtain the profiles of the strain field of the specimens to trace relevant failure modes. Therefore, the robust two-dimensional digital image correlation system of VIC-2D was adopted for image analysis due to its sufficiency for dealing with in-plane deformations and its greater convenience when compared to three-dimensional DIC procedures. The recorded images were uploaded into VIC-2D software, which is installed on the HP-Z600 workstation for image analysis. For experimental purposes, in total, the same experimental protocol was repeated on 12 identical sandstone specimens to secure reliable outcomes.

Figure 4 displays the tested samples and the image recording system. While Figure 4a shows the raw tested specimen, Figure 4b presents the corresponding specimen, which was manually painted to highlight the related surface speckles for better testing effects. First, the surfaces of the sandstone test-piece were thoroughly cleaned. Matte black paint was subsequently sprayed to the surface, resulting in randomly scattered speckles for the sandstone test-piece. After ventilating to dry, the painted specimen was placed at the center of the testing machine. Figure 4c shows an example of specimen failure. The measurement system is shown in Figure 4d. The recording camera was calibrated to be able to view the surface facing the camera and acquire quality images on the piece by adjusting its focal length. Meanwhile, an external lighting source with uniform luminance was added. The testing machine was turned on and then adjusted to enable the machine head to just touch the specimen. Loading rate was first controlled at 0.1 MPa/s by an automating process, and then at 0.3 MPa/s to see if a loading rate change can alter the value of failure strain for sandstone. The CCD-based system was used to record the whole loading process at the image shooting rate of two images per second until the compression process ended.

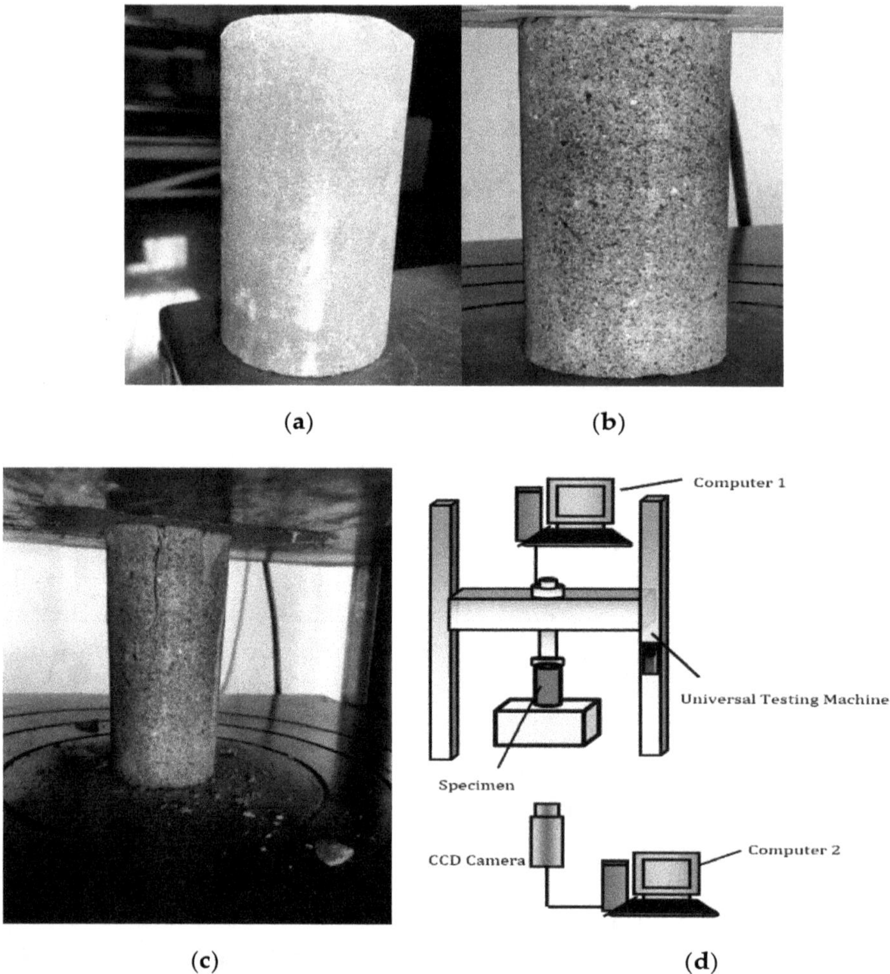

Figure 4. Experiment set-up: (**a**) specimen before painting; (**b**) specimen after painting; (**c**) specimen failure; (**d**) measuring system.

5. Experiment Results and Discussion

As mentioned earlier, the experimental tests were repeated on 12 identical sandstone specimens. Decent consistency was found among all the specimens in terms of failure areas and damage patterns. All the specimens incurred shear failures, with failure angles ranging between 65° and 75°. Some variations were associated with failure angles of specimens, possibly due to random errors and hooping effects from the testing machine on the specimen ends. Limited to the DIC technique, all the following results and discussion correspond to the specimen surface areas captured by the used recording system. The cylinder areas that could not be captured by the system were not considered. Due to unevenness of rock materials, symmetric failures may not have occurred.

5.1. Scenario 1: Loading Rate at 0.1 MPa/s

5.1.1. Results on X-Direction Strain Dynamics

During compression, the dynamic strain change along the x-direction of sandstone is displayed in Figure 5. Figure 5a–d characterizes the compression profiles at the time points of t = 131, 139, 140.5, and 141 s, respectively. They correspond to the failure time at A, B, C, and complete failure. A slight crack can be seen from the right upper side of Figure 5a and the strain value at the crack is relatively large. The strain value at point A reaches a failure threshold value of 0.0044, so failure occurs at point A. The occurrence of failure at point A can be clearly seen from Figure 5a. At this time (the initial stage of compression), point B has not reached the failure value, so failure does not occur, but as compression continues, the strain value at point B begins to increase slowly. When it reaches t = 139 s, the crack suddenly turns much larger with a clear indication of failure. The strain value grows rapidly, and failure occurs suddenly at point B, as shown in Figure 5b. Point C does not show any failure during the initial and middle stages of compression, with the corresponding strain value always fluctuating within a reasonable range. Nevertheless, during the later stage of compression, as shown in Figure 5c, the preceding failure trace further expands. A new curved failure segment with greater failure strain values appears at the upper left. Point C is right on this segment, and then failure occurs at point C. Compression continues to t = 141 s at which the whole sandstone test-piece fails, as shown in Figure 5d.

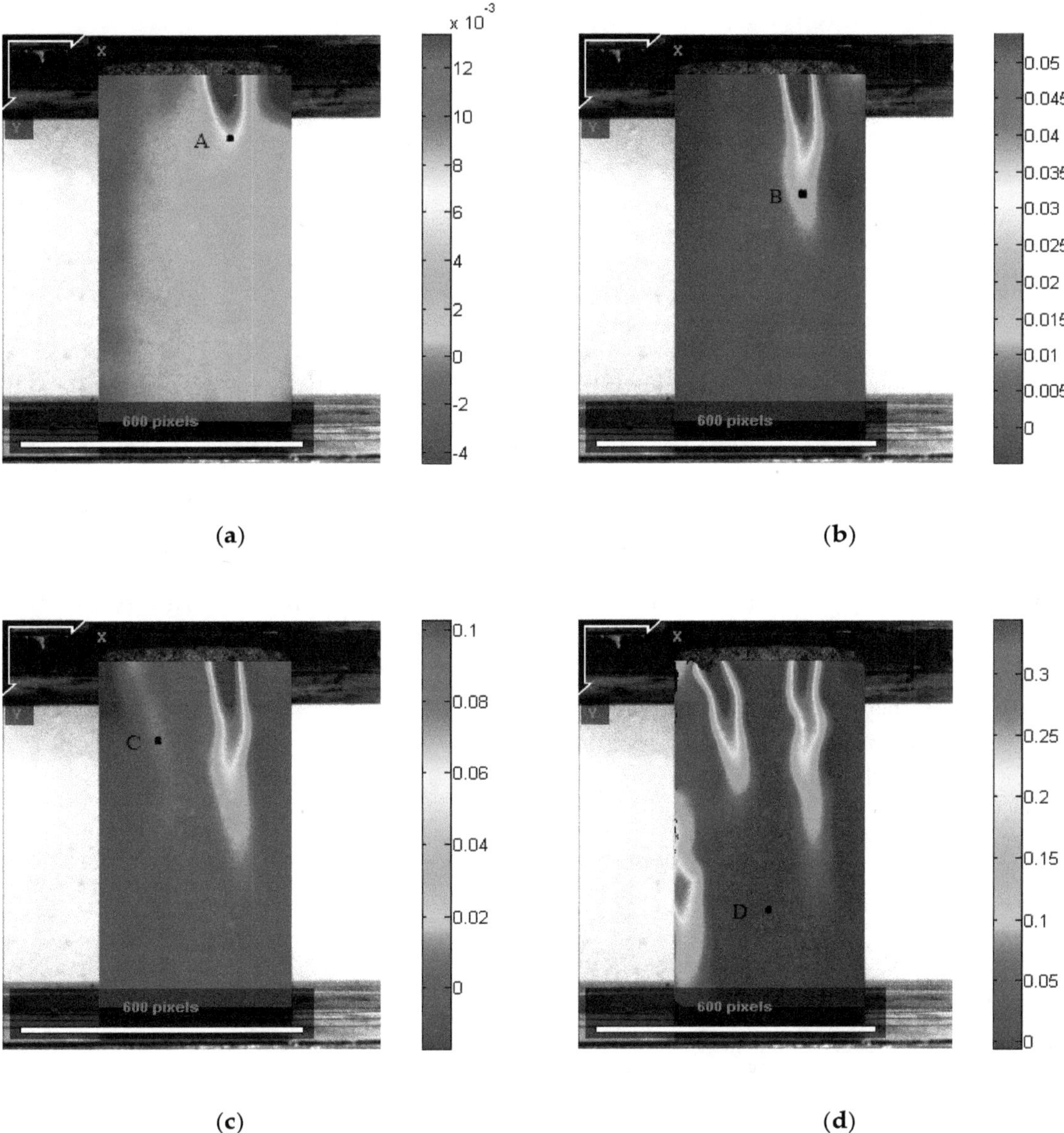

Figure 5. Strain dynamics along the x-direction (loading rate = 0.1 MPa/s): (**a**) t = 131 s; (**b**) t = 139 s; (**c**) t = 140.5 s; (**d**) t = 141 s.

Figure 6 shows the full profiles of strain dynamics at the points A, B, C, and D during the whole compression process, which is in accordance with the phenomenon seen from Figure 5. As seen from Figure 6, the process of sandstone failure is almost instantaneous without a substantial transition course, and therefore, it is a challenge to find the specific features of the failure process, not to mention the accurate prediction of the failure of sandstone. To understand the failure process of sandstone more delicately, this paper conducts more detailed investigations on different individual points to examine and visualize the corresponding failure processes. As mentioned before, these individual points were chosen from the four regions of the full-field in terms of the time sequence of failure occurrence. More clearly, four typical points were selected based on the preliminary results shown in Figure 5. As seen from Figure 5, points A, B, and C indicate the tip locations of the first three failure regions of the tested specimen at the three moments of image collection (t = 131, 139, and 140.5 s). Strain dynamics were analyzed in detail at these three points to identify the failure pattern of the rock specimen. Point D represents a typical location that does not reach failure during the whole failure process. This point was intentionally selected to be compared with the other three points to demonstrate strain pattern differences. These four points expect to represent the corresponding regions and moments of specimen failure.

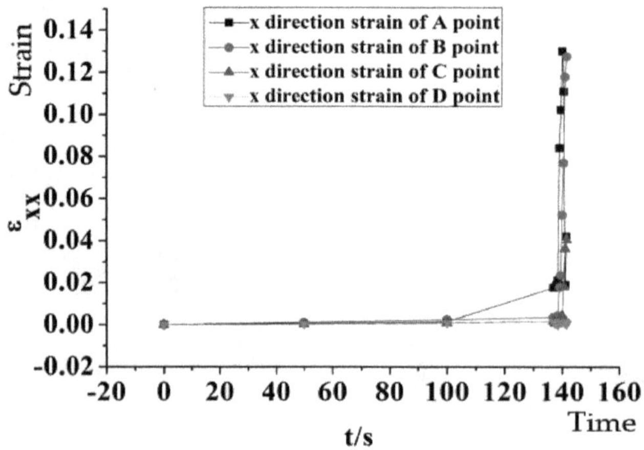

Figure 6. Strain–time profile along the x-direction (loading rate = 0.1 MPa/s).

In general, for elastic materials, during the early period of failure, the strain–time curve shows linearity or near-linearity, with its slope usually low. Once the change in strain value appears nonlinear or shows an exponential growth with time, the material is most likely to be or has already been destructed. In this study, failure at point A occurs at about t = 131 s. From Figure 7, we can see that before t = 131 s, the strain value changes slightly, and shows a near-linear rise. Between t = 131 s and t = 131.5 s, the strain value undergoes a big change from 0.0044 to 0.011. It is then deduced that the strain value of failure is between 0.0044 and 0.011.

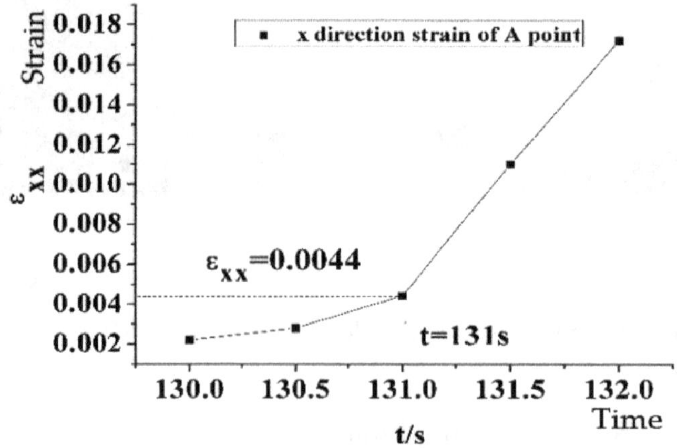

Figure 7. X-direction strain–time relationship at point A (loading rate = 0.1 MPa/s).

It can be seen from Figure 8 that the strain value at point B before t = 137 s generally increases linearly. The strain value is 0.0036 at t = 137 s when strain begins to show nonlinear change and eventually reaches its strain limit. At t = 140 s, the strain value suddenly increases to 0.0177. Within this period, strain value change is dramatic and instantaneous. The strain value of failure is then expected to be between 0.0036 and 0.0177.

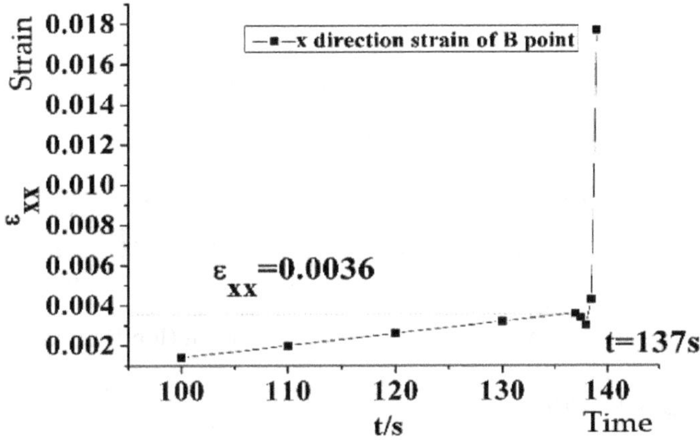

Figure 8. X-direction strain–time relationship at point B (loading rate = 0.1 MPa/s).

As is shown in Figure 9, the strain value before t = 140 s changes very slightly and almost linearly for point C. At t = 140 s, it is 0.0045, while at t = 140.5 s, it is 0.018. During this period of time, similar to those at points A and B, the strain value changes significantly. The strain value of failure is expected to be between 0.0045 and 0.018.

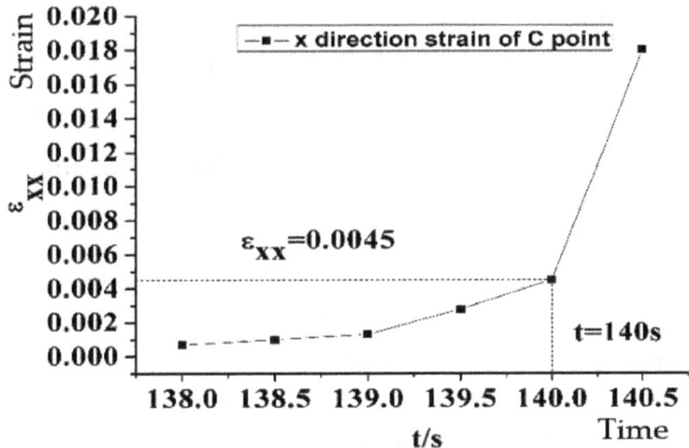

Figure 9. X-direction strain–time relationship at point C (loading rate = 0.1 MPa/s).

From the above, points A, B, and C fail one after another during the loading process. Through the detailed study on the failure processes at each point, the range of the strain values at failure related to points A, B, and C can be revealed. It can be concluded that the strain value of failure for sandstone is approximately between 0.004 and 0.011 when the loading speed lies at 0.1 MPa/s.

5.1.2. Results on Shear Strain and Y-Direction Strain Dynamics

According to the Hill–Tsai criterion [27], the tensile failure strain of the sandstone is about 0.0037 and the shear failure strain is about 0.011, which is three times as large as the tensile strain. The compressive strain is about 0.015–0.1, which is 5 to 50 times as large as the tensile strain.

Based on Figure 6, the time of failure for point A is t = 131 s, for point B is t = 139 s, and for point C is t = 140 s. From Figures 10 and 11, at points A, B, and C, the y-direction strain and shear strain do not reach the failure value at the corresponding time points under study. That is to say, the failure effects caused by the y-direction strain and shear strain are not the main factors affecting the failures at points A, B, and C.

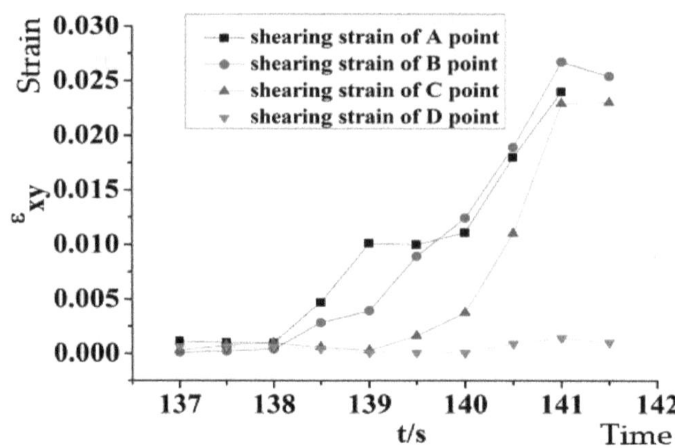

Figure 10. Shear strain–time relationship of sandstone (loading rate = 0.1 MPa/s).

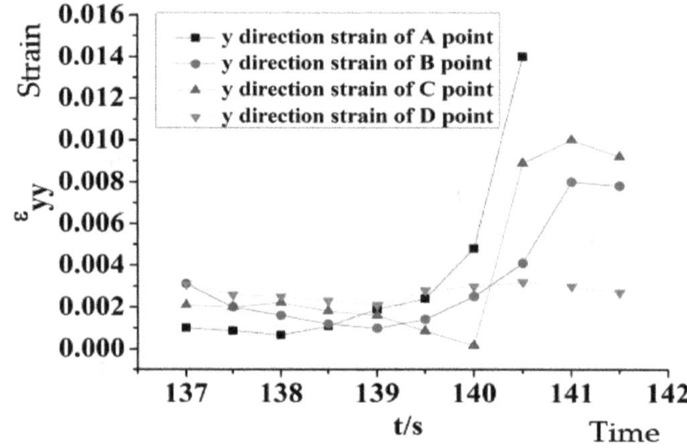

Figure 11. Y-direction strain–time relationship of sandstone (loading rate = 0.1 MPa/s).

5.2. *Scenario 2: Loading Rate at 0.3 MPa/s*

In order to examine the influential effects of different loading speeds on failure strain, another series of experiments with a speed of 0.3 MPa/s were carried out. To ensure an effectual investigation, in the following experiments, the test conditions were controlled to be comparable with that in Scenario 1 (where the loading rate was 0.1 MPa/s), except for loading speed difference. The sandstone test-piece had the equal dimension sizes as the one used above.

5.2.1. Results on X-Direction Strain Dynamics

The same experiment procedure was followed. Four typical points A, B, C, and D were chosen to study the failure mode when the loading speed equals 0.3 MPa/s. Points A, B, and C represent the three point positions where failure occurs at different times, and D indicates the point location where no failure occurs during the whole loading process. Figure 12 shows the strain variations in the x direction throughout the compression process at the loading rate of 0.3 MPa/s. Figure 12a–d plots the strain profiles at t = 58, 71.5, 72.5, and 78 s, respectively, characterizing the failure time of three points of A, B, C, and the compression end event. In the early stage of the compression course, no failure occurs at any point, while local strain values slowly increase but still with linear elastic deformation. At t = 58 s, as shown in Figure 12a, the strain value at point A exceeds the linear elastic strain range extremely and failure occurs. From Figure 12a, it can be clearly seen that the strain at A point is relatively large. As the compression course proceeds further, the crack continues to expand from point A and a new crack occurs at the upper side, as shown in Figure 12b. Point B is right on the new crack, and at t = 71.5 s, the new crack turns significantly larger, and then failure occurs at point B. As shown in Figure 12b, the strain value at point B is relatively high. Quickly, in the following one second, there appears a fresh crack over the old fracture, and as shown in Figure 12c, point C, which is right on the new crack, fails at t = 72.5 s.

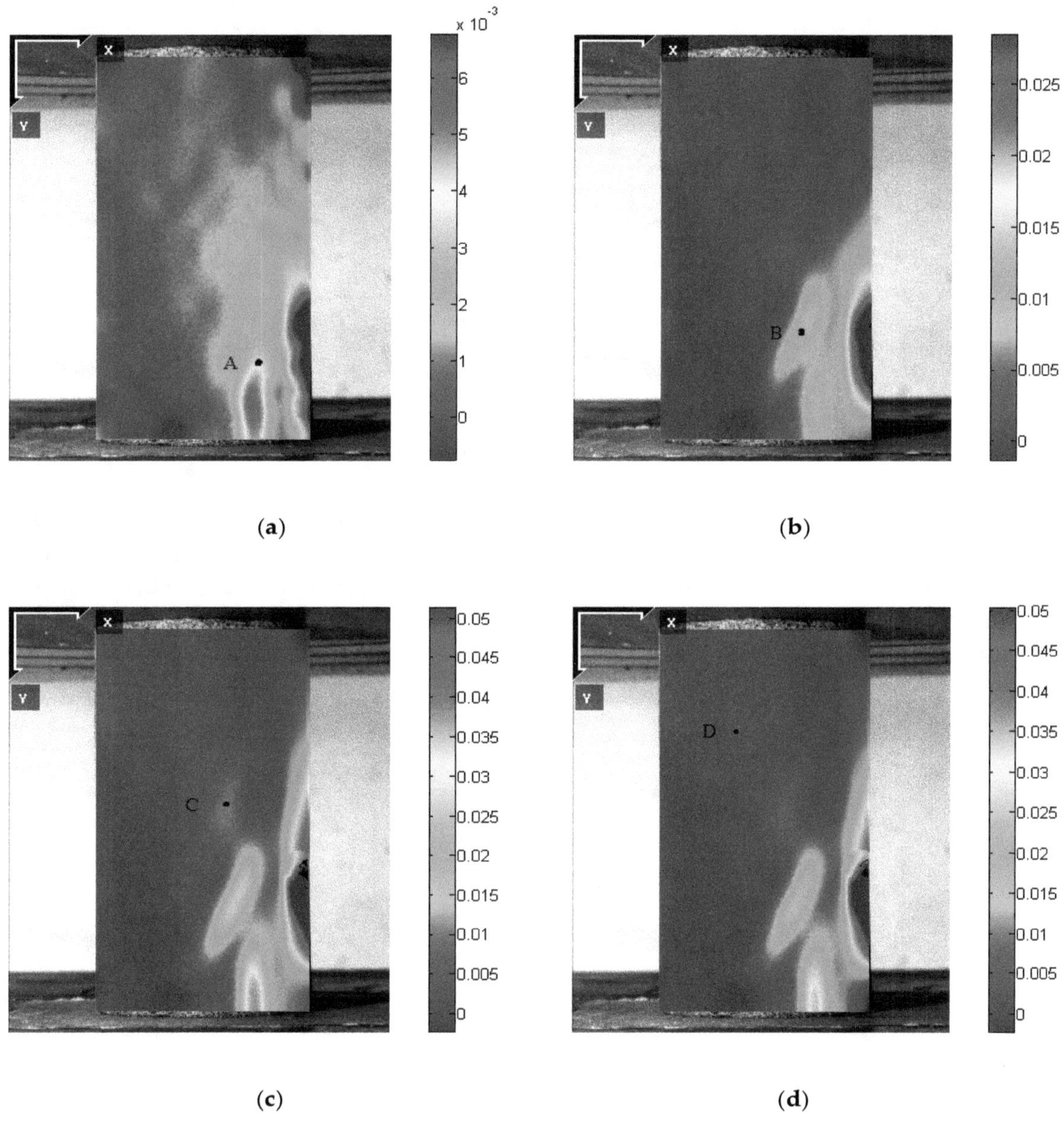

Figure 12. Strain dynamics along the x-direction (loading rate = 0.3 MPa/s): (**a**) 58 s; (**b**) 71.5 s; (**c**) 72.5 s; (**d**) 78 s.

Figure 13 shows the temporal dynamics of strain at points A, B, C, and D with the loading speed of 0.3 MPa/s. It can be learned from Figure 13 that the strain values at points A, B, and C increase over time, displaying a linear or near-linear trend in the early stage. Up to this point, the specimen is undergoing elastic change and the failure does not occur. As the loading process continues, all the three points A, B, and C may be subjected to sharp strain changes where the tested specimen generally goes beyond its elastic range and eventually fails. In order to further uncover the detailed information of failure processes at individual points, individualized deeper analyses were executed to each single point.

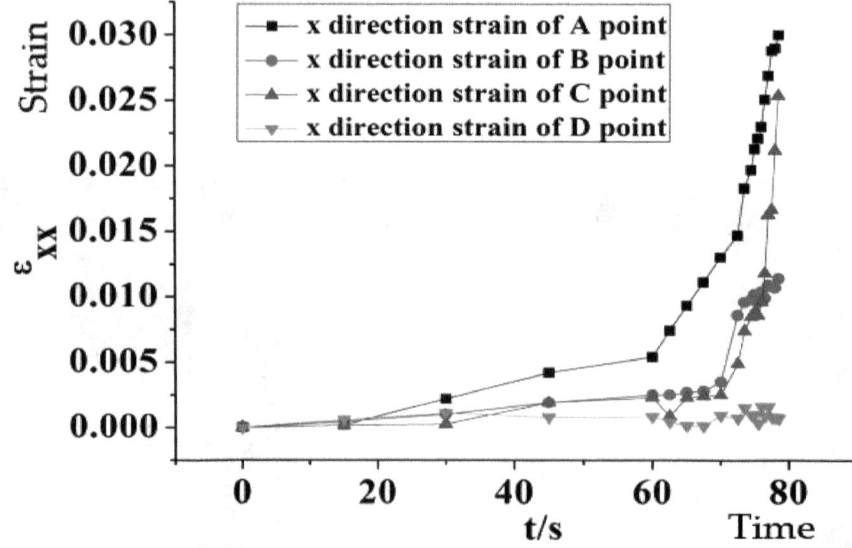

Figure 13. Strain–time profile along the x-direction (loading rate = 0.3 MPa/s).

In Figure 14, before t = 58 s, the variation of strain values at point A is relatively even and in a linear pattern relative to the time dimension without presenting sharp turns. However, starting from t = 58 s, the strain values change nonlinearly, and the strain values increase sharply after t = 60 s. A strain limit may present at t = 58 s. As a result, it can be inferred that the failure strain value for the sandstone sample may lie between 0.0048–0.0093.

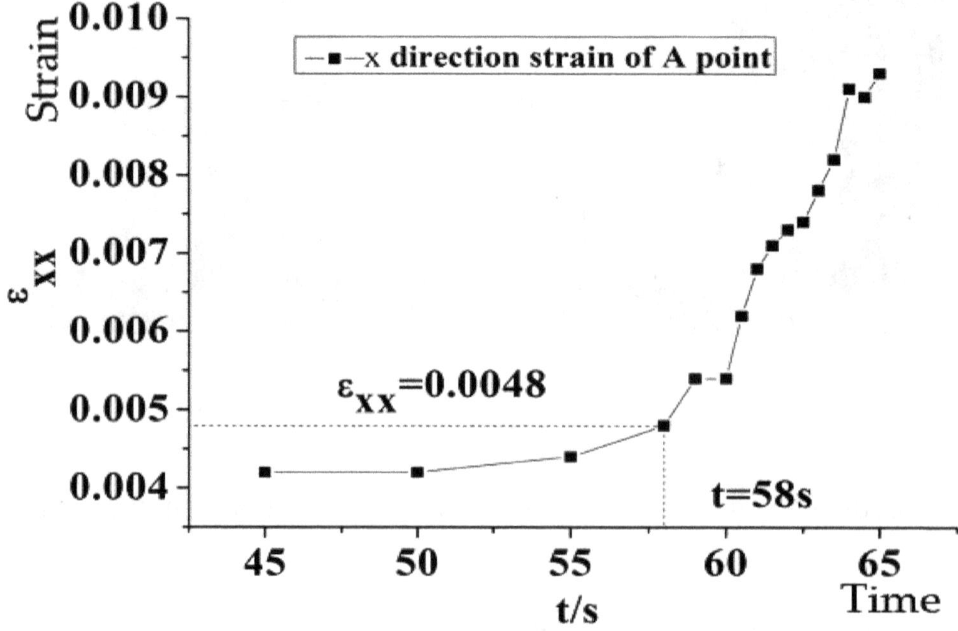

Figure 14. X-direction strain–time relationship at point A (loading rate = 0.3 MPa/s).

As seen from Figure 15, before t = 71.5 s, the strain values of point B vary little and display a near-linear elastic mode. At t = 71.5 s, the strain value of point B is 0.0048, while at t = 72 s, it reaches 0.0086. During this period, the strain values increase more dramatically than before; it is then determined that the failure strain value at B point lies within the range 0.0048–0.0086.

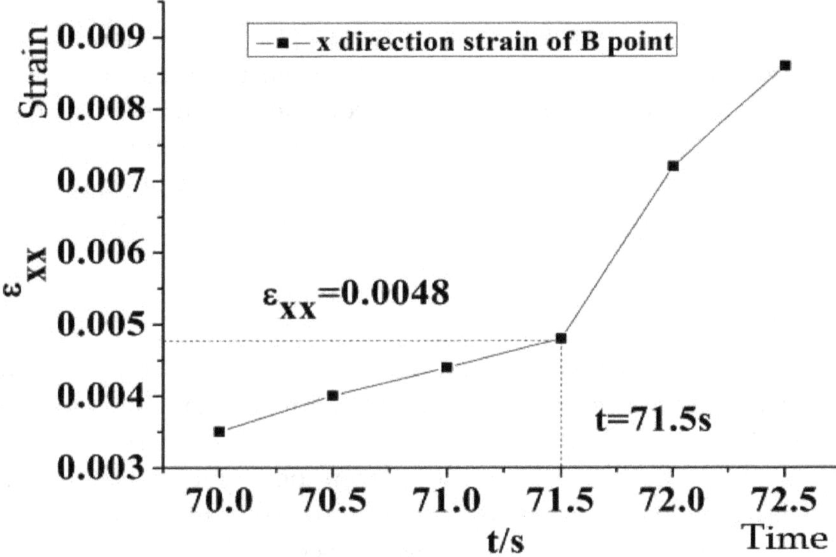

Figure 15. X-direction strain–time relationship at point B (loading rate = 0.3 MPa/s).

From Figure 16, before t = 72.5 s, the strain values of point C vary slightly, showing a near-linear elastic mode. From t = 72.5 s to t = 73 s, the strain values rise significantly, and it is likely that the failure begins at this point. The failure strain at point C is estimated to be within the range 0.0049–0.0071.

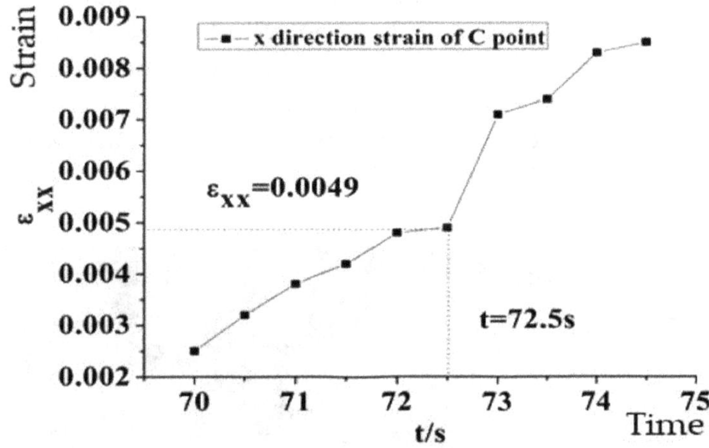

Figure 16. X-direction strain–time relationship at point C (loading rate = 0.3 MPa/s).

The failure occurs successively at points A, B, and C during the loading process. Through the detailed analyses on the failure process of every point, the strain ranges of failure at point A, B, and C were derived. Basically, it can be inferred that, when the loading speed is at 0.3 MPa/s, the failure strain value of sandstone is roughly estimated to be between 0.0048 and 0.093 versus the projection between 0.004 and 0.011 at the loading speed of 0.1 MPa/s (as shown above). It can be seen that little disparity exists between these estimated strain values, which may indicate that the magnitude of the failure strain value of sandstone will not be largely affected when loading speed is changed from 0.1 MPa/s to 0.3 MPa/s (which are at low loading rates).

5.2.2. Results on Shear Strain and Y-Direction Strain Dynamics

Through detailed calculation, it is found that, when the loading speed was at 0.3 MPa/s, the strain values in the shear direction and y-direction were not the main factors affecting sandstone's failure phenomenon, as was found in Scenario 1, where the loading rate was at 0.1 MPa/s.

5.3. *Contrasting Physical Experiment Results against Finite Element Analysis Outcomes*

In order to test the accuracy of the experiment results, a finite element simulation was carried out on the sandstone specimen using the common FEM software Abaqus for the cross-check process. Damage evolution was not considered in the simulation considering the fragility of sandstone materials and the short periods of experiments. The main purpose of the cross-check process was to verify the reliability of strain patterns obtained from DIC by qualitatively examining the consistency between the corresponding strain patterns and trends received from the two methods: DIC analysis and Abaqus simulation.

5.3.1. Model Development

Previous studies showed that sandstone materials involved in the compression experiment share common mechanical properties and constitutive relations with autoclaved aerated concrete (AAC) [28]. Thus, an eight nodded linear brick (C3D8R) element was adopted in this study for the numerical simulation on the sandstone specimen [29]. As seen from the above, sandstone failure occurs at the last moment of the compression process, and before failure, the related deformation shows elastic characteristics. Therefore, only the elastic model was used in the finite element simulation based on Abaqus.

5.3.2. Constraint Definition and Loading Method

For the simulation on the uniaxial compression process of sandstone, the bottom surface was set to be fixed while the loads were exerted to the head surface. A virtual model with the same size as the physical sandstone specimen was built in the Part section in a module of Abaqus. Figure 17a shows the cylinder part (radius = 25 mm, height = 100 mm), while it also presents the upper (Part 2) and lower (Part 3) compression surfaces which are larger than the cylinder base area. In the Property section of Abaqus, the cylinder object was set to be of elastic materials with the elastic modulus of 5×10^3 MPa and Poisson's ratio of 0.25. Part 2 and Part 3 were set to be rigid, without any deformation during the loading process. Finally, by defining the coupling constraints, three parts were coupled together, as is shown in Figure 17a.

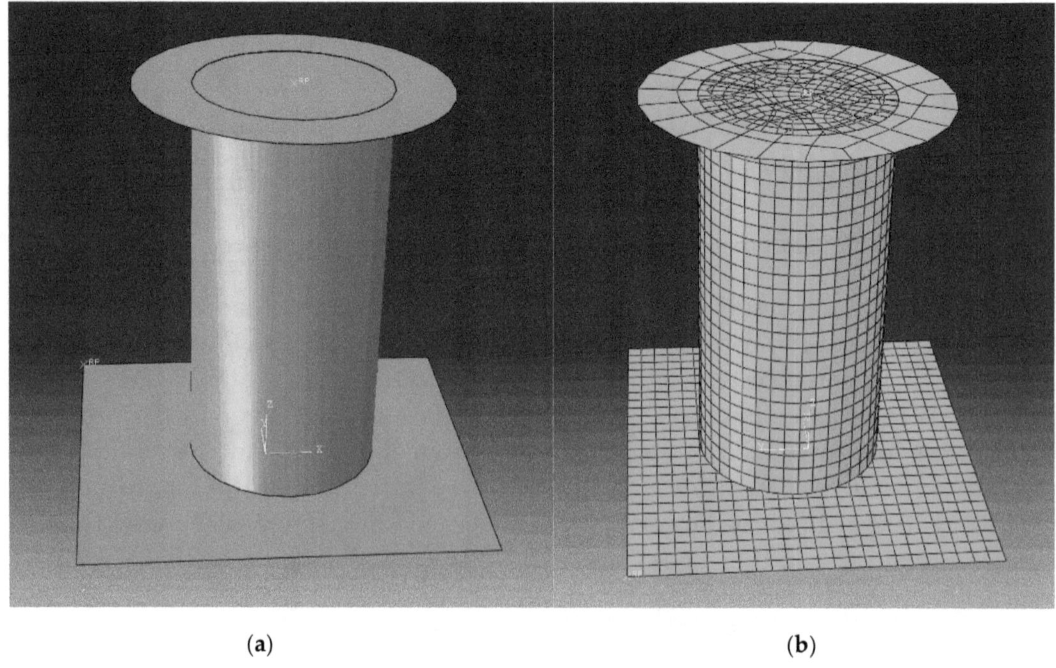

(**a**) (**b**)

Figure 17. Coupling and meshing: (**a**) coupling graph; (**b**) meshing graph.

Meshing (Figure 17b) is critical to the accuracy of the finite element analysis results. In this paper, the three-dimensional cylinder model was divided into grids by gridding seeds with the element division size of 0.01 mm. Then, the axial compressive load was applied to Part 2 by stress with the loading speed at 0.1 MPa/s. Since the above two loading scenarios only have a difference in loading speed, numerical simulation produces similar pattern results. Therefore, only the scenario with the loading speed at 0.1 MPa/s was investigated by the Abaqus simulation.

During the physical experiments, when the loading reaches the compression limit, a principal crack along the vertical direction appears with multiple minor cracks with an axial angle of around 45°. In Figure 18, it is found that during the compression process, larger strain (shown in red color) occurs in some end areas, as well as the central area, with certain angles away from the axle, which is well consistent with the experiment results shown in Figures 5–11. The perceived pattern and trend consistency between the strain profiles obtained from the two methods of the Abaqus simulation and DIC analysis helps to verify the reliability of DIC results in this study.

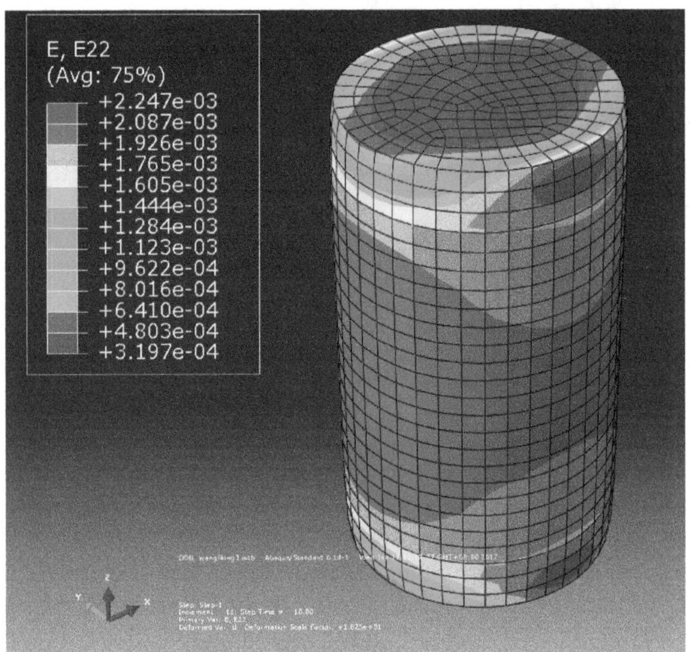

Figure 18. Simulation results.

This study predicts that when the loading speed is 0.1 MPa/s, the failure strain lies between 0.004 and 0.011, while for the loading speed of 0.3 MPa/s, the counterpart strain value is between 0.0048 and 0.0093. The projected information should be useful for the prediction of failure on sandstone in real-world engineering applications.

For the sake of safety, the weakest point of failure should always be given priority. That is to say, for sandstone components under safety monitoring, if the strain value reaches 0.004, there will exist risks of failure which may further lead to accidents. Assuming the safety coefficient is to be 1.5, the calculated allowable strain is then about 0.0027 (=0.004/1.5), and this value is lower than the theoretical result of 0.0037 obtained from the Hill–Tsai criterion. That is to say, when the monitored sandstone strain value reaches 0.0027, caution should be taken, and more attention needs to be paid to strain dynamics to prevent structural damage and engineering failure.

6. Conclusions

Reliable prediction of failure strain for sandstone is critical for engineering security in the construction industry given the wide use of sandstone materials in infrastructure systems. For the first time, through uniaxial compression experiments on sandstone specimens, combining the advanced noncontact digital image correlation (DIC) method, this paper studied the deformation process of sandstone to estimate its full-field failure strain. On classical digital photography principles, digital image correlation contrasts pre- and post-deformation images to obtain numerical strain information by image processing and statistical correlation techniques. A sandstone cylinder 50 mm in diameter and 100 mm in height was adopted as a specimen by considering two loading speed scenarios (loading speeds at 0.1 and 0.3 MPa/s, respectively). Four typical points were selected to obtain detailed investigations, with the strain values along the x-direction, y-direction, and shear direction being analyzed.

It was observed that in both scenarios with the loading speeds of 0.1 and 0.3 MPa/s, the strain–time relationships at all the four selected points displayed linearity before failure occurred. Nevertheless, when strain approaches some critical value of failure, a sharp rise occurs. We defined the identified critical value to be the failure threshold. After comprehensive analysis on the four selected typical points at two different speeds, we concluded that failure may occur if the strain reaches or exceeds 0.004. Sandstone failure mode detected in the physical experiment process was additionally cross-checked by finite element analysis and the Hill–Tsai criterion. In this study, some resulting errors may exist, attributed to the fact that the Cartesian coordinate system was employed to define strain parameters, which may embed uncertainties for cylindrical surfaces subjected to investigation. This study provides a potential basis for future research on mechanical properties of prefabricated rocks associated with the process of failure, and their applications for life cycle deterioration modeling of various structure systems [30].

Author Contributions: Conceptualization, F.C., E.W., B.Z., L.Z., and F.M.; methodology, F.C.; software, F.C.; validation, E.W., B.Z., L.Z., and F.M.; formal analysis, F.C. and E.W.; investigation, F.C.; resources, F.C.; data curation, F.C.; writing—original draft preparation, F.C. and E.W.; writing—review and editing, F.C. and E.W.; visualization, F.C.; supervision, F.C.; project administration, F.C.; funding acquisition, F.C. All authors have read and agreed to the published version of the manuscript.

References

1. Zhou, Z.; Cai, X.; Chen, L.; Cao, W.; Zhao, Y.; Xiong, C. Influence of cyclic wetting and drying on physical and dynamic compressive properties of sandstone. *Eng. Geol.* **2017**, *220*, 1–12. [CrossRef]
2. Verstrynge, E.; Adriaens, R.; Elsen, J.; Van Balen, K. Multi-scale analysis on the influence of moisture on the mechanical behavior of ferruginous sandstone. *Constr. Build. Mater.* **2014**, *54*, 78–90. [CrossRef]
3. Chen, X.; Yu, J.; Li, H.; Wang, S. Experimental and numerical investigation of permeability evolution with damage of sandstone under triaxial compression. *Rock Mech. Rock Eng.* **2017**, *50*, 1529–1549. [CrossRef]
4. Wu, H.; Zhao, G.; Liang, W.; Wang, E.; Ma, S. Experimental investigation on fracture evolution in sandstone containing an intersecting hole under compression using DIC technique. *Adv. Civ. Eng.* **2019**, *2019*, 1–12. [CrossRef]
5. Tating, F.; Hack, R.; Jetten, V. Weathering effects on discontinuity properties in sandstone in a tropical environment: Case study at Kota Kinabalu, Sabah Malaysia. *Bull. Eng. Geol. Environ.* **2015**, *74*, 427–441. [CrossRef]
6. Ma, X.; Haimson, B.C. Failure characteristics of two porous sandstones subjected to true triaxial stresses: Applied through a novel loading path. *J. Geophys. Res. Solid Earth* **2017**, *122*, 2525–2540. [CrossRef]
7. Peng, J.; Han, H.; Xia, Q.; Li, B. Evaluation of the pore structure of tight sandstone reservoirs based on multifractal analysis: A case study from the Kepingtage Formation in the Shuntuoguole uplift, Tarim Basin, NW China. *J. Geophys. Eng.* **2018**, *15*, 1122–1136. [CrossRef]
8. Yue, Z.; Song, Y.; Li, P.; Tian, S.; Ming, X.; Chen, Z. Applications of digital image correlation (DIC) and the strain gage method for measuring dynamic mode I fracture parameters of the white marble specimen. *Rock Mech. Rock Eng.* **2019**. [CrossRef]
9. Lavrov, A.; Gawel, K.; Stroisz, A.; Torsæter, M.; Bakheim, S. Failure modes in three-point bending tests of cement-steel, cement-cement and cement-sandstone bi-material beams. *Constr. Build. Mater.* **2017**, *152*, 880–886. [CrossRef]
10. Wang, H.; Xu, W. Relationship between permeability and strain of sandstone during the process of deformation and failure. *Geotech. Geol. Eng.* **2013**, *31*, 347–353. [CrossRef]

11. Yang, L.; Wang, Y.; Lu, R. Advanced optical methods for whole field displacement and strain measurement. In Proceedings of the 2010 International Symposium on Optomechatronic Technologies, Toronto, ON, Canada, 25–27 October 2010; IEEE: Piscataway, NJ, USA, 2011.
12. Francis, D.; Tatam, R.P.; Groves, R.M. Shearography technology and applications: A review. *Meas. Sci. Technol.* **2010**, *21*, 102001. [CrossRef]
13. Peters, W.H.; Ranson, W.F. Digital imaging techniques in experimental stress analysis. *Opt. Eng.* **1982**, *21*, 427–432. [CrossRef]
14. Chen, F.; Chen, X.; Xie, X.; Feng, X.; Yang, L. Full-field 3D measurement using multi-camera digital image correlation system. *Opt. Lasers Eng.* **2013**, *51*, 1044–1052. [CrossRef]
15. Shao, X.; Dai, X.; Chen, Z.; He, X. Real-time 3D digital image correlation method and its application in human pulse monitoring. *Appl. Opt.* **2016**, *55*, 696–704. [CrossRef]
16. Chen, F.; Zhuang, Q.; Zhang, H. Mechanical analysis and force chain determination in granular materials using digital image correlation. *Appl. Opt.* **2016**, *55*, 4776–4783. [CrossRef] [PubMed]
17. Munoz, H.; Taheri, A. Specimen aspect ratio and progressive field strain development of sandstone under uniaxial compression by three-dimensional digital image correlation. *J. Rock Mech. Geotech. Eng.* **2017**, *9*, 599–610. [CrossRef]
18. Song, H.; Zhang, H.; Kang, Y.; Huang, G.; Fu, D.; Qu, C. Damage evolution study of sandstone by cyclic uniaxial test and digital image correlation. *Tectonophysics* **2013**, *608*, 1343–1348. [CrossRef]
19. Stirling, R.A.; Simpson, D.J.; Davie, C.T. The application of digital image correlation to Brazilian testing of sandstone. *Int. J. Rock Mech. Min. Sci.* **2013**, *60*, 1–11. [CrossRef]
20. Lin, Q.; Labuz, J.F. Fracture of sandstone characterized by digital image correlation. *Int. J. Rock Mech. Min. Sci.* **2013**, *60*, 235–245. [CrossRef]
21. Li, W.; Shaikh, F.; Wang, L.; Lu, Y.; Wang, K.; Li, Z. Microscopic investigation of rate dependence on three-point notched-tip bending sandstone. *Shock Vib.* **2019**, *2019*. [CrossRef]
22. Wang, B.; Pan, B.; Lubineau, G. Some practical considerations in finite element-based digital image correlation. *Opt. Lasers Eng.* **2015**, *73*, 22–32. [CrossRef]
23. Bomarito, G.F.; Hochhalter, J.D.; Ruggles, T.J.; Cannon, A.H. Increasing accuracy and precision of digital image correlation through pattern optimization. *Opt. Lasers Eng.* **2017**, *91*, 73–85. [CrossRef]
24. Pan, B.; Qian, K.; Xie, H.; Asundi, A. Two-dimensional digital image correlation for in-plane displacement and strain measurement: A review. *Meas. Sci. Technol.* **2009**, *20*, 062001. [CrossRef]
25. Su, Y.; Zhang, Q.; Xu, X.; Gao, Z. Quality assessment of speckle patterns for DIC by consideration of both systematic errors and random errors. *Opt. Lasers Eng.* **2016**, *86*, 132–142. [CrossRef]
26. Valliappan, V.; Remmers, J.J.C.; Barnhoorn, A.; Smeulders, D.M.J. A numerical study on the effect of anisotropy on hydraulic fractures. *Rock Mech. Rock Eng.* **2019**, *52*, 591–609. [CrossRef]
27. Hill, R. A theory of the yielding and plastic flow of anisotropic metals. *Proc. R. Soc. A* **1948**, *193*, 281–297.
28. Wang, Q.; Yan, X.; Ding, S.; Huo, Y. Research on the interfacial behaviors of plate-type dispersion nuclear fuel elements. *J. Nucl. Mater.* **2010**, *399*, 41–54. [CrossRef]
29. Das, S.; Shen, L. Experimental and numerical investigation of dynamic failure of sandstone under high strain rates. In Proceedings of the 23rd Australasian Conference on the Mechanics of Structures and Materials (ACMSM23), Byron Bay, Australia, 9–12 December 2014; Southern Cross University: Lismore, Australia, 2014; Volume II, pp. 1057–1062.
30. Wang, E.; Shen, Z. Lifecycle energy consumption prediction of residential buildings by incorporating longitudinal uncertainties. *J. Civ. Eng. Manag.* **2013**, *19*, 161–171. [CrossRef]

11

Feasibility of Low-Cost Thermal Imaging for Monitoring Water Stress in Young and Mature Sweet Cherry Trees

Pedro José Blaya-Ros [1], Víctor Blanco [1], Rafael Domingo [1], Fulgencio Soto-Valles [2] and Roque Torres-Sánchez [2,*]

[1] Dpto Ingeniería Agronómica, Universidad Politécnica de Cartagena (UPCT), Paseo Alfonso XIII, 48, E30203 Cartagena, Spain; pedro.blaya@upct.es (P.J.B.-R.); victor.blanco@upct.es (V.B.); rafael.domingo@upct.es (R.D.)

[2] Dpto Automática, Ingeniería Eléctrica y Tecnología Electrónica, Universidad Politécnica de Cartagena (UPCT), Campus de la Muralla s/n, E30202 Cartagena, Spain; pencho.soto@upct.es

* Correspondence: roque.torres@upct.es

Abstract: Infrared thermography has been introduced as an affordable tool for plant water status monitoring, especially in regions where water availability is the main limiting factor in agricultural production. This paper outlines the potential applications of low-cost thermal imaging devices to evaluate the water status of young and mature sweet cherry trees (*Prunus avium* L.) submitted to water stress. Two treatments per plot were assayed: (i) a control treatment irrigated to ensure non-limiting soil water conditions; and (ii) a water-stress treatment. The seasonal evolution of the temperature of the canopy (Tc) and the difference between Tc and air temperature (ΔT) were compared and three thermal indices were calculated: crop water stress index (CWSI), degrees above control treatment (DAC) and degrees above non-water-stressed baseline (DANS). Midday stem water potential (Ψstem) was used as the reference indicator of water stress and linear relationships of Tc, ΔT, CWSI, DAC and DANS with Ψstem were discussed in order to assess their sensitivity to quantify water stress. CWSI and DANS exhibited strong relationships with Ψstem and two regression lines to young and mature trees were found. The promising results obtained highlight that using low-cost infrared thermal devices can be used to determine the plant water status in sweet cherry trees.

Keywords: water stress; *Prunus avium* L.; stem water potential; low-cost thermography; thermal indexes; canopy temperature; non-water-stressed baselines; non-transpiration baseline

---

1. Introduction

Irrigated agriculture is the largest consumer of fresh water, accounting for 70% of worldwide water use [1]. In this sense, water availability in arid and semi-arid regions is the main factor limiting agricultural production. These regions are subjected to water constraints and are particularly vulnerable to climate change. As a direct result, it is expected that there will be an increase in the mean air temperature with severe drought events occurring during the high evapotranspiration demand periods, accompanied by an irregular rainfall pattern during the wet periods [2].

In addition, Spain—the largest fresh fruit producer in the European Union—has been experiencing severe water supply issues in recent decades, caused mainly by a structural imbalance between water resources and demand [3]. With regards to sweet cherry (*Prunus avium* L.) production, Spain is the seventh-largest producer of cherries in the world and the second-largest producer in Europe [4]. The application of water-saving strategies to this crop, such as deficit irrigation (DI) procedures, should be a priority for their production in areas with water supply issues. Sweet cherry has been described as sensitive to water deficit during the pre-harvest period, when water stress could affect fruit development [5]. However, the application of deficit irrigation in the post-harvest period does not negatively affect yield or fruit size [6,7]. To achieve this, tree water status indicators play the main role and lead to better decisions in DI application strategies, leading to favorable water management at the farm level. These indicators are measured and calculated by sensors which are critical for the correct application of DI. Midday stem water potential (Ψstem) is considered the reference indicator for monitoring plant water status in many woody crops such as sweet cherry trees [8–10]. Even though its measurement is laborious, destructive and cannot be automated, it has been described as the most accurate, reliable and stable water status indicator in fruit trees [11]. In recent years, other water status indicators have increased in popularity due to their consistent, accurate and non-destructive measurements, that enable the implementation of automatic irrigation systems. Moreover, some of them are associated with lower costs and simple management devices.

Infrared thermal sensing has emerged as a powerful technology for monitoring crop water status due to its non-destructive and continuous measurement at an affordable cost and at different scales (from individual plants to complete fields) [12,13]. The principle of infrared thermography is based on leaf energy balance [14]. The transpiration process involves water evaporation through stomata and has a cooling effect, which decreases the crop canopy temperature (Tc) [13]. The degree of canopy cooling can be used as an indicator of stomatal conductance and transpiration rate, and hence, as a measure of plant response to water status, as severe water stress will produce a stomatal closure and the Tc will increase [15]. However, Tc does not only depend on stomatal aperture but is also determined by weather variables such as solar incident angle, solar radiation, air temperature and wind speed [13,16]. To normalize the variation and minimize the effect of environmental factors, several thermal indexes were developed and implemented to monitor and quantify water stress. Idso et al. [17] suggested the first index—the difference between the canopy and air temperature Ta (Tc − Ta = ΔT). ΔT was able to minimize the weather variables; however, it was highly dependent on vapour pressure deficit (VPD). Subsequently, Idso et al. [18] and Jackson et al. [19] developed the crop water stress index (CWSI) for establishing stress for crops by determining non-water-stress baselines (NWSB) and non-transpiration baselines (NTB). NWSB and NTB are the lower and upper limits of temperature that the plant canopy would reach, respectively, related to different VPD values. NWSB refers to a non-limiting water condition when the crop is transpiring at the highest rate and NTB refers to non-transpiration conditions with extreme water stress. Recently, several authors have reported a new index, degrees above the non-stressed canopy (DANS), defined as the difference between the actual temperature of the canopy and the NWSB [14,15]. It is much simpler than CWSI and has been successfully used as the water status indicator in different crops. It is yet to have been used for woody crops; thus, it is important to evaluate the feasibility of using DANS for sweet cherry trees.

Thermal and multispectral cameras have been used over recent years for water stress monitoring with unmanned aerial vehicles (UAVs). However, the difficulty and high cost of using UAVs regularly has meant that their use is reduced to specific events in the crop phenology. Conesa et al. [20] recommended that care should be taken when using instantaneous remote sensing indicators to evaluate moderate water deficits in deciduous fruit trees, and more severe/longer water stress conditions are probably needed to detect significant differences.

Low-cost thermal cameras could be an alternative and robust means of obtaining satisfactory thermal information instead of high-resolution cameras, due to their price (around 20-fold cheaper), user familiarity and ease of implementation in the farm context as a precision irrigation tool [21,22]. Furthermore, this technology can be integrated into intelligent sensor systems to use appropriate image-segmentation algorithms, which are capable of identifying regions of interest [23]. However, the lower sensor resolution must be an impediment for remote acquisition or establishing plant water status at larger scales, such as row-level, due to the pixel size [24,25].

The objectives of the present study were (i) to test the feasibility of low-cost thermal imaging using several thermal indicators (Tc, ΔT, CWSI and DANS) to detect and quantify the water status of young and mature sweet cherry trees subjected to water stress; (ii) to define the non-water-stressed baseline (NWSB) and non-transpiration baseline (NTB) for both cultivations; and (iii) to assess the relationship between thermal indicators and midday stem water potential by linear correlation analysis.

2. Materials and Methods

2.1. Study Site

Two experiments were carried out during 2018 in Murcia (SE Spain). Plot 1 (from 29 June to 1 October, 180–274 DOY) located at the "Tomás Ferro" Experimental Agro-food Station of the Technical University of Cartagena (37°41′ N, 0°57′ W, 32 m elevation, La Palma). The plant material consisted of three-year-old sweet cherry trees (*P. avium* L.), 'Lapins' grafted on 'Mirabolano' rootstock. The trees, planted at a spacing of 3.5 m × 2.25 m, were drip-irrigated by three on-line pressure-compensated emitters per tree, each with a discharge of 2.2 L h^{-1} and fitted on a single lateral per tree row. The irrigation water, with an electrical conductivity ($EC_{25\,°C}$) of 1.1 dS m^{-1} and pH of 8, was from the Tajo-Segura Water Transfer System. The soil was deep and well-drained, had a sandy-clay-loam texture (34.5% clay, 21.3% silt and 44.2% sand), with an available water capacity of about 0.18 mm^{-1} and bulk density of 1.4 ± 0.1 Mg m^{-3} and a low organic matter content (1.5%). Plot 2 (from 27 April to 7 November, 117–311 DOY) is located in a commercial orchard (38°8′ N; 1°22′ W, 680 m elevation, Jumilla) and consisted of sweet cherry trees (*P. avium* L.) 'Prime Giant' that were fifteen years old grafted onto 'SL64' rootstock, and with 'Brooks' and 'Early Lory' as pollinizers. The tree spacing was 5 m between rows and 3 m within rows. The soil was moderately stony and presented a sandy loam texture (67.5% clay, 17.5% silt and 15% sand), with high available phosphorus (108.67 mg kg^{-1}), low potassium (0.32 meq 100 g^{-1}) and a normal active limestone (2.7%) content. The irrigation water was drawn from a well and it had an average electrical conductivity $EC_{25\,°C}$ of 0.8 dS m^{-1}. Water was applied using a single lateral with three pressure-compensated emitters (4 L h^{-1}) per tree.

2.2. Treatments

Plot 1: the young sweet cherry trees were irrigated to satisfy the full crop water requirements from the beginning of the irrigation season until July 5 2019. From that date, two irrigation treatments were imposed: (i) a control, YCTL, irrigated daily at 115% of the crop water requirements (ETc) to guarantee the trees were under non-limiting soil water conditions; and (ii) severe deficit irrigation, YS, in which the trees were submitted to two drought cycles that reached a midday stem water potential (Ψstem) of −1.6 MPa and −2.2 MPa in the first and second drought cycle, respectively. After each drought period, a recovery period was applied in which YS trees were irrigated until their Ψstem values reached similar values to the YCTL trees.

Plot 2: In the orchard of mature sweet cherry trees, we applied two irrigation treatments: (i) a control, MCTL, irrigated daily at 110% ETc during all irrigation season to maintain the trees under non-limiting soil water conditions; and (ii) a regulated deficit irrigation, MS, irrigated at 100% of ETc during pre-harvest and the first days of flower differentiation (from April until the end of June) and 55% of ETc post-harvest, from the end of June to November (see Blanco et al. [26] for details). The irrigation doses for both Plot 1 and Plot 2 were calculated using the methodology proposed by Allen et al. [27]: $ETc = ET_0 \times Kc \times Kr$, where ET_0 is reference evapotranspiration, Kc is a crop-specific coefficient for sweet cherry reported by Marsal [28], and Kr is a factor of localization related to the percentage of ground covered by the crop [29].

Treatments were distributed according to a completely randomized block design in both Plot 1 and Plot 2. In Plot 1, each treatment consisted of three replicates and each replicate had a row of four trees. The two central trees (6 per treatment) were used to measure stem water potential and canopy temperature. In Plot 2, each treatment had three blocks and each replicate consisted of seven adjacent trees. The measurements were taken in the two central trees per replicate, with the other trees serving as guard trees.

2.3. Field Data

Meteorological variables were collected by two weather stations of the Agricultural Information System of Murcia (CA52 for Plot 1 and JU42 for Plot 2; SIAM, http://siam.imida.es/). Daily reference crop evapotranspiration (ET_0) was estimated using the Penman–Monteith equation and daily mean air vapour pressure deficit (VPD) using air temperature and relative humidity data [27]. Additionally, in Plot 1, three microclimate sensors (ATMOS-14, METER Group Inc., Pullman, WA, USA) were installed. The ATMOS-14 sensors were connected to a datalogger (CR1000 with AM16/32B multiplexer, Campbell Scientific Ltd., Logan, UT, USA), programmed to take measurements every 30 s and report mean values every 10 min.

In both experiments, every 2–5 days in Plot 1 and 10–15 days in Plot 2, midday stem water potential (Ψstem) was measured at solar noon (12:00 to 13:00 UT) with a Scholander-type pressure bomb (mod. SF-PRES-70, SolFranc Tecnologías, S.L., 43480 Tarragona Spain) following the recommendations of McCutchan and Shackel [30]. Ψstem was measured in 2 mature leaves per replicate (6 leaves per treatment). The mature and healthy leaves, close to the trunk, were enclosed in small black plastic bags and covered with aluminium foil for 2 h before the measurement.

The canopy temperature (Tc) was measured at the same time as Ψstem with a low-cost thermal camera (ThermalCam Flir One, Flir Systems, Wilsonville, OR, USA) connected to a smartphone. Two images per replicate ($n = 6$) were taken at 1.5 m from the sunny side of the trees in order to identify the highest differences between irrigation treatments, according to Costa et al. [31] and Jones [13]. The camera uses a thermal sensor with a spectral range of 8–14 μm and 80 × 60 pixels, and a visible-light sensor of 1440 × 1080 pixels with ±2% precision. The emissivity, ε, was set at matt ($\varepsilon = 0.95$), as suggested by Stoll and Jones [32] and Costa et al. [31]. The images were analyzed using the Flir Tools application (Flir One, Flir Systems, Wilsonville, OR, USA). The Tc average of four sunny areas was selected within the same image (24 areas per treatment; Figure 1). The distance of the camera from the canopy, the background temperature, relative humidity and air temperature were used as input to discard the effect of reflection by the object's surface and the radiation emitted by the object's surroundings, according to the methodology proposed by Gómez-Bellot [33] and García-Tejero [22].

Three thermal indices were calculated to mitigate the effect of meteorological variables: (i) The difference between the canopy and air temperature (ΔT); (ii) crop water stress index (CWSI), calculated following the recommendation by Jackson et al. [19]; and (iii) the degree above control treatment (DAC) and degree above non-water-stressed baseline (DANS) were calculated according to Taghvaeian et al. [15]:

$$\Delta T = Tc - Tair, \tag{1}$$

$$CWSI = \frac{\Delta Tc - \Delta T_{wet}}{\Delta T_{dry} - \Delta T_{wet}}, \tag{2}$$

$$DAC = T_S - T_{CTL}, \tag{3}$$

$$DANS = T_c - (Tair + \Delta T_{wet}), \tag{4}$$

where Tc is the canopy temperature; Tair is the air temperature at the moment of the measurement; T_S is the canopy temperature of the water-stress treatment; T_{CTL} is the canopy temperature of the control treatment; ΔT_{wet} and ΔT_{dry} are the differences between canopy and air temperature when the crop has the stomata fully transpiring and fully closed, respectively. According to Idso et al. [18] ΔT_{wet} was calculated from non-water-stress baselines (NWSB; $\Delta T_{wet} = a + b \cdot VPD$). As stated by Jones [34], NWSB was obtained by spraying a thin layer of water on leaves 15 to 30 s before images were taken and ΔT_{dry} was estimated by covering two leaves with a layer of petroleum-jelly (Vaseline) on both sides, blocking all transpiration flows. In this regard, several authors do not empirically calculate ΔT_{dry}, and they work with a value set to 5 °C [22,35,36]. Consequently, with the aim of testing whether ΔT_{dry} can always be taken as 5 °C or should be measured every day, CWSI was calculated from the two different methods depending on ΔT_{dry}.

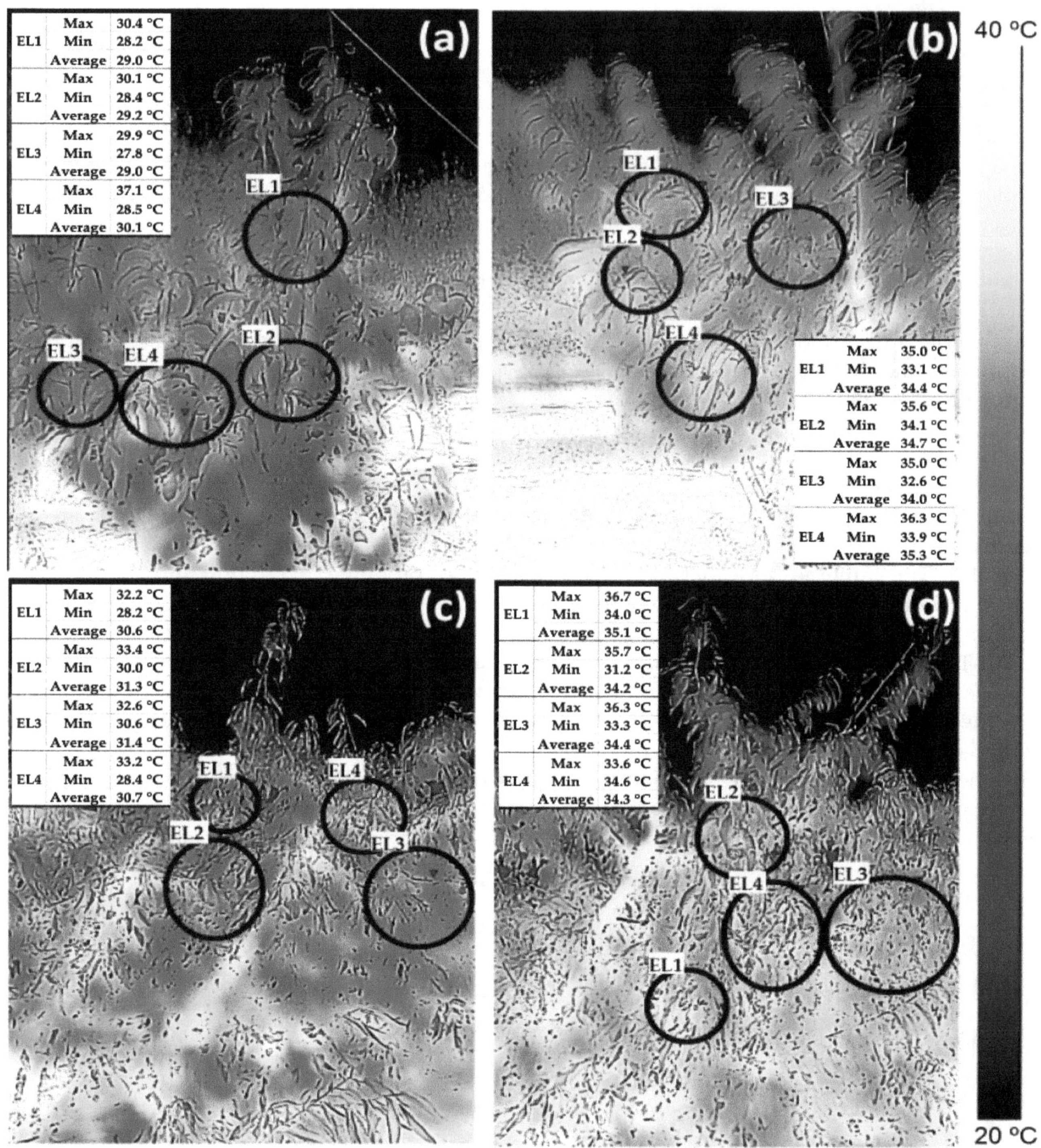

Figure 1. Example of thermal images at plant level taken using Flir One (Flir Systems, Wilsonville, OR, USA) connected to a smartphone for young (**a**,**b**) and mature (**c**,**d**) sweet cherry trees.

2.4. Statistical Analysis

Data were analyzed using statistical software Statgraphics Centurion XVI (StatPoint Technologies Inc., The Plains, VA, USA) and IBM SPSS Statistics (SPSS Inc., 24.0 Statistical package; Chicago, IL, USA). Statistically significant differences among treatments and water stress indicators were determined using analysis of variance (ANOVA) with a significance level of $p < 0.05$. Linear and nonlinear regression analysis among water indicators were determined using Sigmaplot Plus for Windows v.12.5 (Systat Software, San Jose, CA, USA).

3. Results and Discussion

3.1. Environmental Conditions

Environmental conditions at both locations during the experimental period were characteristic of areas with a Mediterranean climate (Table 1). All climatic parameters showed a similar trend with values that increased during spring and early summer and dropped in autumn. Mean temperatures in Plot 1 were generally 3 °C higher than Plot 2. This could be due to the lower daily minimum temperatures recorded in Plot 2 compared to Plot 1. The highest differences in VPD values were recorded during early summer (July) when VPD values in Plot 1 were double those measured in Plot 2.

Table 1. Environmental conditions of Plot 1 (La Palma) and Plot 2 (Jumilla) during the experimental period.

Location	Parameter	May. (121–151) [1]	Jun. (152–181)	Jul. (182–212)	Aug. (213–243)	Sep. (244–273)	Oct. (274–304)	Nov. (305–334)
Plot 1	VPD (kPa)	0.88	1.16	1.10	1.42	0.99	0.71	0.51
	ET_0 (mm d^{-1})	4.99	5.78	6.11	5.34	3.81	2.54	1.60
	P (mm)	3.60	14.00	0.00	0.00	70.20	42.60	106.60
	T (°C)	18.69	22.79	25.62	26.89	24.42	18.91	14.62
Plot 2	VPD (kPa)	0.76	1.27	2.07	1.44	0.83	0.61	0.34
	ET_0 (mm d^{-1})	4.25	5.21	6.07	4.81	3.09	2.09	1.25
	P (mm)	22.95	35.27	0.00	21.17	35.21	22.06	27.95
	T (°C)	15.44	20.44	24.97	23.99	20.61	14.19	9.65

VPD: vapour pressure deficit; ET_0: crop reference evapotranspiration; P: accumulated rainfall; T: mean air temperature. [1]: Day of year.

The highest difference between both experimental sites occurred in late summer. In late August a considerable decline of both air temperature and ET0 occurred in Plot 2, while in Plot 1 the decrease in both parameters was observed in late September.

3.2. Midday Stem Water Potential

Midday stem water potential, Ψstem, accurately reflected the tree water status in both young and mature sweet cherry trees (Figure 2). Ψstem has been reported as a sensitive water stress indicator in mature sweet cherry trees [6,8]; however, there is scarce information about the use of this indicator in young sweet cherry trees, for which pre-dawn stem water potential and midday leaf water potential have been reported as robust water status indicators [37,38].

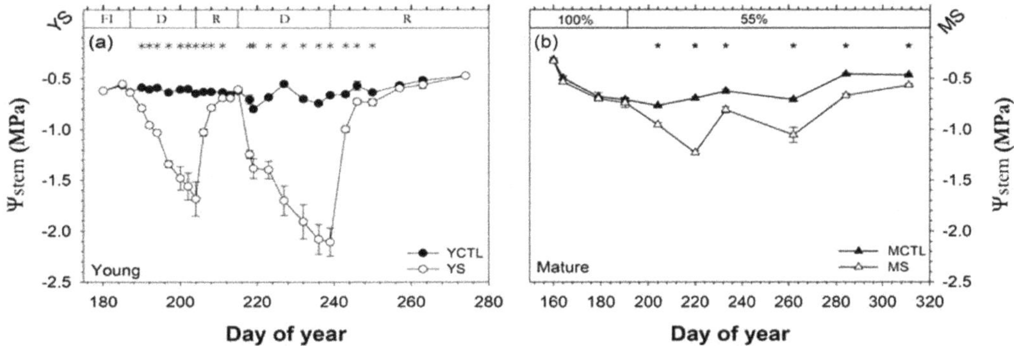

Figure 2. Seasonal evolution of the midday stem water potential (Ψstem) in young (**a**) and mature (**b**) sweet cherry trees during the study period. Each point corresponds to the mean ± standard error of the mean for six measurements per treatment. Asterisks indicate statistically significant differences between treatments by ANOVA ($p < 0.05$). CTL and S correspond to control and deficit irrigation treatment for young (Y) and mature (M) sweet cherry trees, respectively. FI is full irrigation period, D is drought period and R is recovery period in young sweet cherry trees (Plot 1), and 100% and 55% are the percentages of crop water requirements (ETc) applied to mature sweet cherry trees (Plot 2).

The mean Ψstem measured in young and mature CTL trees was between −0.5 and −0.7 MPa, values typical of well-watered trees. These differences in water potential of control trees were due to changes in the climatic demand. Regarding the water stress treatments, the lowest Ψstem values were measured in young trees which were submitted to two drought and recovery cycles, with minimum values that fell below −1.7 and −2.1 MPa for the first and second cycle, respectively. After irrigation was resumed, recovery of Ψstem in young sweet cherry trees was rapid in both cycles. The Ψstem values measured in the young trees showed that they were submitted to severe water stress. During the first drought period, Ψstem in young sweet cherry trees continuously declined from values similar to those of CTL trees down to −1.7 MPa in 16 days, and needed eight days of full irrigation to exhibit similar values to CTL trees. During the second drought cycle, a steeper drop of Ψstem was observed, and the minimum value reached −2.1 MPa (Figure 2a). Ψstem values measured were lower than those reported by Higgs et al. [39] for unirrigated young sweet cherry trees.

In the mature trees (Figure 2b), deficit irrigation trees resulted in Ψstem values that remained above −1.5 MPa, which could be considered a mild–severe water stress that would not compromise the tree's yield the following year [5,40]. Water stress in mature trees resulted in different rates depending on the evaporative demand. Thus, in mid-August (DOY 229, 230), as a result of several rainy episodes in Plot 2, the ET_0 decreased from 6 mm day^{-1} to 3 mm day^{-1} and consequently, mature trees exhibited higher Ψstem values. Similarly, at the end of the season, the evaporative demand decreased and the trees of the deficit treatment resulted in Ψstem values similar to those measured in control trees.

3.3. Canopy Temperature

The pattern of Tc was in accordance with the evolution of Ψstem in young sweet cherry trees (Figure 3); however, in mature trees it was not possible to differentiate between the control and water-stressed trees at the end of the season (September, DOY 270 onwards) using the temperature of the canopy, when the air temperature significantly decreased from 24 to 13 °C (Figure 3a,b).

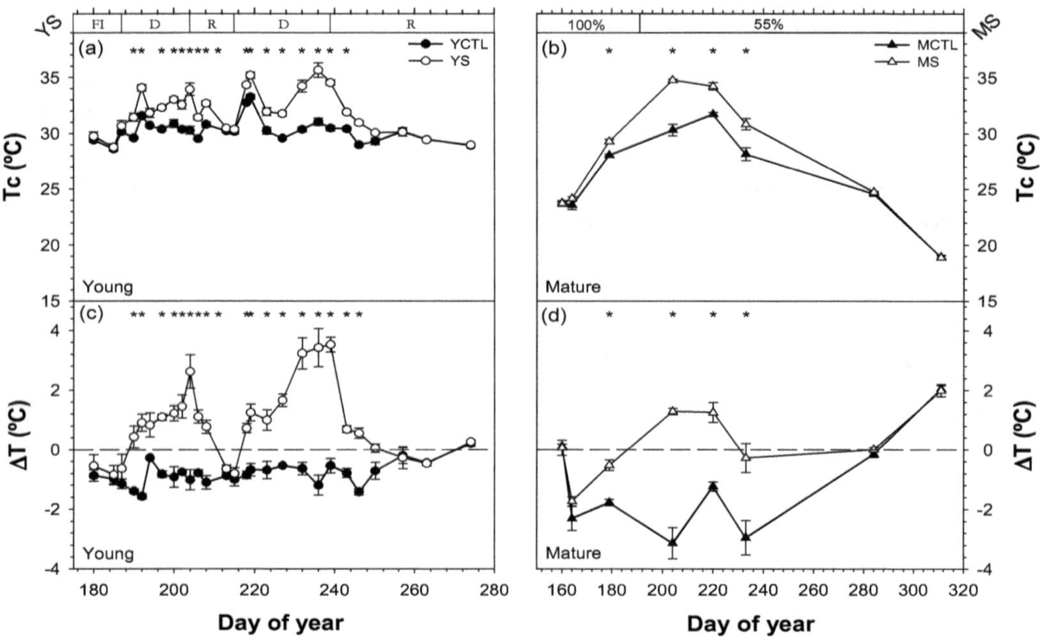

Figure 3. Seasonal evolution of the canopy temperature (Tc) and the difference between canopy and air temperature (ΔT) in young (**a**,**c**) and mature (**b**,**d**) sweet cherry trees during the study period. Each point corresponds to the mean ± standard error of the mean for six images per treatment. CTL and S correspond to control and deficit irrigation treatment for young (Y) and mature (M) sweet cherry trees, respectively. Asterisks indicate statistically significant differences between treatments by ANOVA ($p < 0.05$). FI is full irrigation period, D is drought period and R is recovery period in young sweet cherry trees (Plot 1), and 100% and 55% are the percentages of ETc applied in mature sweet cherry trees (Plot 2).

As expected, young and mature control trees had lower values of canopy temperature minus air temperature than water-stressed trees during the period of water restriction (Figure 3c,d). Regarding the control trees, it was observed that mature control trees had a canopy temperature on average 2.5 °C below the temperature of the air, while in the same period the young trees had a temperature of the canopy only 1 °C below the air temperature. This difference in ΔT of control trees depended on their age, according to Taghvaeian et al. [15], who related the influence of leaf area on the temperature of the plants. Thus, mature trees with greater canopy volume exhibited lower canopy temperatures than young trees with lower canopy volume.

The maximum ΔT was measured on DOY 239 in stressed young trees (3.5 °C), which was the day with the lowest Ψstem (−2.1 MPa, Figure 2a). The difference in canopy temperature between stressed and control young trees was higher than 4 °C on that day. These results indicated a smaller difference than that reported by Ballester et al. [41] and Wang and Gartung [42] in non-irrigated citrus (ΔT = 5.0 °C) and peach trees (ΔT = 6.5 °C) under similar values of Ψstem (<−2.0 MPa). Similarly, the maximum difference of ΔT observed between water-stressed and control mature trees was 4.4 °C (DOY 204, Figure 3d). The difference of 4.4 °C between treatments was mainly due to the contribution of the control trees (ΔT_{MCTL} = −3.1 °C) rather than the high value of the temperature of the canopy of water-stressed trees above the air temperature (ΔT_{MS} = 1.3 °C). These values of canopy temperatures that were lower than the air temperature in control sweet cherry trees are similar to those reported in almond [43] and peach trees [42], but are contrary to those recorded for orange trees [44]. This difference with citrus trees might be due to the stomatal closure of citrus trees at midday, which increases the leaf temperature even though the tree has no soil water restrictions, while in well-watered *Prunus* trees this does not occur [45,46].

Data from control and water-stressed trees were pooled to determine the upper (non-transpiration) and lower (non-water-stress) baselines for the mature and young sweet trees (Figure 4). All the obtained equations for the non-water-stress baselines showed a strong linear relationship between VPD and canopy temperature of sunny leaves (Table 2). Regardless of the different location and age of trees, the non-water-stress baseline did not differ among them, and fitted in the linear regression: ΔT = 3.87 − 2.62·VPD (R^2 = 0.91). Mature trees overestimated ΔT by 1 °C compared to young trees for the lowest VPD value (1 kPa), and underestimated by 1.3 °C for the highest value (4 kPa). The non-transpiration baseline obtained for both young and mature trees achieved 6 °C, a similar value to that reported in peach trees under semiarid climate conditions by Paltineanu et al. [47] and 1 °C above the stated value of 5 °C reported by Jones et al. [35].

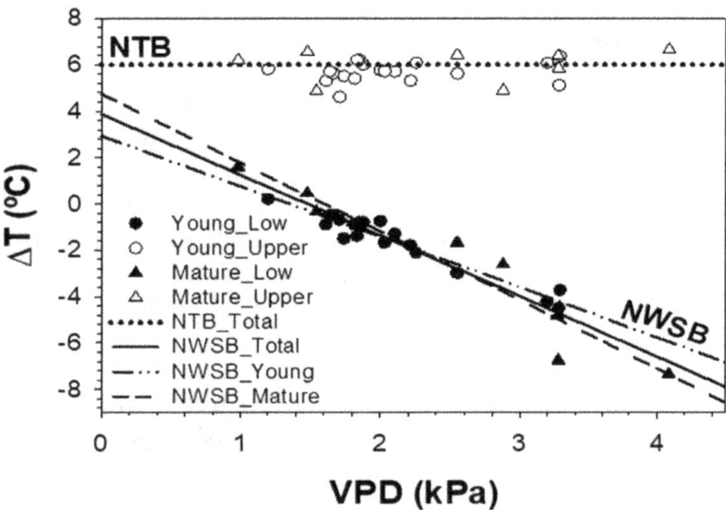

Figure 4. Non-water-stress baselines (NWSB) and non-transpiration baselines (NTB) for young and mature sweet cherry trees. VPD is vapour pressure deficit and ΔT is the difference between canopy and air temperature.

Table 2. Fitted parameters for the non-water-stress baselines ($\Delta T_{wet} = a + b \cdot VPD$) for young and mature sweet cherry trees.

Treatment	Slope (°C kPa^{-1})	Intercept (°C)	R^2
Young sweet cherry trees	−2.174	2.936	0.93
Mature sweet cherry trees	−2.962	4.738	0.92
Global relationship	−2.618	3.868	0.91

3.4. Crop Water Stress Index and Degrees above Non-Stress

CWSI was calculated based on the methodology proposed by Idso et al. [19], which uses a water stress baseline of 5 °C, and with the baselines we obtained from our measures in non-transpiring leaves (Figure 5). In accordance with the results obtained, both methodologies showed similar results; however, the method of Idso et al. [19] led to slightly higher CWSI maximum values.

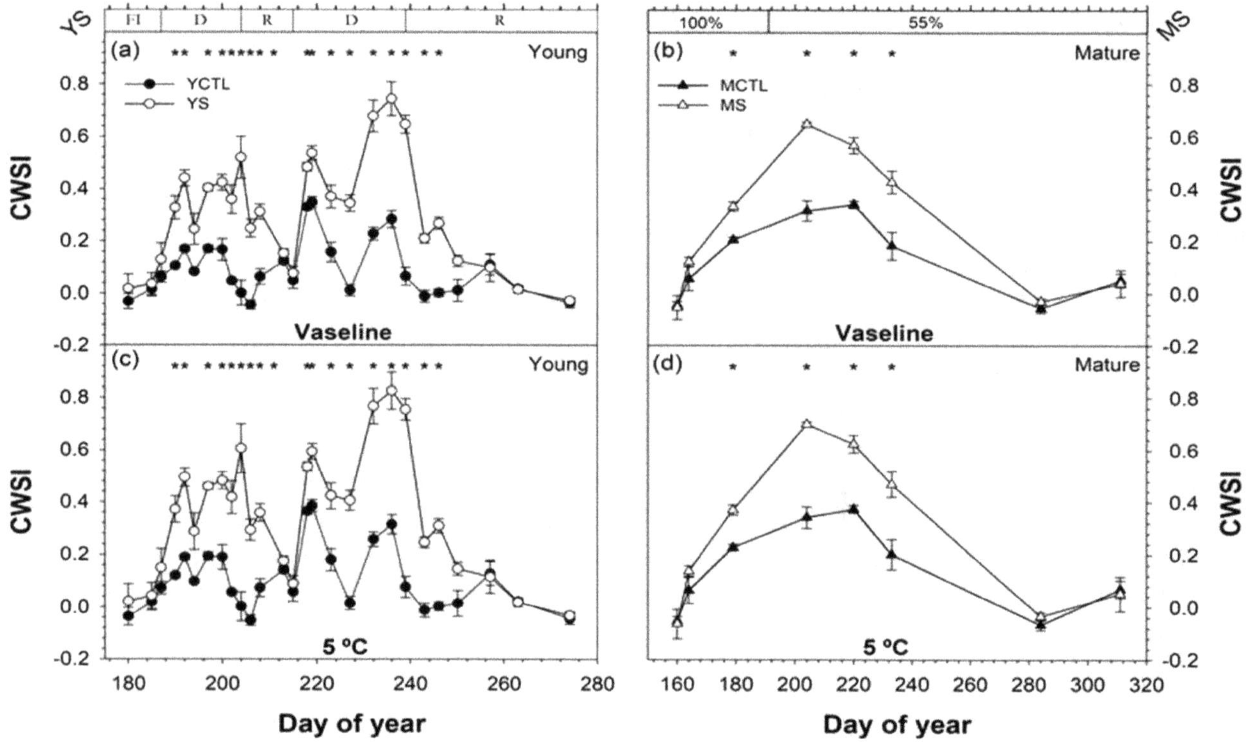

Figure 5. Seasonal evolution of the crop water stress index (CWSI) calculated using a transpiration inhibitor [22] (**a**,**b**) and ΔT_{dry} equal to 5 °C (**c**,**d**) in young (**a**,**c**) and mature (**b**,**d**) sweet cherry trees. Each point corresponds to the mean ± standard error of the mean for six images per treatment. Asterisks indicate statistically significant differences between treatments by ANOVA ($p < 0.05$). CTL and S correspond to control and deficit irrigation treatment for young (Y) and mature (M) sweet cherry trees, respectively. FI is full irrigation period, D is drought period and R is recovery period in young sweet cherry trees (Plot 1), and 100% and 55% are the percentages of ETc applied in mature sweet cherry trees (Plot 2).

In general, the control treatment in both young and mature trees exhibited CWSI values significantly lower than those of water-stressed trees. The CWSI values of control trees ranged from −0.05 to 0.35 (Figure 5). Negative CWSI values were measured on days of low evaporative demand and high Ψstem (−0.5 MPa, Figure 2), and have been related to increased transpiration in almond trees [48]. The water -stressed treatment exhibited CWSI values that achieved 0.75 and 0.65 for young and mature sweet cherry trees, respectively, calculated with the upper baseline of 6 °C (Figure 5a,b). These CWSI values obtained in water-stressed trees were similar to those reported in nectarine trees [49] but are lower than those described in almond trees [48,50], which reached values close to 1 on dates with similar values of ΔT (4.0 °C). When the evolutions throughout the experiment of CWSI and ΔT were compared, a trend that CWSI values presented sharper peaks and troughs and greater oscillations than the evolution of ΔT values was observed, particularly in young trees. However, CWSI as a water stress indicator showed significant differences between treatments on the same days that ΔT showed differences, and the absolute minimum and maximum values occurred on the same days in both water stress indicators.

The DANS index followed the same pattern of significance as the CWSI, with significant differences between treatments on the same dates. The DANS values of young and mature water-stressed trees ranged from slightly below 0.0 °C when they were irrigated as control trees to over 8 °C at the time with the highest difference (DOY 236 and 207 for young and mature trees, respectively; Figure 6). Contrary to CWSI, the DANS index exhibited higher values in mature trees than young trees (Figure 6c,d), despite the young trees being submitted to greater water stress. Regarding the DAC index, in young trees the seasonal evolution was barely higher than results obtained by ΔT; on the other hand, in mature trees, the DAC index resulted in values which achieved a 4.4 °C difference between control and water-stressed trees, while on the same dates ΔT did not achieve values higher than 2.0 °C (Figure 6a,b).

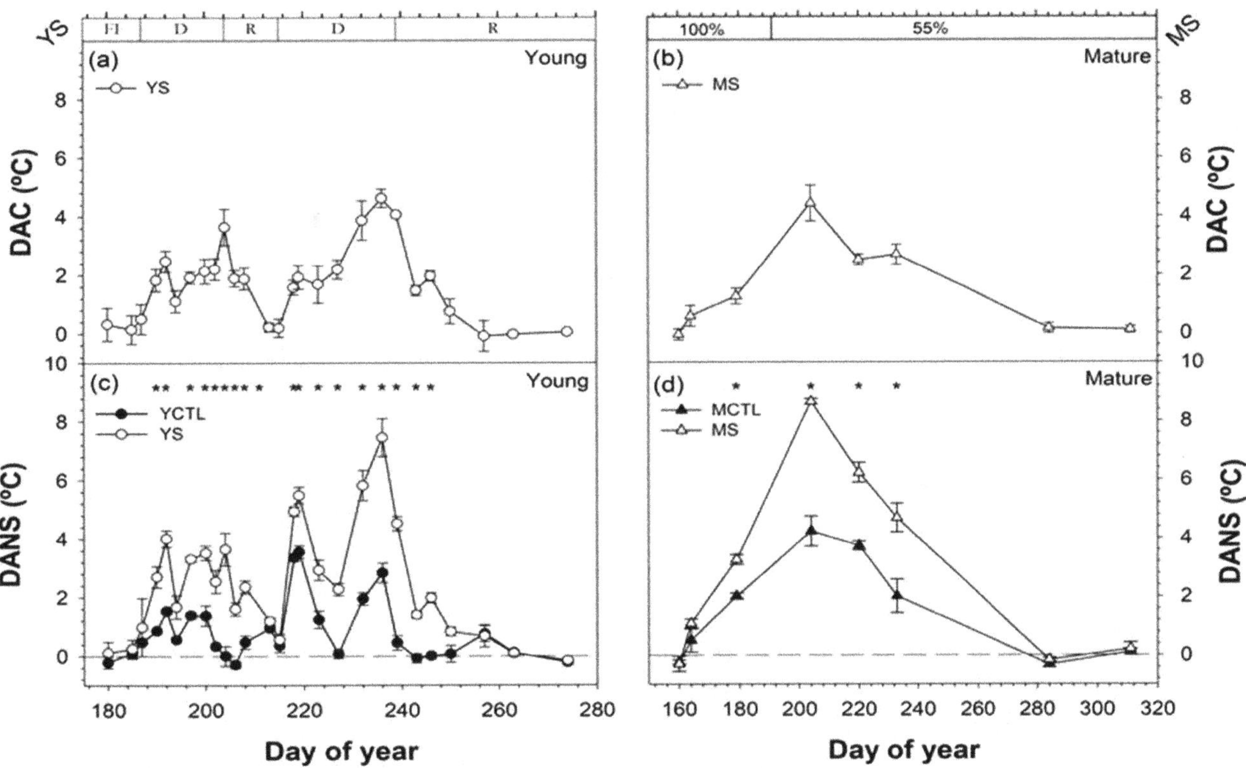

Figure 6. Seasonal evolution of degrees above control (**a**,**b**) and non-stressed (**c**,**d**) in young (**a**,**c**) and mature (**b**,**d**) sweet cherry trees during the study period. Each point corresponds to the mean ± standard error of the mean for six images per treatment. Asterisks indicate statistically significant differences between treatments by ANOVA ($p < 0.05$). CTL and S correspond to control and deficit irrigation treatment for young (Y) and mature (M) sweet cherry trees, respectively. FI is full irrigation period, D is drought period and R is recovery period in young sweet cherry trees (Plot 1), and 100% and 55% are the percentages of ETc applied in mature sweet cherry trees (Plot 2).

A linear relationship between the thermal indicators and Ψstem was calculated. The Tc showed a non-linear relationship with Ψstem (Figure 7a). As expected, higher Tc values were related to trees submitted to water stress (MS and YS). Although the coefficient of correlation obtained between Ψstem and Tc for all the trees exhibited a strong relationship (r = 0.73), Tc as a water stress indicator showed important limitations. Thus, the second-grade polynomial relationship obtained showed two different relationships. At first, Tc increased linearly as Ψstem fell from −0.5 MPa to a threshold value close to −1.0 MPa, which corresponded to 33 °C. From that value onwards, Ψstem values below −1.0 MPa were not related to higher values of Tc. It was observed that Tc did not exceed values above 36 °C regardless of the intensity of the water deficit applied. Consequently, within the Tc range between 33 and 36 °C, similar values were measured in CTL trees on a hot day of high evaporative demand (Ψstem = −0.8 MPa) and in sweet cherry trees under severe water stress (Ψstem = −2.0 MPa). Therefore, while it is known that in sweet cherry trees water deficit induces stomatal closure and increases leaf temperature [6,8], it is also well known that Tc is highly dependent on tree density, canopy architecture, tree phenological stage and environmental conditions [14,51]. In light of this, the use of absolute values of Tc cannot be recommended as a water stress indicator.

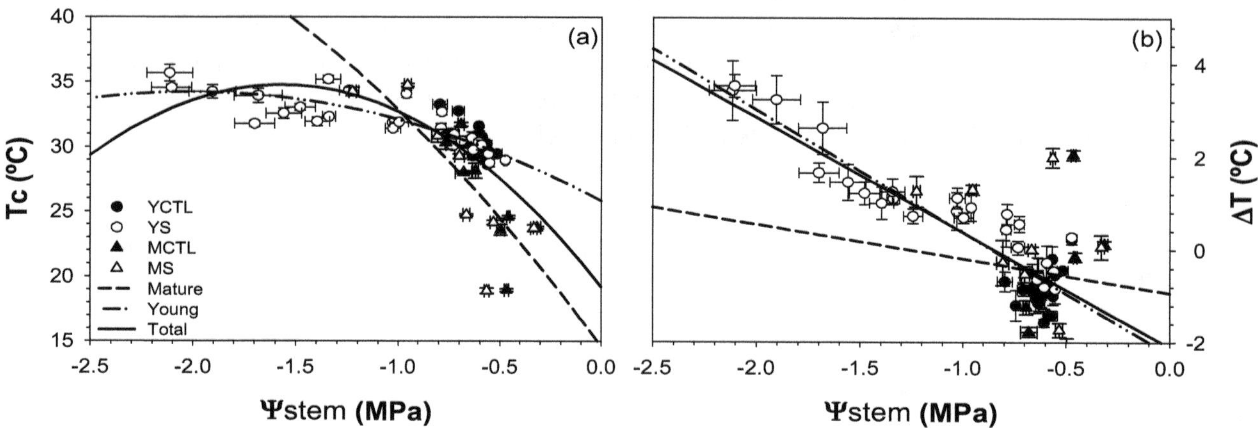

Figure 7. (**a**) Relationship between midday stem water potential (Ψstem) and canopy temperature (Tc). (**b**) Relationship between Ψstem and the difference between canopy and air temperature (ΔT). Each point corresponds to the mean ± standard error of the mean of six measurements per treatment. CTL and S correspond to control and deficit irrigation treatment for young (Y) and mature (M) sweet cherry trees, respectively.

The ΔT exhibited a linear relationship with Ψstem (Figure 7b). The negative ΔT values obtained by CTL trees (young and mature) were related to Ψstem values below −0.8 MPa, which corresponded to trees under non-limiting soil water conditions. In sweet cherry trees under post-harvest deficit irrigation, −1.5 MPa is generally considered a threshold value for irrigation management and higher values have been reported not to negatively affect the yield in the following year and reduce excessive vegetative growth [5]. In this sense, 1.6 °C has been suggested as the ΔT corresponding value to −1.5 MPa. The relationship between Ψsteam and ΔT was significantly different in young and mature trees. The weaker relationship found in mature trees is due to the fact that MS trees did not reach Ψstem values below −1.3 MPa (Figure 2b). The consistent relationship found in the young sweet cherry trees (r = 0.91) was similar to that reported in peach trees by Wang and Gartung [42] and higher than that reported in almond trees by García-Tejero et al. [52] with similar ΔT maximum values at 3.6 °C and Ψstem values below −2.0 MPa. According to the results obtained, ΔT was less dependent than Tc on weather conditions, clearly identified control and stressed trees, and did not show any inflexion point in its relationship with Ψstem. Consequently, these advantages of ΔT over Tc highlight its utility as a water stress indicator.

Similar to ΔT, CWSI showed a strong linear relationship with Ψstem (Figure 8). Young and mature trees resulted in high correlation coefficients (r = 0.89 and 0.88, respectively). These results are similar to those reported in sweet cherry trees by Köksal et al. [53] on the relationship between CWSI and leaf water potential. The correlation between Ψstem and CWSI was identified as CWSI = −0.44 Ψstem = −0.17 in young sweet cherry trees and as CWSI = −0.86 Ψstem = −0.36 in mature sweet cherry trees. Regarding the different regression lines found in young and mature trees, Oberhuber et al. [54] reported that young trees have a greater capacity to extract water from water reserves in their organs (water storage tissues) than mature trees, and quickly transport it through the plant with the aim of sustaining leaf transpiration. Mature trees require a larger amount of water for their transpiration process because they have a greater leaf area, release more water to the atmosphere and have a proportionally smaller water reserve. Consequently, the mechanism used by mature trees to face water stress does not only consist of recruiting water from the water storage tissues but to promote stomatal closure. Stomatal closure avoids plant water release, decreases tree transpiration, and leads to an increase in leaf temperature [8]. These increments in leaf temperature of mature sweet cherry trees are referred to against the same baselines for young sweet cherry trees (Figure 4). Therefore, for a similar value of Ψstem, mature trees exhibit higher CWSI values. However, despite the difference in results for young and mature trees, it can be stated for both of them that CWSI values lower than 0.2 match with Ψstem values of well-watered trees.

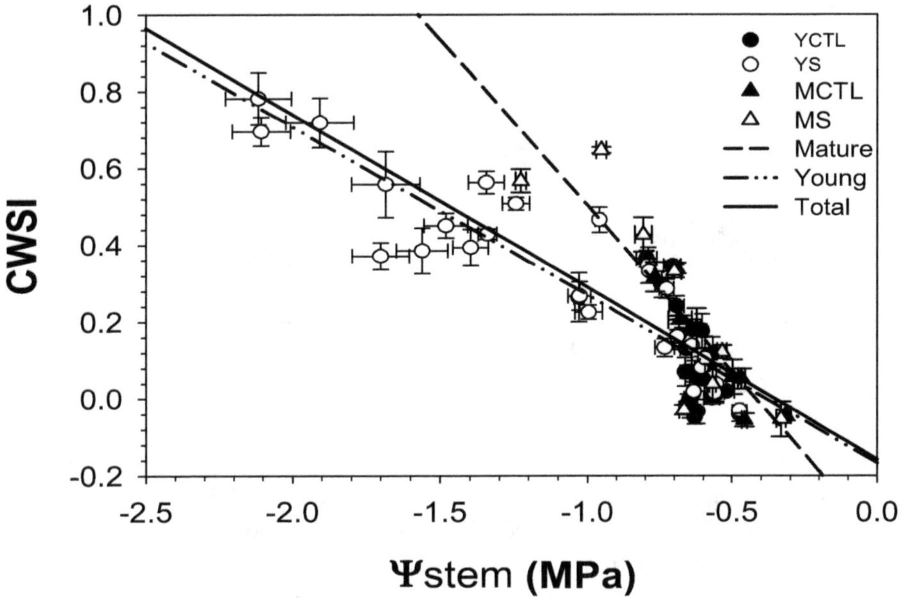

Figure 8. Relationship between midday stem water potential (Ψstem) and crop water stress index (CWSI). Each point corresponds to the mean ± standard error of the mean of six measurements per treatment. CTL and S correspond to control and deficit irrigation treatment for young (Y) and mature (M) sweet cherry trees, respectively.

Similar to CWSI, when DAC and DANS indices were compared to Ψstem, young and mature trees, they showed significantly different linear regressions (Figure 9a,b). In the case of the DAC index, young trees were more closely related to Ψstem than mature trees (r = 0.9 and 0.76, respectively), with maximum values of 4.5 °C. In the case of the DANS index, mature and young trees were closely related (r = 0.84), and the maximum value (8.6 °C) was found in mature trees at −0.95 MPa. As expected according to the results reported by Taghvaeian et al. [15], DAC and DANS were strongly related to CWSI, especially DANS (Figure 9c,d).

In general, water stress indicators derived from thermal imaging evaluated in the present work were not sensitive enough to detect slight plant water stress in sweet cherry trees, due to Tc strongly depending on both stomatal conductance and transpiration rate, which are physiological processes that are less sensitive than other plant water indicators such as micrometric fluctuation of the different plant organs (trunk, branch, fruit, etc.) [8,11,55]. This limitation has been observed in all indices and has been reported in several fruit trees such as apple, citrus and nectarine [56–58]. This is because water status indicators based on leaf temperature when the soil water deficit is not moderate or severe are highly dependent on weather conditions. However, when trees were submitted to moderate water stress, CWSI, ΔT and DANS were robust water indicators able to assess young and mature sweet cherry tree water statuses.

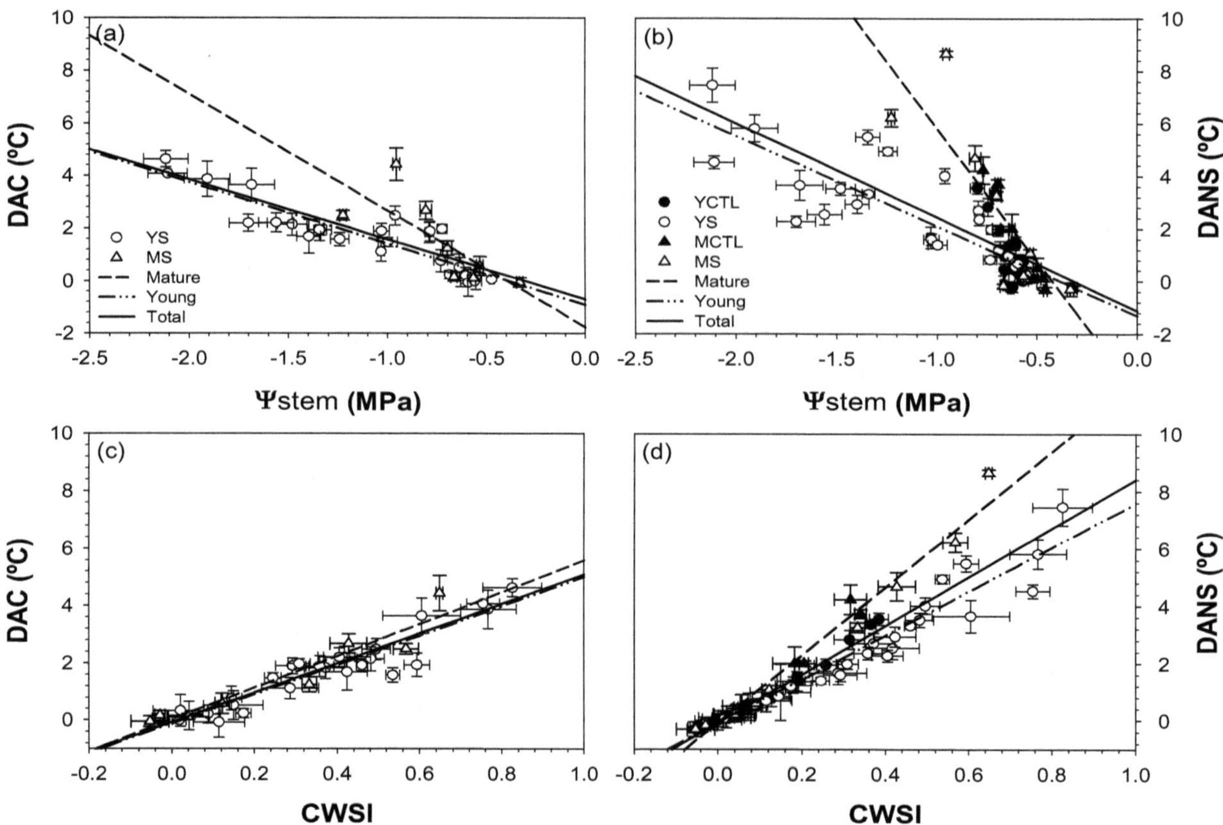

Figure 9. Relationship between midday stem water potential (Ψstem) and (**a**) degree above control treatment (DAC) and (**b**) degree above non-stress baseline (DANS). Relationship between crop water stress index (CWSI) and (**c**) DAC and (**d**) DANS. Each point corresponds to the mean ± standard error of the mean of 6 measurements per treatment. CTL and S correspond to control and deficit irrigation treatment for young (Y) and mature (M) sweet cherry trees, respectively.

4. Conclusions

The use of thermal imaging obtained from low-cost devices provided reliable data which were used to obtain the thermal indicator Tc and to calculate the thermal indices ΔT, CWSI, DAC and DANS to assess the response of young and mature sweet cherry trees submitted to water stress. Our results revealed that Tc was highly dependent on weather conditions, while the thermal indices mitigated this dependency, so the use of Tc in water stress detection is not recommended. ΔT was highly influenced by VPD, and when upper and lower baselines were obtained there were no differences found either between young and mature sweet cherry trees or between plots, which supports the use of the proposed baselines. CWSI and DANS were strongly related to Ψstem and were calculated on the basis of the experimental non-water-stress baseline and water stress baseline, over arange of VPD values between 1 and 4 kPa. The DANS index differentiated between irrigation treatments as well as CWSI, despite being much easier to calculate than CWSI, and exhibited a strong relationship with Ψstem. These results indicate that the DANS index is a promising thermal index which can be used in fruit tree water assessment. It must also be added that CWSI and DANS resulted in different regression lines with Ψstem, depending on the plot studied. These differences might not be solely attributable to the different age of the trees but also to the different soil and weather conditions of each plot. When thermal indices were compared with Ψstem, it was observed that, under non-limiting soil water conditions (values below −0.7 MPa), all indices were highly influenced by climatic conditions. Moreover, despite thermal indices being a non-invasive and fast means with which to assess tree water status, Tc strongly depends on crop transpiration rate. This is a limiting factor in the interpretation of thermography data for the early detection of water stress, so in phenological stages when even slight water stress must be avoided, its use should be coupled with other water status indicators. However, when deficit irrigation was applied, CWSI and DANS could be considered reliable water stress indicators. The results of this study could help improve sweet cherry cultivation, as well as other *Prunus* fruit trees with similar phenology and water stress behavior such as extra early plum trees, and not only in areas where water is scarce, but in regions where water availability is not currently a problem and sweet cherry trees are mainly rainfed, to assess the tree water status.

Author Contributions: Conceptualization, P.J.B.-R., V.B., R.T.-S. and R.D.; methodology, P.J.B.-R., V.B. and R.D.; software, R.T.-S. and F.S.-V.; validation, V.B., P.J.B.-R. and R.D.; formal analysis, V.B. and P.J.B.-R.; investigation, P.J.B.-R., V.B., F.S.-V., R.T.-S. and R.D.; resources, F.S.-V., R.T.-S. and R.D.; data curation, P.J.B.-R. and V.B.; writing—original draft preparation, V.B. and P.J.B.-R.; writing—review and editing, R.D. and R.T.-S.; visualization, P.J.B.-R. and V.B.; supervision, P.J.B.-R., V.B., R.D. and R.T.-S.; project administration, R.D. and R.T.-S.; funding acquisition, R.D. and R.T.-S. All authors have read and agreed to the published version of the manuscript.

Acknowledgments: The authors are grateful to Pedro and Agustín Carrión-Guardiola, "Finca Toli" farm owners, and the staff of the "Tomás Ferro" Experimental Agro-food Station of the Technical University of Cartagena for letting them use their facilities to carry out the study.

References

1. Fereres, E.; Soriano, M.A. Deficit irrigation for reducing agricultural water use. *J. Exp. Bot.* **2007**, *58*, 147–159. [CrossRef] [PubMed]
2. García-Tejero, I.F.; Gutiérrez-Gordillo, S.; Ortega-Arévalo, C.; Iglesias-Contreras, M.; Moreno, J.M.; Souza-Ferreira, L.; Durán-Zuazo, V.H. Thermal imaging to monitor the crop-water status in almonds by using the non-water stress baselines. *Sci. Hortic.* **2018**, *238*, 91–97. [CrossRef]
3. Ruiz-Sánchez, M.C.; Domingo, R.; Castel, J. Review. Deficit irrigation in fruit trees and vines in Spain. *Span. J. Agric. Res.* **2010**, *8*, 5–20. [CrossRef]
4. EUROSTAT. *Structure of Orchards in 2017*; Newsrelease 32/2019; Eurostat Press Office: Luxembourg, 2019.
5. Blanco, V.; Torres-Sánchez, R.; Blaya-Ros, P.J.; Pérez-Pastor, A.; Domingo, R. Vegetative and reproductive response of 'Prime Giant' sweet cherry trees to regulated deficit irrigation. *Sci. Hortic.* **2019**, *249*, 478–489. [CrossRef]
6. Marsal, J.; Lopez, G.; del Campo, J.; Mata, M.; Arbones, A.; Girona, J. Postharvest regulated deficit irrigation in "Summit" sweet cherry: Fruit yield and quality in the following season. *Irrig. Sci.* **2010**, *28*, 181–189. [CrossRef]
7. Blanco, V.; Martínez-Hernández, G.B.; Artés-Hernández, F.; Blaya-Ros, P.J.; Torres-Sánchez, R.; Domingo, R. Water relations and quality changes throughout fruit development and shelf life of sweet cherry grown under regulated deficit irrigation. *Agric. Water Manag.* **2019**, *217*, 243–254. [CrossRef]
8. Blanco, V.; Domingo, R.; Pérez-Pastor, A.; Blaya-Ros, P.J.; Torres-Sánchez, R. Soil and plant water indicators for deficit irrigation management of field-grown sweet cherry trees. *Agric. Water Manag.* **2018**, *208*, 83–94. [CrossRef]
9. Shackel, K.A.; Ahmadi, H.; Biasi, W.; Buchner, R.; Goldhamer, D.; Gurusinghe, S.; Hasey, J.; Kester, D.; Krueger, B.; Lampinen, B.; et al. Plant water status as an index of irrigation need in deciduous fruit trees. *Horttechnology* **1997**, *7*, 23–29. [CrossRef]
10. Naor, A. Irrigation scheduling and evaluation of tree water status in deciduous orchards. *Hortic. Rev.* **2010**, *32*, 111–165.
11. Puerto, P.; Domingo, R.; Torres, R.; Pérez-Pastor, A.; García-Riquelme, M. Remote management of deficit irrigation in almond trees based on maximum daily trunk shrinkage. Water relations and yield. *Agric. Water Manag.* **2013**, *126*, 33–45. [CrossRef]
12. Jones, H.G.; Stoll, M.; Santos, T.; de Sousa, C.; Chaves, M.M.; Grant, O.M. Use of infrared thermography for monitoring stomatal closure in the field: Application to grapevine. *J. Exp. Bot.* **2002**, *53*, 2249–2260. [CrossRef] [PubMed]
13. Jones, H.G. Thermal imaging and infrared sensing in plant ecophysiology. In *Advances in Plant Ecophysiology Techniques*, 1st ed.; Sánchez-Moreiras, A.M., Reigosa, M.J., Eds.; Springer: Cham, Switzerland, 2018; pp. 135–151.

14. DeJonge, K.C.; Taghvaeian, S.; Trout, T.J.; Comas, L.H. Comparison of canopy temperature-based water stress indices for maize. *Agric. Water Manag.* **2015**, *156*, 51–62. [CrossRef]
15. Taghvaeian, S.; Comas, L.; DeJonge, K.C.; Trout, T.J. Conventional and simplified canopy temperature indices predict water stress in sunflower. *Agric. Water Manag.* **2014**, *144*, 69–80. [CrossRef]
16. García-Tejero, I.F.; Rubio, A.E.; Viñuela, I.; Hernández, A.; Gutiérrez-Gordillo, S.; Rodríguez-Pleguezuelo, C.R.; Durán-Zuazo, V.H. Thermal imaging at plant level to assess the crop-water status in almond trees (cv. Guara) under deficit irrigation strategies. *Agric. Water Manag.* **2018**, *208*, 176–186. [CrossRef]
17. Idso, S.B.; Jackson, R.D.; Reginato, R.J. Remote-Sensing of Crop Yields. *Science* **1977**, *196*, 19–25. [CrossRef] [PubMed]
18. Idso, S.B.; Jackson, R.; Pinter, P.J.; Reginato, R.J.; Hatfield, J.L. Normalizing the stress-degree-day parameter for environmental variability. *Agric. Meteorol.* **1981**, *24*, 45–55. [CrossRef]
19. Jackson, R.D.; Idso, S.B.; Reginato, R.J.; Pinter, P.J. Canopy temperature as a Crop Water Stress Indicator. *Water Resour. Res.* **1981**, *17*, 1133–1138. [CrossRef]
20. Conesa, M.R.; Conejero, W.; Vera, J.; Ramírez-Cuesta, J.M.; Ruiz-Sánchez, M.C. Terrestrial and remote indexes to assess moderate deficit irrigation in early-maturing nectarine trees. *Agronomy* **2019**, *9*, 630. [CrossRef]
21. Puértolas, J.; Johnson, D.; Dodd, I.C.; Rothwell, S.A. Can we water crops with our phones? Smartphone technology application to infrared thermography for use in irrigation management. *Acta Hortic.* **2019**, *1253*, 443–448. [CrossRef]
22. García-Tejero, I.F.; Ortega-Arévalo, C.J.; Iglesias-Contreras, M.; Moreno, J.M.; Souza, L.; Tavira, S.C.; Durán-Zuazo, V.H. Assessing the crop-water status in almond (*Prunus dulcis* Mill.) trees via thermal imaging camera connected to smartphone. *Sensors* **2018**, *18*, 1050. [CrossRef]
23. Giménez-Gallego, J.; González-Teruel, J.D.; Jiménez-Buendía, M.; Toledo-Moreo, A.B.; Soto-Valles, F.; Torres-Sánchez, R. Segmentation of multiple tree leaves pictures with natural backgrounds using deep learning for image-based agriculture applications. *Appl. Sci.* **2020**, *10*, 202. [CrossRef]
24. Noguera, M.; Millán, B.; Pérez-Paredes, J.J.; Ponce, J.M.; Aquino, A.; Andújar, J.M. A new low-cost device based on thermal infrared sensors for olive tree canopy temperature measurement and water status monitoring. *Remote Sens.* **2020**, *12*, 723. [CrossRef]
25. Jones, H.G.; Sirault, X.R.R. Scaling of thermal images at different spatial resolution: The mixed pixel problem. *Agronomy* **2014**, *4*, 380–396. [CrossRef]
26. Blanco, V.; Blaya-Ros, P.J.; Torres-Sanchez, R.; Domingo, R. Influence of regulated deficit Irrigation and environmental conditions on reproductive response of sweet cherry trees. *Plants* **2020**, *9*, 94. [CrossRef] [PubMed]
27. Allen, R.G.; Pereira, L.S.; Raes, D.; Smith, M. Irrigation and drainage paper 56. In *Crop Evapotranspiration. Guidelines for Computing Crop Water Requirements*; FAO: Rome, Italy, 1998.
28. Marsal, J. FAO irrigation and drainage paper 66. In *Crop Yield Response Water. Sweet Cherry*; FAO: Rome, Italy, 2012; pp. 449–457.
29. Fereres, E.; Castel, J.R. Drip irrigation saves money in young almond orchards. *Calif. Agric.* **1982**, *36*, 12–13.
30. McCutchan, H.; Shackel, K.A. Stem-water Potential as a Sensitive Indicator of Water Stress in Prune Trees (*Prunus domestica* L. cv. French). *J. Am. Soc. Hortic. Sci.* **1992**, *117*, 607–611. [CrossRef]
31. Costa, J.M.; Grant, O.M.; Chaves, M.M. Thermography to explore plant-environment interactions. *J. Exp. Bot.* **2013**, *64*, 3937–3949. [CrossRef]
32. Stoll, M.; Jones, H.G. Thermal imaging as a viable tool for monitoring plant stress. *J. Int. Sci. Vigne Vin* **2007**, *41*, 77–84. [CrossRef]
33. Gómez-Bellot, M.J.; Nortes, P.A.; Sánchez-Blanco, M.J.; Ortuño, M.F. Sensitivity of thermal imaging and infrared thermometry to detect water status changes in Euonymus japonica plants irrigated with saline reclaimed water. *Biosyst. Eng.* **2015**, *133*, 21–32. [CrossRef]
34. Jones, H.G. Use of thermography for quantitative studies of spatial and temporal variation of stomatal conductance over leaf surfaces. *Plant Cell Environ.* **1999**, *22*, 1043–1055. [CrossRef]
35. Jones, H.G.; Hutchinson, P.A.; May, T.; Jamali, H.; Deery, D.M. A practical method using a network of fixed infrared sensors for estimating crop canopy conductance and evaporation rate. *Biosyst. Eng.* **2018**, *165*, 59–69. [CrossRef]
36. Jackson, R.D. Canopy temperature and crop water stress. *Adv. Irrig.* **1982** *1*, 43–85.

37. Abdelfatah, A.; Aranda, X.; Savé, R.; de Herralde, F.; Biel, C. Evaluation of the response of maximum daily shrinkage in young cherry trees submitted to water stress cycles in a greenhouse. *Agric. Water Manag.* **2013**, *118*, 150–158. [CrossRef]
38. Livellara, N.; Saavedra, F.; Salgado, E. Plant based indicators for irrigation scheduling in young cherry trees. *Agric. Water Manag.* **2011**, *98*, 684–690. [CrossRef]
39. Higgs, K.H.; Hipps, N.A.; Collard, L.G. Effects of irrigation and nitrogen fertilization on the water relations of *Prunus avium* and Colt' (*P. avium* L. x *P. pseudocerasus* Lind.) in the nursery, and residual effects after outplanting. *J. Hortic. Sci.* **1995**, *70*, 235–243. [CrossRef]
40. Carrasco-Benavides, M.; Espinoza Meza, S.; Olguín-Cáceres, J.; Muñoz-Concha, D.; von Bennewitz, E.; Ávila-Sánchez, C.; Ortega-Farías, S. Effects of regulated post-harvest irrigation strategies on yield, fruit quality and water productivity in a drip-irrigated cherry orchard. *N. Z. J. Crop Hortic. Sci.* **2020**, *48*, 97–116. [CrossRef]
41. Ballester, C.; Jiménez-Bello, M.A.; Castel, J.R.; Intrigliolo, D.S. Usefulness of thermography for plant water stress detection in citrus and persimmon trees. *Agric. For. Meteorol.* **2013**, *168*, 120–129. [CrossRef]
42. Wang, D.; Gartung, J. Infrared canopy temperature of early-ripening peach trees under postharvest deficit irrigation. *Agric. Water Manag.* **2010**, *97*, 1787–1794. [CrossRef]
43. Gonzalez-Dugo, V.; Zarco-Tejada, P.; Berni, J.A.J.; Suárez, L.; Goldhamer, D.; Fereres, E. Almond tree canopy temperature reveals intra-crown variability that is water stress-dependent. *Agric. For. Meteorol.* **2012**, *154*, 156–165. [CrossRef]
44. Ribeiro, R.V.; Eduardo, C.M.; Santos, M.G. Leaf temperature in sweet orange plants under field condition: Influence of meteorlogical elements. *Rev. Bras. Agrometeorol.* **2005**, *13*, 353–368.
45. Nicolás, E.; Barradas, V.L.; Ortuño, M.F.; Navarro, A.; Torrecillas, A.; Alarcón, J.J. Environmental and stomatal control of transpiration, canopy conductance and decoupling coefficient in young lemon trees under shading net. *Environ. Exp. Bot.* **2008**, *63*, 200–206. [CrossRef]
46. Mira-García, A.B.; Conejero, W.; Vera, J.; Ruiz-Sánchez, M.C. Leaf water relations in lime trees grown under shade netting and open-air. *Plants* **2020**, *9*, 510. [CrossRef]
47. Paltineanu, C.; Septar, L.; Moale, C. Crop water stress in peach orchards and relationships with soil moisture content in a chernozem of dobrogea. *J. Irrig. Drain. Eng.* **2013**, *139*, 20–25. [CrossRef]
48. González-Dugo, V.; López-López, M.; Espadafor, M.; Orgaz, F.; Testi, L.; Zarco-Tejada, P.; Lorite, I.J.; Fereres, E. Transpiration from canopy temperature: Implications for the assessment of crop yield in almond orchards. *Eur. J. Agron.* **2019**, *105*, 78–85. [CrossRef]
49. Park, S.; Ryu, D.; Fuentes, S.; Chung, H.; Hernández-Montes, E.; O'Connell, M. Adaptive estimation of crop water stress in nectarine and peach orchards using high-resolution imagery from an unmanned aerial vehicle (UAV). *Remote Sens.* **2017**, *9*, 828. [CrossRef]
50. Bellvert, J.; Adeline, K.; Baram, S.; Pierce, L.; Sanden, B.L.; Smart, D.R. Monitoring crop evapotranspiration and crop coefficients over an almond and pistachio orchard throughout remote sensing. *Remote Sens.* **2018**, *10*, 2001. [CrossRef]
51. Leuzinger, S.; Körner, C. Tree species diversity affects canopy leaf temperatures in a mature temperate forest. *Agric. For. Meteorol.* **2007**, *146*, 29–37. [CrossRef]
52. García-Tejero, I.; Durán-Zuazo, V.H.; Arriaga, J.; Hernández, A.; Vélez, L.M.; Muriel-Fernández, J.L. Approach to assess infrared thermal imaging of almond trees under water-stress conditions. *Fruits* **2012**, *67*, 463–474. [CrossRef]
53. Köksal, E.S.; Candogan, B.N.; Yazgan, S.; Yildirim, Y.E. Determination of water use and water stress of cherry trees based on canopy temperature, leaf water potential and resistance. *Zemdirbyste* **2010**, *97*, 57–64.
54. Oberhuber, W.; Hammerle, A.; Kofler, W. Tree water status and growth of saplings and mature Norway spruce (*Picea abies*) at a dry distribution limit. *Front. Plant Sci.* **2015**, *6*, 703. [CrossRef]
55. Ortuño, M.F.; Conejero, W.; Moreno, F.; Moriana, A.; Intrigliolo, D.S.; Biel, C.; Mellisho, C.D.; Pérez-Pastor, A.; Domingo, R.; Ruiz-Sánchez, M.C.; et al. Could trunk diameter sensors be used in woody crops for irrigation scheduling? A review of current knowledge and future perspectives. *Agric. Water Manag.* **2010**, *97*, 1–11. [CrossRef]
56. Bellvert, J.; Marsal, J.; Girona, J.; Gonzalez-Dugo, V.; Fereres, E.; Ustin, S.L.; Zarco-Tejada, P.J. Airborne thermal imagery to detect the seasonal evolution of crop water status in peach, nectarine and Saturn peach orchards. *Remote Sens.* **2016**, *8*, 39. [CrossRef]

57. Ballester, C.; Zarco-Tejada, P.J.; Nicolás, E.; Alarcón, J.J.; Fereres, E.; Intrigliolo, D.S.; Gonzalez-Dugo, V. Evaluating the performance of xanthophyll, chlorophyll and structure-sensitive spectral indices to detect water stress in five fruit tree species. *Precis. Agric.* **2017**, *19*, 178–193. [CrossRef]
58. González-Dugo, V.; Zarco-Tejada, P.J.; Fereres, E. Applicability and limitations of using the crop water stress index as an indicator of water deficits in citrus orchards. *Agric. For. Meteorol.* **2014**, *198*, 94–104. [CrossRef]

12

A Method for Detecting Coffee Leaf Rust through Wireless Sensor Networks, Remote Sensing and Deep Learning

David Velásquez [1,2,3,*], Alejandro Sánchez [1], Sebastian Sarmiento [1], Mauricio Toro [1], Mikel Maiza [2] and Basilio Sierra [3]

1 I+D+i on Information Technologies and Communications Research Group, Universidad EAFIT, Carrera 49 No. 7 Sur - 50, Medellín 050022, Colombia; asanch41@eafit.edu.co (A.S.); ssarmien@eafit.edu.co (S.S.); mtorobe@eafit.edu.co (M.T.)

2 Department of Data Intelligence for Energy and Industrial Processes, Vicomtech Foundation, Basque Research and Technology Alliance (BRTA), 20014 Donostia-San Sebastián, Spain; mmaiza@vicomtech.org

3 Department of Computer Science and Artificial Intelligence, University of Basque Country, Manuel Lardizabal Ibilbidea, 1, 20018 Donostia/San Sebastián, Spain; b.sierra@ehu.eus

* Correspondence: dvelas25@eafit.edu.co

Abstract: Agricultural activity has always been threatened by the presence of pests and diseases that prevent the proper development of crops and negatively affect the economy of farmers. One of these pests is Coffee Leaf Rust (CLR), which is a fungal epidemic disease that affects coffee trees and causes massive defoliation. As an example, this disease has been affecting coffee trees in Colombia (the third largest producer of coffee worldwide) since the 1980s, leading to devastating losses between 70% and 80% of the harvest. Failure to detect pathogens at an early stage can result in infestations that cause massive destruction of plantations and significantly damage the commercial value of the products. The most common way to detect this disease is by walking through the crop and performing a human visual inspection. As a result of this problem, different research studies have proven that technological methods can help to identify these pathogens. Our contribution is an experiment that includes a CLR development stage diagnostic model in the *Coffea arabica*, Caturra variety,scale crop through the technological integration of remote sensing (through drone capable multispectral cameras), wireless sensor networks (multisensor approach), and Deep Learning (DL) techniques. Our diagnostic model achieved an F_1-score of 0.775. The analysis of the results revealed a p-value of 0.231, which indicated that the difference between the disease diagnosis made employing a visual inspection and through the proposed technological integration was not statistically significant. The above shows that both methods were significantly similar to diagnose the disease.

Keywords: coffee leaf rust; machine learning; deep learning; remote sensing; Fourth Industrial Revolution; Agriculture 4.0

---

1. Introduction

The food and beverage industry is characterized by a relatively small number of multinational companies that link small producers around the world with consumers. A development analysis conducted by the World Economic Forum and Accenture, in 2018 [1], focused, predominantly, on upstream value chain segments due to the low tech nature of food and beverage processing and production and the substantial potential for improving efficiency in agrifood activities.

According to the Organisation for Economic Co-operation and Development (OECD), the food and beverage industry is classified as a low tech industry, so it can add innovation without significant social disadvantages [2]. According to the OECD, each opportunity presented by the Fourth Industrial Revolution must be used to realize a global food production system that can address challenges with limited environmental impact while taking advantage of opportunities for growth, innovation, and development [2].

The developments of the Fourth Industrial Revolution will change production systems in the food and beverage industry through innovation in digital, physical, and biological technologies [1]; for instance, vertical agriculture, advanced wastewater treatment, advanced packaging, precision agriculture [3], advanced organic agriculture, supply chain traceability [4], genome editing, cell and tissue engineering, automated agriculture [5], remote sensing [6], 3D food printing, and Agriculture 4.0.

The three main developments with the most significant growth potential for value creation in the food and beverage industry are: precision agriculture, advanced organic agriculture, and genome publishing [1]. In particular, precision agriculture integrates data analysis processes with crop science and technologies such as GPS, soil sensors, meteorological data, and the Internet of Things (IoT) for decisions related to fertilizer, irrigation, harvest time, and seed spacing, among others. Precision agriculture is applicable to the entire agricultural production system and drives substantial yield increases while optimizing for resource use [1]. The goal of precision agriculture is to enable scientific decisions in agriculture to improve value creation.

One industry in which precision agriculture can improve value creation is the coffee industry; in particular, the specialty market. Coffee is one of the world's most popular drinks and merchantable commodities. Every year, over 500,000 million cups are consumed, and over 158 million bags of 60 kg are produced. Coffee is grown in around 70 countries around the world in a region known as the Bean Belt. This region is located between the Tropics of Cancer and Capricorn, and the world's primary producers are Brazil (2720 million kg/year), Vietnam (1650 million kg/year), and Colombia (810 million kg/year). Furthermore, the social impact of the coffee growing industry is very significant because the people who depend on this activity for all or most of their living exceeds 100,000,000 worldwide [7].

The market is divided into two groups, known as the standard and specialty markets, according to the quality of the final product, which depends on the cultivated coffee variety, the environmental conditions, and the post-harvest process. This quality is measured with a score between zero and 100 and is known as the cup quality. When the cup quality is less than 80 points, the coffee belongs to the standard market, and its selling price depends primarily on the New York Commodity Exchange. On the other hand, when the coffee has a cup quality greater than or equal to 80 points, it belongs to the specialty market, and its selling price is at least twice the standard coffee price [8]. Nevertheless, it is a fact that coffee, which is cultivated with a view toward the specialty market, needs a more careful and judicious agronomic management.

Regardless of the product's target market, coffee growers around the world face three significant challenges currently to preserve quality: (i) unpredictable climate variations, (ii) the presence of nutritional deficiencies, and (iii) attacks of pests and diseases. Concerning the latter, for instance, Coffee Leaf Rust (CLR), which is a disease considered to be the main phytosanitary problem for coffee crops, causes in Latin America losses of 30% of the efficiency of each harvest [9].

The fungus *Hemileia vastatrix* is the cause of the CLR disease, which is the major phytosanitary problem for coffee crops. Once high levels of severity are reached, the corrective actions can be minimal. Inappropriate management of the disease can harshly compromise the coffee plants, as seen in Figure 1a, resulting in only a few leaves remaining on the trees, which has a direct negative impact on the quantity and quality of the harvest [10].

(a)

(b)

Figure 1. Coffee Leaf Rust (CLR) effects: (**a**) on the Caturra variety crops; (**b**) on a leaf at the disease's highest development stage [11].

CLR progresses gradually in time and reaches three noticeably phases. The first one, called the "slow phase" (severity $\leq$5%), is where the first structures responsible for the production of spores emerge and low levels of infection are evident. The second one, which is named the "fast or explosive phase" (5% $<$ severity $\leq$ 30%), starts with the fungus sporulation and is represented by more plants getting sick in a short period. The final phase is called the "maximum or terminal phase" (severity $>$30%) and occurs when most of the leaves are severely attacked and a small amount of healthy leaves remains. At that moment, the epidemic stops in the host due to the lack of biological matter to continue the infection. When the CLR is not controlled and the climatic conditions are favorable, the disease can develop at a daily rate of 0.19–0.38%, reduce the impact of the chemical controls, and cause significant economic damage [10].

1.1. Context

In the Colombian context, coffee is the most exported agricultural product, followed by cut flowers, bananas, cocoa, and sugarcane [12]. In the country, there are more than 903,000 hectares dedicated to it, and approximately 563,000 families depend directly on this economic activity. Colombian coffee has been considered one of the best soft coffees in the world, and this product has traditionally been of great importance for Colombian exports. Currently, 14,000,000 bags of 60 kg are exported every year to the USA, Japan, and Germany, among other countries [13].

In terms of employment generation and income distribution, coffee growing is a sector with superlative relevance for local economies and the maintenance of the social fabric in many regions of the country. For this reason, it is justified to contribute by solutions that strengthen the profitability of families engaged in this activity and improve their life quality, either by increasing the selling price of the product, reducing production costs, or increasing the number of units produced per unit of cultivated area.

Among the main threats for strengthening the coffee growing families' profitability, nutritional deficiencies and phytosanitary problems stand out. Phytosanitary problems are caused by pests such as the coffee borer beetle and diseases such as CLR, whose proliferation increases due to the drastic climate changes (from long drought periods to extended rainy seasons) that occur in Colombia. In the case of CLR, when the climatic conditions are unfavorable and the agronomic management deficient, at least 20% of the total expected harvest is not able to be collected. Additionally, the quality of coffee deteriorates dramatically, reducing the marketing price and increasing the costs associated with its control [10]. In extreme cases of CLR, the disease has caused devastating losses that have represented between 70% and 80% of the total harvest.

Although it is a disease with vertiginous spread and highly negative repercussions for the coffee farmers' economy, its detection and diagnosis are carried out using visual inspection while walking through the crops. This method refers to the recognition of plant diseases using visual inspections, development scales, and standard severity diagrams for their measurement [14]. People in charge of the crops walk through them, watching and touching the plants to identify symptoms associated with the particular disease that produces them and calculate infection levels [15].

Unfortunately, because the process consists of a visual inspection, which is not done with enough regularity, most of the time, the detection of the development stage of the disease is late, its control becomes more difficult, and considerable economic losses are inevitable.

1.2. State-of-the-Art

Plenty of research has been done on applying technological methods and strategies to diagnose diseases [16], to detect pests [17], and to obtain nutritional information [18], among other objectives, for different types of crops. The phytosanitary status of the plantations is closely related to different crucial factors in their ecosystem, such as weather, altitude, and type of soil, among others. Therefore, several biological and engineering studies aim to implement practical solutions based on these factors to improve farming techniques to preserve healthy crops.

The most commonly used methods for monitoring the phytosanitary status efficiently, including those that make use of technology, are: (i) Remote Sensing (RS), (ii) visual detection, (iii) biological intervention, (iv) Wireless Sensor Networks (WSN), and (v) Machine Learning (ML) supported on a source of data. Thus, this work is intended to present recent relevant studies based on the mentioned methods for detecting anomalies on the plantations.

(i) Remote Sensing (RS)

RS is based on the interaction of electromagnetic radiation with any material. In the case of agriculture, it involves the non-contact measurement of the reflected radiation from soil or plants to assess different attributes such as the Leaf Area Index (LAI), chlorophyll content, water stress, weed density, and crop nutrients, among others. Those measurements can be made using satellites, aircraft, drones, tractors, and hand held sensors [19]. In addition to measuring reflected radiation, there are two other RS techniques that analyze fluorescent and thermal energy emitted by the leaves. However, the most common technique is reflectance, because the amount of reflected radiation from the plants is inversely related to the radiation absorbed by their pigments, and this can serve as an indicator of their health status [19]. RS helps the indirect detection of problems in agricultural fields since this method captures unusual behaviors in crops' reflectance, which can be caused by factors like nutritional deficiencies, pests and diseases, and water stress. In 2017, Calvario et al. [20] monitored agave crops using Unmanned Aerial Vehicles (UAVs) and integrating RS with unsupervised machine learning (*k*-means) to classify agave plants and weed. In 2003, Goel et al. [21] studied the detection of changes in the spectral response in corn (*Zea mays*) due to nitrogen application rates and weed control. For that purpose, the researchers employed a hyperspectral sensor called the Compact Airborne Spectrographic Imager (CASI) and analyzed the reflectance values of 72 bands with a wavelength between 409 and 947 nm, which comprise part of the visible and Near-Infrared (NIR) regions of the electromagnetic spectrum. The obtained results demonstrated the potential of detecting weed infestations and nitrogen stress using the hyperspectral sensor CASI. Specifically, the researchers found that the best fitting bands for the detection were the wavelength regions near 498 nm and 671 nm, respectively, as seen in Figure 2.

It has been shown that using satellites' multispectral images, it is possible to detect the location of crops [22], but the resolution of satellites images does not allow early detection of the phytosanitary of individual lots of plants. Regarding the phytosanitary status of the plants, the water and the type of soil are two components that play an essential role in their health. In 2017, Bolaños et al. [23] proposed a characterization method using the visible and infrared spectrum to identify these components, through low cost cameras with two different filters, Roscolux #19 and Roscolux #2007, and a multi-rotor air vehicle. Through this method and using portable and highly qualified devices, those hard-to-reach places were monitored and analyzed to detect anomalies that may cause diseases in the crop. This monitored phase provided a characterization of the Normalized Difference Vegetation Index (NDVI), as seen in the example of Figure 3, which was used to categorize essential characteristics of the crop, such as crop health, diseased plants or soil, and water or others.

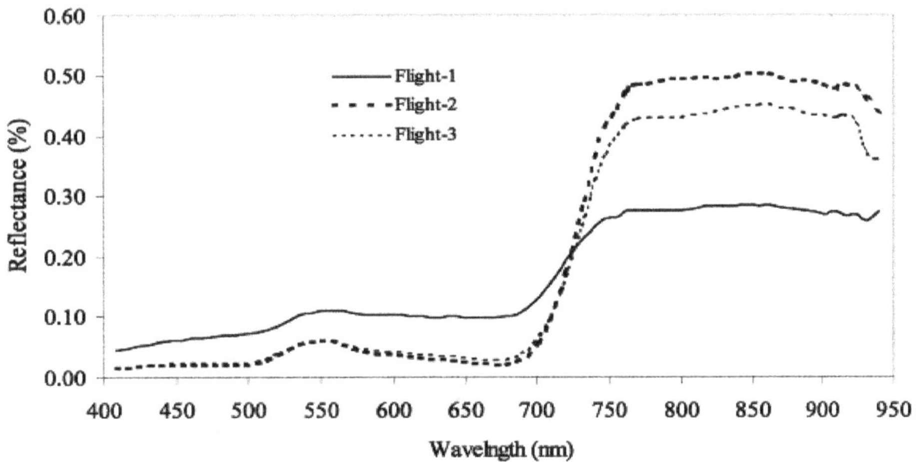

Figure 2. Reflectance (%) of the corn response during different flights under normal nitrogen rates and no weed control [21], Copyright Elsevier, 2003.

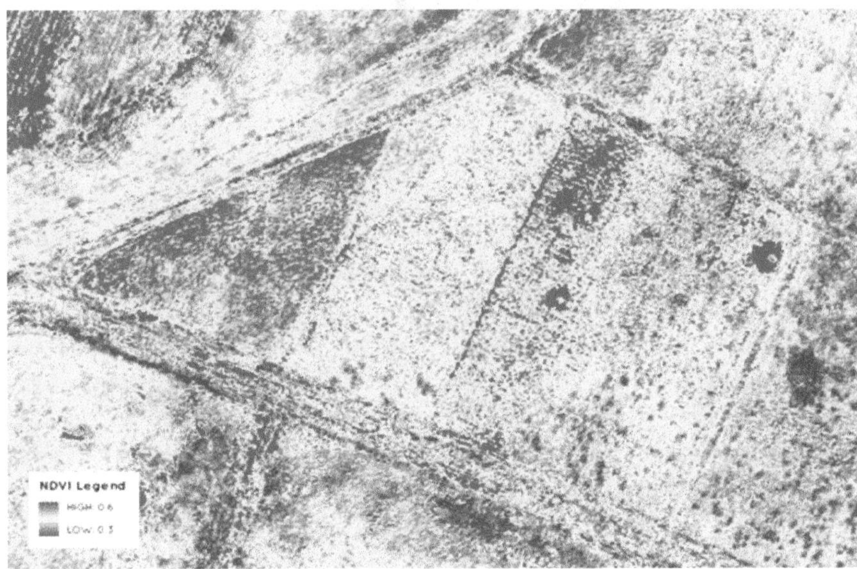

Figure 3. Characterization of the NDVI with low cost solutions [24].

In 2017, Chemura et al. proposed a method to predict the presence of diseases and pests early among coffee trees based on unnoticeable water stress. For that purpose, multispectral scanners with filters with wavebands from the visual spectrum and near infrared region were placed on a UAV [25]. The wavebands scanner results showed inflections points between the regions 430 nm and 705–735 nm due to the water content in coffee trees.These results underlined the importance of a suitable irrigation plan according to the water requirements of the trees, causing an improvement in productivity. Although the later region indicated relevant values, the waveband of 430 nm was the most relevant band of remote sensing for predicting the water plant content directly related to its stress. However, in [25], the authors remarked that although the results were promising, there were some missing valid components that could allow the model to be suitable and testable in real conditions. For that purpose, they recommended using hyperspectral cameras, which provide more precise measured waveband results.

(ii) Visual Detection

The detection of visual symptoms uses the changes in the plant's appearance (colors, forms, lesions, spots) as an indicator of it being attacked by a disease or pest [15]. In the survey of Hamuda et al. [26], image based plant segmentation, which is the process of classifying an image into plant and non-plant, was used for detecting diseases in plants [27]. For instance, for the evaluation of the CLR's infection percentage in a specific lot, the number of diseased leaves in 60 random trees had to be divided into the total number of leaves in those trees and multiplied by 100 (see Equation (1)). A leaf is considered diseased with CLR when chlorotic spots or orange dust are observed on it. The severity of the disease can be divided into five categories depending on the number and diameter of rust orange spots, as seen in Figure 4.

$$Average\ infection\ \%\ in\ the\ lot = \frac{Number\ of\ diseased\ leaves\ in\ the\ 60\ trees}{Total\ number\ of\ leaves\ in\ the\ 60\ trees} \times 100 \quad (1)$$

Figure 4. CLR development stages [28].

A visual inspection can be carried out to detect the presence of chlorotic spots on the leaves, which are then used for measuring the incidence and severity of the disease [10].

To understand the conditions conducive to the development of CLR and, subsequently, refine the disease control, Avelino et al. [29] monitored such development on 73 coffee crops in Honduras for 1–3 years. Thereby, through the analysis of production situation variables such as climate, soil components, coffee tree productive characteristics, and crop management patterns, the researchers aimed to establish a relationship with the presence of rust. The result of this research indicated that CLR epidemics depend on the diverse production situations based on Table 1, linked as well to the local conditions of the plantation. Due to the above, these results reflect the need for the consideration of a certified growing system that aims for sustainability, taking into consideration production situations and, thus, preventing the development of pests and diseases.

Table 1. Kinds of variables that describe the importance of coffee plots in the presence of CLR [29].

Kind of Variable	Relevance
Climate variation (Altitude and rainfall)	High
Soil components	Medium–low
Cropping practices	Medium
Coffee tree productive characteristics	High

(iii) Biological Intervention

Several authors stated the importance of the relationship between living beings sharing the same environment. One of them was Haddad et al. [30], who in 2009, proposed a study to determine if seven selected isolated bacteria under greenhouse conditions would efficiently detect and control CLR. For the development of this research, they inoculated these bacteria: six *Bacillus* sp., B10, B25, B157, B175, B205, and B281, and one *Pseudomonas* sp., P286, which help to detect and control CLR in the early development stages, according to a preliminary result presented by Haddad et al. (2007). For the experiment, two important coffee varieties, Mundo Novo and Catuai, were selected due to the high susceptibility to CLR. Therefore, for three years, the varieties with the disease interacted with different treatments (bacteria) to analyze the behavior evolution between them. Based on the results of the treatments, the isolates P286 and B157 were as efficient as the copper fungicide in controlling the rust. Hence, considering the harmful effects due to the copper fungicide, the application of biological control with the B157 isolate of *Bacillus* sp. may be a reliable alternative solution to CLR management. That is why this research displayed the opportunity to successfully biocontrol CLR, for specialty coffee growers.

Jackson et al. [31], in 2012, proposed as well a biological detection and control based on a fungus, *Lecanicillium lecanii*. Their primary interest in the crops, in general, was the analogy of the coexistence of organisms in a specific environment with defined conditions that encounter a perfect balance. Given the above, the biological control system of the *A. instabilis* ants were mutualistically associated with the white halos of the fungus, *Lecanicillium lecanii*, based on the CLR effect.

However, the hypothesis stated the possibility that spores from *Lecanicillium lecanii* help to attack the *Hemileia vastatrix* before the rainy season. The effect of the time delay of *Lecanicillium lecanii* in *Hemileia vastatrix* resulted in a relationship between the two fungi and the ants not to be demonstrated, in spite of the control experiment resembling the real world. In conclusion, the restriction of biotic factors directly affects the development of CLR; therefore, for future work, it is important to consider the climate variation of an ecosystem to be able to predict such development [31].

(iv) Wireless Sensor Networks (WSN)

Wireless Sensor Networks (WSN) are a technology that is being used in many countries worldwide to monitor different agricultural characteristics in real time and remotely. It consists of multiple non-assisted embedded devices, called sensor nodes, that collect data in the field and communicate them wirelessly to a centralized processing station, which is known as the Base Station (BS). The BS has data storage, data processing, and data fusion capabilities, and it is in charge of transmitting the received data to the Internet to present them to an end-user [32]. Once the collected data are stored on a central server on the Internet, further analysis, processing, and visualization techniques are applied to extract valuable information and hidden correlations, which can help to detect changes in crop characteristics. These changes could be used as indicators of phytosanitary problems such as nutritional deficiencies, pests, diseases, and water stress. WSN is a powerful technology since the information of remote and inaccessible physical environments can be easily accessed through the Internet, with the help of the cooperative and constant monitoring of multiple sensors [33]. The sensor nodes in a WSN setup can vary in terms of their functions. Some of them can serve as simple data collectors that monitor a single physical phenomenon, while more powerful nodes may also perform more complex processing and aggregation operations. Some sensors can even have GPS modules that help them determine their particular location with high accuracy [33]. The most common sensors used in WSN for agriculture are the ones that collect climate data, images, and frequencies. Chaudhary et al. [34] emphasized in 2011 the importance of WSN in the field of PA by monitoring and controlling different critical parameters in a greenhouse through a microcontroller technology called Programmable System on a Chip (PSoC). As a consequence of the disproportionate rainfall dynamics, the need for controlling a suitable water distribution meeting those parameters inside the greenhouse arises. Thereby, the study tested the integration of wireless sensor node structures., with high bandwidth spectrum telecommunication technology. Mainly, it was proven that the integration was useful to determine an ideal irrigation plan that met the specific needs of a crop based on the interaction of the nodes within the greenhouse. Furthermore, the researchers recommended using reliable hardware with low current consumption to develop WSN projects, because it generates more confidence for the farmers concerning its incorporation with their crops and provides a longer battery life.

Besides, Piamonte et al. [35] proposed in 2017 a WSN prototype for monitoring the bud rot of the African oil palm. With the use of pH, humidity, temperature, and luminosity sensors, they aimed to measure climate variations and edaphic (related to the soil) factors to detect the presence of the fungus that causes the disease indirectly.

(v) Machine Learning

The domain concerned with building intelligent machines that can perform specific tasks just like a human is called Artificial Intelligence (AI) [36]. One of the main subareas of AI is Machine Learning (ML), which aims to extract complex patterns from large amounts of raw data automatically to predict future behaviors. When the extracting process of those patterns is taken to a more detailed level, where computers learn complicated real-world concepts by building them out of simpler ones in a hierarchical way, ML enters one of its most relevant subsets: Deep Learning (DL) [37]. The functionality of DL is an attempt to mimic the activity in layers of neurons in the human brain. The central structure that DL uses is called an Artificial Neural Network (ANN), which is composed of multiple layers of neurons and weighted connections between them. The neurons are excitable units that transform information, whereas the connections are in charge of rescaling the output of one layer of neurons and transmitting it to the next one to serve as its input [38]. Inputting data such as images, videos, sound, and text through the ANN, DL builds hierarchical structures and levels of representation and abstraction that enable the identification of underlying patterns [36]. One application of finding patterns through DL can be for estimating plant characteristics using non-invasive methodologies by means of digital images and machine learning. Sulistyo et al. [39] presented a computational intelligence vision sensing approach that estimated nutrient content in wheat leaves. This approach analyzed color features of the leaves' images captured in the field with different lighting conditions to estimate nitrogen content in wheat leaves. Another work of Sulistyo et al. [40] proposed a method to detect nitrogen content in wheat leaves by using color constancy with neural networks' fusion and a genetic algorithm that normalized plant images due to different sunlight intensities. Sulistyo et al. [41] also developed a method for extracting statistical features from wheat plant images, more specifically to estimate the nitrogen content in real context environments that can have variations in light intensities. This work provided a robust method for image segmentation using deep layer multilayer perceptron to remove complex backgrounds and used genetic algorithms to fine tune the color normalization. The output of the system after image segmentation and color normalization was then used as an input to several standard multi-layer perceptrons with different hidden layer nodes, which then combined their outputs using a simple and weighted averaging method. Fuentes et al. [42] presented a robust deep learning based detector to classify in real-time different types of diseases and pests in tomatoes. For such a task, the detector used images from RGB cameras (multiple resolutions and different devices such as mobile phones or digital cameras). This method detected if the crop had a disease or pest and which type it was. Similarly, Picon et al. [43] developed an automatic deep residual neural network algorithm to detect multiple plant diseases in real time, using mobile devices' cameras as the input source. The algorithm was capable of detecting three types of diseases on wheat crops: (i) *Septoria* (*Septoria tritici*), (ii) tan spot (*Drechslera tritici-repentis*), and (iii) rust (*Puccinia striiformis* and *Puccinia recondita*). Related to CLR, research has been done, such as that by Chemura et al. [44], who evaluated the potential of Sentinel-2 bands to detect the CLR infection levels early due to its devastating rates. Through the employment of the Random Forest (RF) and Partial Least Squares Discriminant Analysis (PLS-DA) algorithms, such levels could be identified for early CLR management. The researchers employed the variety of Yellow Catuai, which was chosen due to its CLR susceptibility. In this matter, Chemura et al. considered only seven Sentinel-2 Multispectral Instrument (MSI) bands due to the high resolution stated by previous works in biological studies. Based on the selected bands, the research results determined that the CLR reflectance was higher in NIR regions of the spectrum, as could be seen in leaves from the bands B4 (665 nm), B5 (705 nm), and B6 (740 nm). These bands achieved a high overall CLR discrimination of 28.5% and 71.4% using the RF and PLS-DA algorithms respectively.

Thus, the band and vegetation indices derived from the MSI of Sentinel-2 achieved the detection of the disease and an evaluation of CLR in the early stages, avoiding unnecessary chemical protection in healthy trees.

In 2017, Chemura et al. [45] studied the detection of CLR through the reflectance of the leaves at specific electromagnetic wavelengths. The objective of their investigation was to assess the utility of the wavebands used by the Sentinel-2 Multispectral Imager in detection models. The models were created using Partial Least Squares Regression (PLSR) and the non-linear Radial Basis Function partial Least Squares Regression (RBF-PLS) machine learning algorithm. Then, both models were compared, resulting in a low accuracy prediction of the state of the disease for the PLSR, due to its over-fitting, and a high accuracy prediction for the RBS-PLS model. Additionally, Chemura et al., through weighting of the importance of the variables, found that the blue, red, and RE1 bands had a high model correlation, but the implementation excluding the remaining four bands led to lower accuracy in both models. On the other hand, if more than one NIR or red edge (RE) band were used, then the RBS-PLS model developed would over-fit, resulting in a non-transferable model. However, Chemura et al. emphasized the utilization of the RBS-PLS model due to its machine learning advantage and its excellent adaptation to possible model over-fitting.

1.3. Conclusions of the Literature Review

The presented state-of-the-art showed that several researchers sought the detection of any vital element like water stress, nitrogen levels, and vegetation indexes that could lead to an improvement of production and quality in crops, which translated to an increase in profitability. However, most of the research did not integrate different means of detecting CLR to have more insights and better accuracy in predicting this disease. Furthermore, the determination of the infection percentage of the crop through visual inspection is a tedious task, which is also laborious, time consuming, and subject to human error and inconsistency [46]. For this reason, the objective of this research is to evaluate to what extent it is possible to diagnose the CLR development stage in the Colombian Caturra variety (the most susceptible to the disease) through a technological integration system that involves Remote Sensing (RS), Wireless Sensor Networks (WSN), and Deep Learning (DL). Adequate management of CLR could preserve the quality and selling price of the final product, reduce production costs by rationing control costs, and protect productivity. The present research aims to facilitate the management of the most dangerous disease in the Colombian Caturra variety's coffee production to strengthen the profitability of the rural inhabitants.

The present work provides empirical evidence of a novel diagnostic method for the classification of the development stage of CLR in coffee crops, by means of a technological integration of image data (RS), WSN, and DL. This contribution allows coffee growers to detect CLR disease automatically, thus optimizing the production and maintenance of their crops and replacing the task of manual inspection. Through this method, the performance evaluation is done, and the results are presented to conclude to what extent it is possible to diagnose CLR disease. Thus, this information can be useful for coffee growers to determine if the integration of RS, WSN, and DL in our method could positively impact their profitability.

2. Proposed Method

The design of experiment implemented in this research was a Completely Randomized Design (CRD). It was used to compare two or more treatments considering only two sources of variability: treatments and random error. The objective of using this design of experiment in this project was to analyze whether the diagnosis of the CLR development stage through the integration of RS, WSN, and DL was similar to the one made with a traditional visual inspection. A summarized diagram of this process is shown in Figure 5.

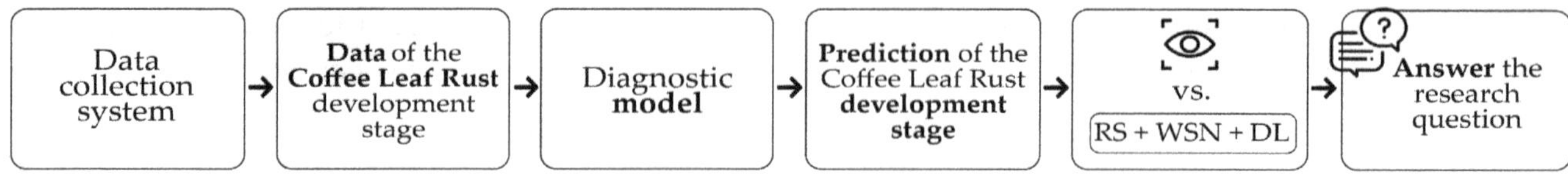

Figure 5. Proposed methodology flowchart (based on [44]).

In that sense, the study factor was the type of inspection, which had two levels ("visual inspection" and "technological integration"), and the response variable was the development stage of the disease, which was a whole number between 0 and 4. Thereby, the fundamental hypothesis to prove, presented in Equation (2), helped by deciding whether Treatments 1 ("visual inspection") and 2 ("technological integration") were statistically equivalent with respect to their means [47].

$$\begin{aligned} H_0 &: \mu_1 = \mu_2 \\ H_A &: \mu_1 \neq \mu_2 \end{aligned} \quad (2)$$

The procedure for proving the mentioned hypothesis is called Analysis Of Variance (ANOVA) and required a data table containing a row for each observation and a column for each treatment indicating the measurements of the response variable. This procedure separated the variability due to the treatments from the one attributed to the random error and compared them. If the former was higher than the latter, the means of the treatments were different, and thus, the type of diagnosis influenced the determined CLR development stage. Otherwise, the means were statistically equivalent, and it was possible to conclude that the visual inspection and the technological integration were similar for diagnosing the disease. Lastly, it is essential to mention that the significance level that was used for proving the hypothesis was 10% ($\alpha = 0.1$), since the problem at hand was related to agriculture, where many noise factors associated with the variation of environmental conditions were involved [47].

For the data collection experiment, 16 six month old, healthy coffee plants coming from Jardín, Antioquia, were used. Those plants were stored in a Universidad EAFIT's greenhouse. A biology team was in charge of their transplantation, agronomic management (elimination of weeds, fertilization, and fumigation), inoculation, and supervision. For the inoculation, the biology team followed the process described in Chemura et al. [44]. It is relevant to clarify that a new group of diseased plants was held as a reserve in case the inoculation of the healthy plants did not take effect over time.

Furthermore, an engineering team was dedicated to the design and assembly of a system, in the same greenhouse, that integrated RS and WSN. It allowed building a scale crop, recording different characteristics of it regularly, and storing them on a remote server to analyze its phytosanitary status later using DL. In that way, once the plants were inoculated and the system was verified, they were transplanted to it so that the data collection may begin. For that purpose, the scale crop was divided into four lots with certain differences in their agronomic management, which sought to recreate various circumstances of a real coffee crop. Thereby, a greater number of scenarios were covered, and the false positive rate regarding the diagnosis was reduced. *LOT 1* contained four non-inoculated plants, and they were neither fertilized nor fumigated; *LOT 2* had four non-inoculated plants and was fertilized but not fumigated; *LOT 3* had four inoculated plants, and they were also fertilized but not fumigated; and *LOT 4* had four inoculated plants, and they were neither fertilized nor fumigated. The previous distribution can be seen in Figure 6.

Finally, the visual inspections for diagnosis of the CLR development stage were carried out by the biology team for three months. Once per day, one of them examined the severity of the disease for each lot and indicated the value of the response variable for each observation; this measure corresponded to the ground truth. Similarly, the technological system automatically recorded the scale crop's characteristics from each lot seven times per day at different moments (with and without sunlight, because the field sensors and cameras had different illuminance requirements), assigning to each of these samples the above mentioned daily ground truth. After the data collection phase finished, the diagnostic model using DL was generated, and a comparative data table for the statistical analysis was produced, based on its predictions and the results of the visual inspections. As it was expected that a considerable amount of observations would be made, only 25% of all collected data were used for the statistical study. It should also be noted that, as was recommended, the order of the table's entries were randomized before executing the analysis in order to minimize bias.

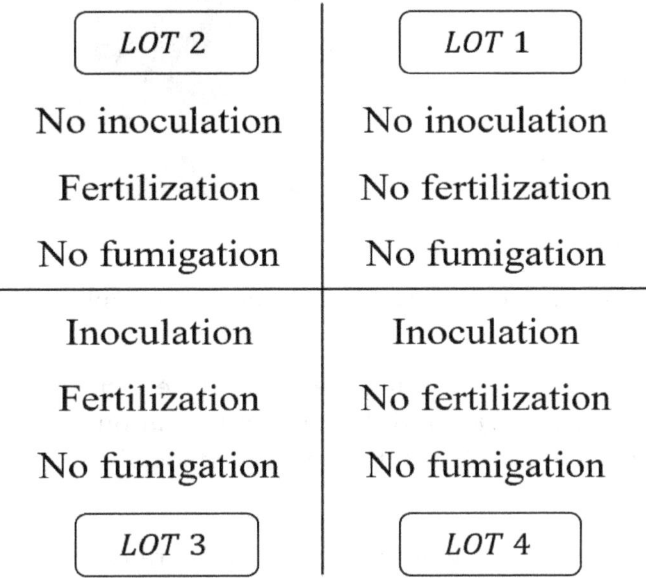

Figure 6. Data collection distribution.

2.1. Experimental Testbed

To evaluate to what extent it was possible to diagnose the CLR development stage in the Colombian Caturra variety through the integration of RS, WSN, and DL, it was necessary to obtain empirical evidence employing an experiment. Therefore, an experimental testbed prototype was built, which included a scale coffee crop. This testbed was capable of simulating different agronomic conditions and allowed capturing data for diagnosing the disease. The experimental testbed consisted of a data collection system prototype that integrated remote sensing and wireless sensor networks. In this testbed, the coffee plants were grouped, combined with the soil, and then divided into four lots. Furthermore, they were separated to inoculate CLR in half of them, and after that, the four lots were assembled again. For their agronomic management, fertilizer and fungicide were distributed and incorporated. Then, each lot was isolated from the others to make the four lots independent, and the whole scale crop was combined with a rain emulation system and a wind system. Both rainfall and wind speed for the whole crop were perceived. Furthermore, using sensors in each lot, pH, illuminance, temperature, humidity, and electrical conductivity were perceived, which will be further called "sensor data", and RGB and multispectral images were captured. RGB pictures were acquired through a regular RGB camera with a resolution of 720 p. These cameras were positioned on the bottom of the plants since CLR was commonly visible at the underside of the leaf [10]. Regarding the multispectral cameras, which allowed capturing the reflected radiation of wavelengths that were not perceptible to the human eye, two cameras from MAPIR®, called Survey3, were used. Based on the information cited in the state-of-the-art [21,23,25], the Red + Green + NIR (RGN) and Red Edge (RE) camera filters were chosen as being suitable to identify crop diseases, including CLR. Thus, one camera centered in the wavelengths 660 nm–550 nm–850 nm and another one centered in the 735 nm wavelength were selected to capture images from the top of the plants. The Survey3 incorporated a Sony® Exmor R IMX117 12MP sensor and a sharp non-fish eye lens for perceiving light in specific wavelengths. The created experimental testbed is shown in Figure 7.

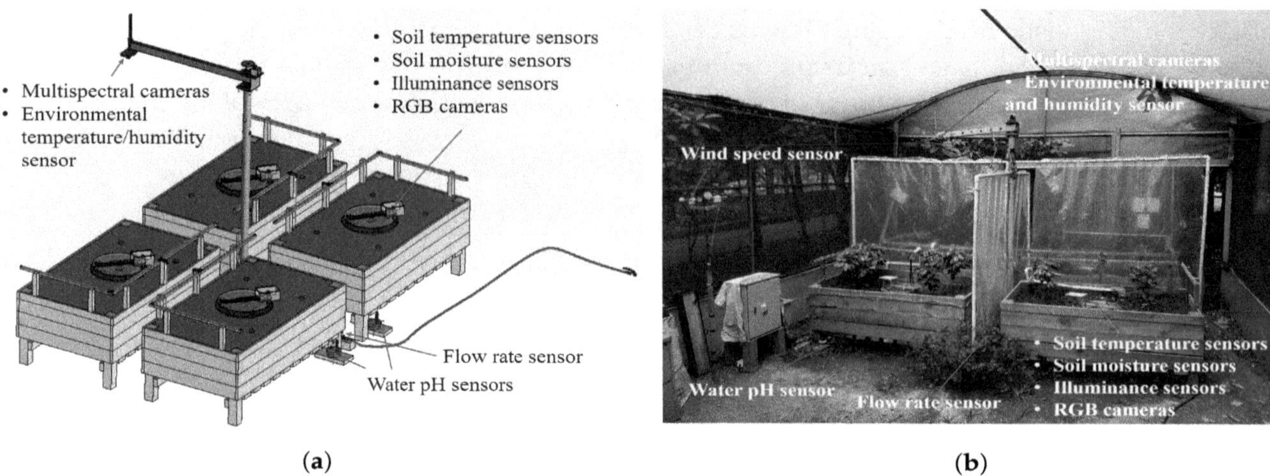

Figure 7. Experimental testbed: (**a**) 3D CAD model; (**b**) implemented prototype.

Afterwards, the state of each lot was integrated with the expert's visual inspection information to diagnose the CLR development stage, and then, this information was clustered with the collected data. To finish the data collection process, data were stored locally and sent to a remote server over the Internet.

On the other hand, the data that were received on the remote server were preprocessed for cleaning purposes and stored in a remote database. An example of the *LOT 3* directory's content on the remote server after one data collection routine was concluded is presented in Figure 8.

Figure 8. *LOT 3* directory's content after a data collection routine.

To clarify how a data collection routine worked, Figure 9 details the whole pipeline from the sensor readings and image captures until the remote storage. The data from sensors were gathered and smoothed by a microcontroller. RGB and multispectral images were captured by the cameras. The totality of the data was collected by a Single Board Computer (SBC), which continually notified the progress to the Internet of Things (IoT) platform (see Figure A3 inside Appendix C for the IoT platform dashboard) while it created a single data package. The package containing the documents with the lots and general data, as well as the images was stored locally. Furthermore, the documents were inserted into the remote MongoDB®, which resided in the data center, and the entire data package was uploaded via Secure File Transfer Protocol (SFTP) to the data center's file system. At that point, the data collection routine finished.

Finally, it is also relevant to mention how the collected data can be reviewed so that the process can be verified. Using a personal computer, the IoT platform, the single board computer, and the data center could be accessed over the Internet. The access to the IoT platform required a web browser, while the single board computer and the data center could be remotely inspected through the graphical desktop sharing system Virtual Network Computing (VNC) or the cryptographic network protocol Secure Shell (SSH).

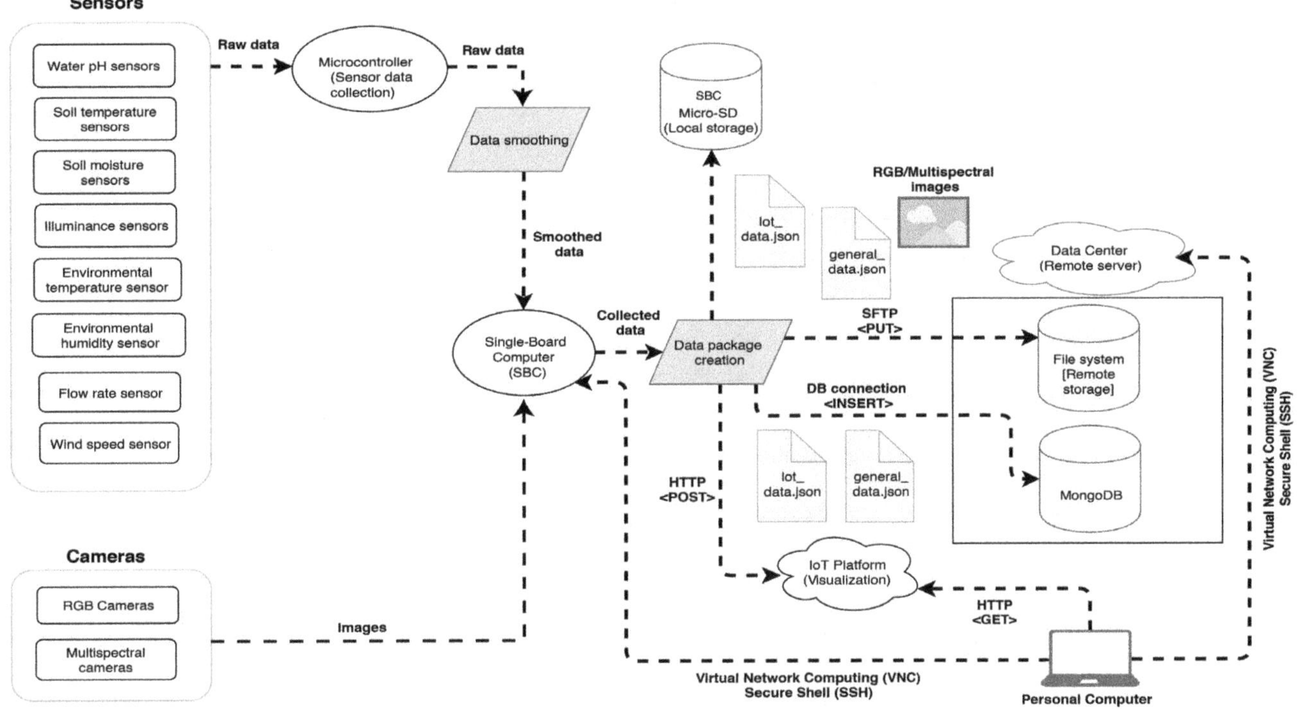

Figure 9. Data collection pipeline.

2.2. Machine Learning Pipeline

To create an adequate model for diagnosing the CLR development stage, the stored data were first divided into two sets, namely training (with cross-validation) and test. The training set was processed to build the diagnostic model with cross-validation, which served to assess its intermediate performance and tune it. Once the diagnostic model was generated, the test set was used for evaluating its final performance. All the developed models and cloud storage were implemented using an academic data center.

Within the framework of this project, the data center was used to store the data collected remotely on the physical part of the prototype. Both the MongoDB® instance in it, as well as its file system made the replication of single-board computer's local storage possible and facilitated the ubiquitous access to that information. Furthermore, the data center was the place where the data preprocessing, model generation and CLR development stage diagnosis occurred. It is also relevant to mention that the software libraries used for the implementation were Python 3.6.0, NumPy 1.16.0 [48], Pandas 0.24.0 [49], Scikit-learn 0.20.2 [50], and Keras 2.2.4 [51] running on top of TensorFlow 1.12 [52].

The machine learning pipeline model to show how the collected data were manipulated to extract the model that was used to diagnose the development stage of the disease in question is shown in Figure 10.

This pipeline model initially consisted of four sub-directories ranging from *LOT 1* to *LOT 4* where each lot's data would be correspondingly labeled later on. For that purpose, the biology team determined the labels by carrying out visual inspections in the field on all plants once a day during the whole data collection phase. In that sense, it assigned a whole number between 0 and 4 to each plant on each lot, evaluating the plant leaves' severity level, and calculated the specific lot's label as the rounded average of its four plants' disease development stages. All data directories of the current day and corresponding lot were labeled with the value of the last visual inspection, which was determined in the most recent checkup.

Subsequently, a new `rgb_images` directory containing five sub-directories (ranging from 0 to 4) representing the diseases' five stages was created. In these five sub-directories, RGB images coming from all lots (*LOT* 1 to *LOT* 4) were correspondingly stored according to their label. Similarly, the sensor data, which were stored as a JavaScript Object Notation File (JSON), and the multispectral images had the same label as the RGB images belonging to the same lot. Furthermore, in the case of images in general, they were visually checked one-by-one to keep only the ones with valuable content (focus, brightness level) and remove the others. In addition to this, a script was executed to eliminate the irrelevant JSON files (those with missing values and outliers), as well as the sub-directories that ended up with no content. The last two actions were part of the depuration stage. In the end, five sub-directories would exist containing the data from all lots (*LOT 1* to *LOT 4*) adequately labeled. Those sub-directories were the ones that were used for the generation and final evaluation of the diagnostic model, taking into account that the diagnosis occurred at the lot level.

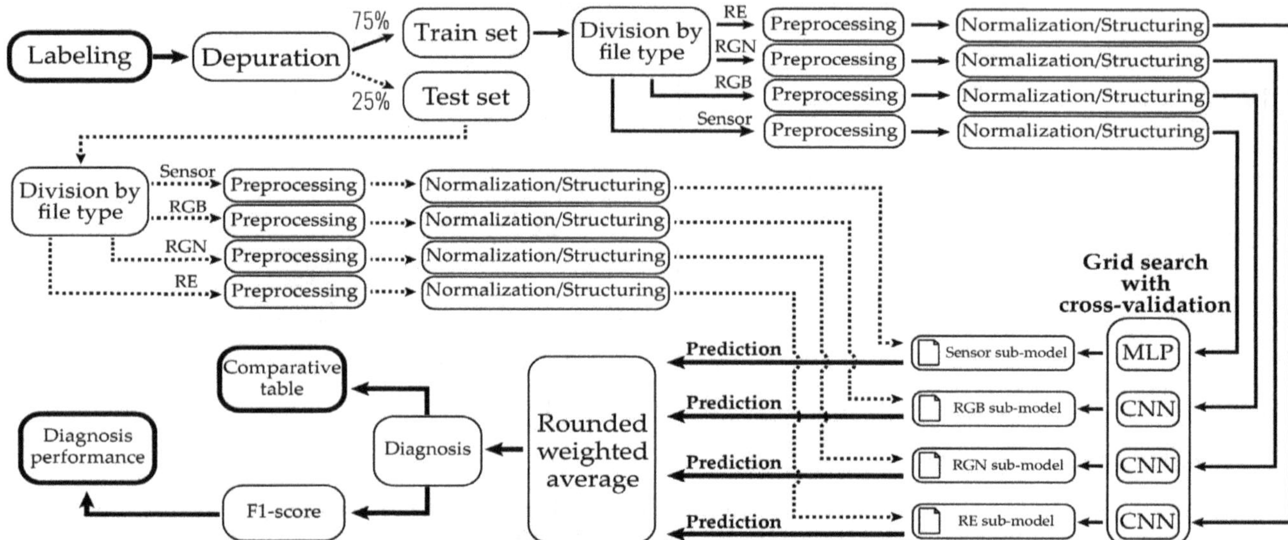

Figure 10. Machine learning summarized pipeline model.

Once all the data were correctly distributed, the content of each of the five sub-directories was virtually shuffled, and the elements per file type were counted for every sub-directory. Then, per label sub-directory, the minimum of those values was found. Twenty-five percent was calculated, and the file type associated with that minimum was determined. The resulting numbers indicated the number of files per respective determined file type and per label that could be used, at most, for testing the diagnostic model. Taking those threshold numbers into account, the shuffled lot data directories within each label subdirectory were individually analyzed to split them into two groups, namely training and test sets. If a particular lot;s data directory was considered as complete (i.e., it had a JSON file, the two multispectral images, and at least one RGB image) and supposing that using its files for testing did not exceed the corresponding threshold, then it was copied under the same structure to another location in order to feed the test set. Otherwise, the lot data directory was also copied, but to grow the training set. Thereby, the training set (∼75% of all data) was used to train and tweak the model, while the test set (∼25% of all data), with no overlapping with training set, was only incorporated at the time of the diagnosis evaluation. The data distribution after the above mentioned process was importantly imbalanced, as seen in Table 2. It can be noted that Stage 1 was not included in the table. This was due to the fact that only one sample was identified in that stage. Consequently, it could not be used for the model construction, and therefore, it could be considered as not relevant.

Table 2. Data distribution between the training and test sets by each CLR development stage.

# of Samples	Stage 0	Stage 2	Stage 3	Stage 4	Total
Training	711	55	90	112	968
Test	149	12	18	23	202

After the two sets were correctly obtained, one submodel was generated for each file type, i.e., sensor data (JSON), RGB, RGN, and RE. For the JSON files, Multi-Layer Perceptron (MLP) was used, whereas Convolutional Neural Networks (CNNs) were implemented to classify the RGB, RGN, and RE images. For that purpose, the data in the training set were first divided into four subdirectories according to the file type, while preserving the same structure. Then, each of them was preprocessed so that the noise was removed from the images, and the irrelevant keys in the documents were also identified and eliminated. Figure 11 illustrates an example of preprocessed image files.

(a) (b) (c)

Figure 11. Example of preprocessed image files: (**a**) RGB image; (**b**) RGN image; (**c**) RE image.

After that, the corresponding data were loaded within each submodel's generation, divided into feature data (the files themselves) and label data (the names of the label subdirectories that contained the files), and permuted. Thereby, the data were randomly mixed while it was still possible to know each feature's respective label unequivocally. Then, if applicable, the data were normalized and structured to scale the input and format, as was recommended when using deep ANNs. The normalization used for this experiment was the z-score (subtracting the mean of the feature and dividing by its standard deviation), which scaled the data to have the properties of a standard normal distribution [53]. Upon having the data prepared, different architectures and hyperparameter values were tried to train the submodel to tune it to reach higher performance values on the predictions.

The technique used for tuning the submodel is called grid search with cross-validation. It consisted of executing an exhaustive search over specified hyperparameter values for an estimator to find out which combination achieved the best performance, which was by default the higher accuracy, but different metrics could be chosen. One candidate estimator for each combination of hyperparameters was built and evaluated, so that the best estimator, its attributes, and its average performance could be extracted once the search was complete [54]. Furthermore, the procedure for measuring the average performance of each candidate estimator during the generation of the submodel is called *k*-Fold Cross-Validation, where *k* separate learning experiments are run on the the same estimator to calculate *k* performance values and average them. To achieve this, the feature and label data were split at the beginning into *k* non-overlapping subsets (also known as "folds"), so that for every experiment, one different fold was kept for measuring the performance, whereas the remaining $k-1$ were put together to form the training set to fit the estimator [55]. Finally, when the grid search processes concluded, the four submodels were extracted and saved for the definitive diagnosis about the CLR development stage.

To select the best estimator during the grid search with cross-validation, the chosen metric was the F_1-score, which, in the multi-label case, was the weighted average of the labels' F_1-scores. This metric was used due to the importantly imbalanced dataset (skewed classes) between the development stages of the CLR, as seen in Table 2. The F_1-score of the label L is a value in the [0, 1] range, and it was calculated as the harmonic mean of the estimator's precision and recall with respect to L (see

Equation (3)). The precision with respect to L is the ratio of the number of times that L was correctly predicted to the overall number of times that L was predicted. Furthermore, the recall with respect to L is the ratio of the number of times that L was correctly predicted to the overall number of times that L should have been predicted. Thereby, the general F_1-score reaches its best value at 1, indicating that the estimator perfectly matched reality, and its worst at 0, showing that the estimator never coincided with reality [53].

$$F_1\text{−}score_{\mathrm{L}} = \frac{2 * precision_{\mathrm{L}} * recall_{\mathrm{L}}}{precision_{\mathrm{L}} + recall_{\mathrm{L}}} \tag{3}$$

At this point, the data that were kept to be only incorporated at the time of the diagnosis evaluation were brought up. First, the submodels were loaded. Then, each lot's data directory contained in the test set was submitted to the following process. At the beginning, its data were divided according to the file type. After that, each type was sent to its corresponding submodel, where it was first cleaned, normalized, and structured, applying the same particular procedures that were used to prepare the data for the submodel generation. Subsequently, the submodel made its prediction based on the trained model output. It is also relevant to mention that, considering that the diagnosis was made at the lot level, the RGB submodel could be used up to four times per lot data directory before retrieving its result (which was the rounded average of its predictions). The final step consisted of combining the outcomes of the four submodels and calculating their rounded weighted average, the weights being the respective F_1-scores. Thereby, the definitive lot's CLR diagnosis was obtained, and it was recorded along with the processed lot's data directory label. Once the whole test set was covered, a table showing comparative results was generated for the statistical analysis, and the performance reached by the composite model was assessed with the calculation of the F_1-score. Figure A2 from Appendix B illustrates the above machine learning pipeline in a detailed manner, and Table 3 shows the selected hyperparameters and obtained F_1-score for each of them. Tables A1, A2 and A3 from Appendix D details the architectures of the submodels.

Table 3. Hyperparameters and F_1-score for each generated submodel.

Submodel	Batch Size	Epochs	Kernel Initializer	Activation	Rate	Optimizer	F_1-Score (Cross-Val Set)
Sensor data	16	20	normal	ReLU	0.4	Adam	0.651
RGB	16	6	glorot_uniform	ReLU	0.4	Adam	0.949
RGN	32	9	glorot_uniform	elu	0.3	Adam	0.928
RE	16	6	normal	ReLU	0.4	Adam	0.878

The last step of the proposed ML pipeline consisted of integrating the four presented submodels and evaluating the composite model, i.e., diagnosing the CLR development stage through it, creating a comparative table with the results achieved and calculating the model's performance. For that purpose, a model evaluator script was implemented. This script was in charge of loading the submodels into memory, iterating over the whole test set, taking each lot data directory within it, dividing the contained files according to their type and preprocessing them, resizing them to reduce the spatial complexity (in the case of images), normalizing and structuring each file according to the submodels' expected input, and sending them to their corresponding submodel to get a prediction. In addition, the script allowed gathering the four predicted labels and calculating their rounded weighted average, since the generated submodels presented different performances for diagnosing the CLR development stage. Table 4 shows the weights for the predictions of each submodel, which were determined as the ratio of each F_1-score in Table 3 with respect to the sum of all F_1-scores.

Table 4. Weights for the predictions of each submodel.

Submodel	Weight for Predictions
Sensor Data (JSON)	0.191
RGB	0.279
RGN	0.272
RE	0.258

To further explain the weighted average, let us assume that a sample folder with all the collected data (sensor data, RGB, RGN, and RE images) was labeled as CLR Development Stage 2 ($Label = 2$). Then, these data inside this folder were fed into the developed submodels (sensor data, RGB, RGN, and RE submodels) which produced an output based on their trained model. Let us assume that the sensor data submodel classified this as 0, the RGB submodel as 3, the RGN submodel as 2, and the RE submodel as 2. Then, considering the weights from Table 4, the averaged development stage would be approximately 1.90. Then, rounding this value up, the final output of the ML pipeline would be $DevelopmentStage = 2$. This example is shown in Figure 12.

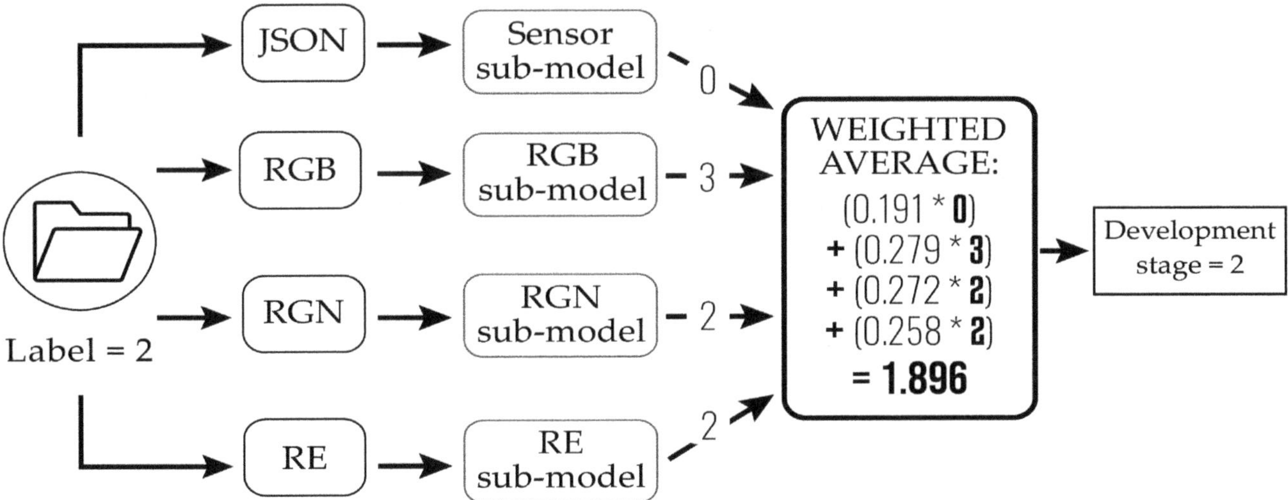

Figure 12. Machine learning classification example.

3. Results

The results of this experiment were a composite trained model with an F_1-score of 0.775. This model was tested using ANOVA to prove the validity of the hypothesis presented in Section 2, with respect to the visual inspection and our proposal using the technological integration methods. The p-value obtained was 0.231, which was greater than the significance $\alpha = 0.1$. This result indicated that the proposed method for automatically detecting the CLR disease presented an equivalent performance compared to the manual/visual inspection method (the ANOVA test will be further discussed in Section 3.1). All the inputs for the obtained results are detailed below.

On the one hand, it must be mentioned that, during the data collection phase, the biology team had to replace 12 coffee plants of the scale crop with external diseased ones because the inoculation did not take effect after two months (all plants stayed in Development Stage 0).

On the other hand, the training set used for fitting the submodels was composed of 968 directories. In total, they contained 672 sensor data (JSON) files, 2192 RGB files, 603 RGN files, and 641 RE files. In addition, the test set employed for the composite model evaluation comprised 202 lot data directories, with 224 sensor data (JSON), 730 RGB files, 202 RGN files, and 202 RE files. Finally, after evaluating the diagnosis of the CLR development stage in the Colombian Caturra variety employing the created DL model, a comparative table, along with a performance table, was successfully generated. Figure A1 from Appendix A shows the comparative table for the statistical analysis. Table 5 presents the definitive F_1-score reached by each submodel and the composite model.

Table 5. F_1-score reached by the individual submodels and the composite model.

Model	F_1-score (Test Set)
Sensor Data (JSON)	0.570
RGB	0.920
RGN	0.946
RE	0.944
Composite	**0.775**

3.1. Analysis of the Results

Statistical analysis of the results regarding the performance evaluation of the diagnostic model was carried out using the comparative table found in Figure A1 from Appendix A. The purpose of the analysis was to determine whether there was a significant difference in the mean CLR development stage diagnosed with a visual inspection and using the proposed technological integration. The outcome was relevant to get the necessary statistical support for answering the research question.

The comparative table contained 202 observations with the corresponding diagnosed development stage for both treatments. Figure 13 shows the box plot chart describing the measurements. The x-axis contains the two treatments ("visual inspection" and "technological integration"), whereas the CLR development stage is presented on the y-axis. The graphical similarity of the data distribution of each treatment suggested a possible similarity to the means of the response variable. To assess this condition and make a decision based on the hypothesis, an ANOVA was executed.

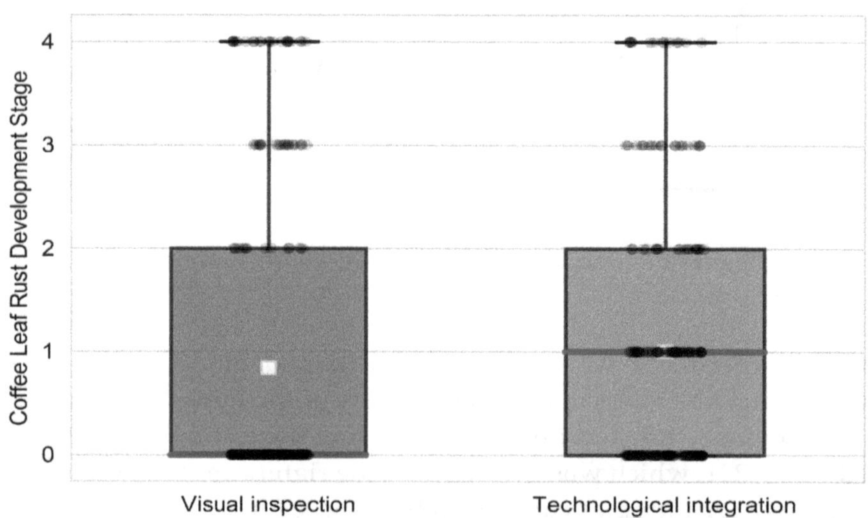

Figure 13. Data distribution of the observations for both treatments.

The results of the ANOVA can be seen in Table 6. The obtained p-value for the treatments factor was 0.231. This value was greater than the set significance ($\alpha = 0.1$), which meant that there was not sufficient evidence for rejecting the null hypothesis. Thus, it was concluded, with 90% confidence, that there was no statistically significant difference between the diagnosis of the CLR development stage made by using visual inspection and the technological integration. This result indicated that both methods were significantly similar to diagnose the disease.

This research demonstrated the feasibility of diagnosing the CLR development stage in the Colombian Caturra variety, with significant performance, through the integration of RS, WSN, and DL. The analysis of the results allowed obtaining statistical evidence for supporting the research hypothesis. In that sense, the outcome suggested that a technological integration could contribute to the protection of the phytosanitary status of coffee crops since it showed potential for complementing the traditional visual inspections towards the diagnosis of the most economically limiting disease for Colombian coffee production.

Table 6. ANOVA table of the statistical analysis.

	Df	**Sum Sq**	**Mean Sq**	**F Value**	***Pr(>F)***
Treatments	1	2.7	2.696	1.437	**0.231**
Residuals	402	753.9	1.875		

4. Conclusions

The integration of RS, WSN, and DL within the framework of this study successfully allowed evaluating to what extent it was possible to diagnose the CLR development stage in the Colombian Caturra variety. To this end, the most relevant information obtained was consolidated, the knowledge about the study context and CLR was detailed, and the repercussions of the disease in the Colombian coffee growing industry were identified. Furthermore, the state-of-the-art methods were reviewed and used for the current research. Creative design sessions were carried out to define the most useful technological integration of RS and WSN. Afterward, a functional prototype that automatically collected data in the field and transferred them to a remote server over the Internet was built. Besides, a diagnostic model using DL was implemented based on the stored data, and it successfully allowed evaluating the CLR development stage with unknown field data.

The motivation of this research project was to contribute to rural development through technological innovation to strengthen the profitability of Colombian coffee growers. Considering that the country has the potential, in terms of environmental conditions and diverse ecosystems, to generate a giant portfolio of exotic products that would be better valued in the specialty coffee market, this research evaluated, with empirical evidence, a technological approach that attempted to facilitate the diagnosis and mitigate the risks of one of the most economically limiting diseases for coffee production. In that sense, the proposed technological integration could positively impact the rural sector since those innovations promote investments in infrastructure, which are crucial to empower the rural community and improve the living standards and activities concerning progress, productivity, and income generation.

The obtained p-value in the analysis of the results was 0.231, which helped to determine, with 90% confidence, that the visual inspection and the technological integration did not present a statistically significant difference regarding the diagnosis of the CLR development stage. Thus, it could be said that the assessment of the disease led to a similar outcome using either method, which suggested that the obtained results supported the research hypothesis. Finally, it could be asserted that through the integration of RS, WSN, and DL, it was possible to diagnose the CLR development stage in the Colombian Caturra variety with a F_1-score of 0.775. This value indicated that, on average, the diagnostic model was excellent in terms of the certainty and usefulness of its diagnosis.

Regarding the data processing phase, a further extension of this research could include the implementation of a simple user interface for visualizing the diagnosis of the CLR development stage through the generated DL model to better illustrate the results to a coffee grower. Additionally, the proposed technological integration could be scaled to a real context by using drones with one or both of the two multispectral cameras used in the experiment presented by this work (depending on the project budget) as a possible approach, knowing that the identification of the CLR could be done with just one camera, e.g., RGN (F_1-score of 0.946), due to its high score. Another real context approach could be further explored using a mobile autonomous robot with a single RGB camera. Finally, the F_1-score values achieved on the test set, which showed that the submodels based on images presented a higher performance than the JSON submodel (sensor data model), suggested reconsidering the composite model for future work and focusing all efforts on improving the collection and processing of just RGB and multispectral data or using more robust sensors when the technology allows it; by using just the three submodels (RGB, RGN, and RE), we computed an average F_1-score of 0.93, which clearly showed that an improved composite F_1-score could be surely achieved, but a real context commercial application may only implement one of the best three previous submodels due to both implementation and maintenance costs.

Author Contributions: A.S. and S.S. designed and implemented the experimental testbed and algorithms for integrating WSN, RS, and DL for CLR detection. D.V. and M.T. supervised the experimental design and managed the project. M.M. and B.S. reviewed the machine learning and deep learning parts of this research. All authors contributed to the writing and reviewing of the present manuscript. All authors read and agreed to the published version of the manuscript.

Acknowledgments: We would like to thank Engineer Carlos Mario Ospina for his support during the visit to CENICAFÉ's Experimental Station "El Rosario", as well as coffee grower Nabor Giraldo, Diego Miguel Sierra, and Dentist Samuel Roldán for their support in the procurement of the coffee plants and the biological matter containing CLR. Additionally, we would like to thank Alejandro Marulanda, Edwin Nelson Montoya, Engineer Hugo Murillo, Universidad EAFIT, and Colciencias for the constant academic, infrastructural, and financial support. It is also important to highlight the enormous work and dedication of Project Supervisors Engineer Camilo Velásquez, Engineer Felipe Gutiérrez, Engineer Sebastián Osorio, Engineer Juan Diego Zuluaga, and physics engineering undergraduate student Sergio Jurado. In addition, we want to acknowledge the biology team, formed of undergraduate students Alisson Martínez and Laura Cristina Moreno and led by Luisa Fernanda Posada. We would also like to mention computing-systems engineering undergraduate students Alejandro Cano, Luis Javier Palacio, and Sebastián Giraldo, and we also thank the Engineers Ricardo Urrego and Luis Felipe Machado and Sebastián Rodríguez.

Abbreviations

The following abbreviations are used in this manuscript:

CLR	Coffee Leaf Rust
FCP	Fondo Colombia en Paz
RS	Remote Sensing
WSN	Wireless Sensor Networks
ML	Machine Learning
DL	Deep Learning
LAI	Leaf Area Index
CASI	Compact Airborne Spectrographic Imager
NIR	Near-Infrared
RGB	Red Green Blue
RE	Red Edge
RGN	Red Green Near-Infrared
NDVI	Normalized Difference Vegetation Index
UAV	Unmanned Aerial Vehicle
BS	Base Station
PSoC	Programmable System on a Chip
AI	Artificial Intelligence
ANN	Artificial Neural Network
CNN	Convolutional Neural Network
MSI	Multispectral Instrument
CRD	Completely Randomized Design
ANOVA	Analysis Of Variance
SBC	Single Board COmputer
SFTP	Secure File Transfer Protocol
VNC	Virtual Network Computing
SSH	Secure Shell
IoT	Internet of Things
JSON	JavaScript Object Notation

Appendix A. Comparative Statistical Analysis Table

ID	Visual inspection	Technological integration	ID	Visual inspection	Technological integration	ID	Visual inspection	Technological integration	ID	Visual inspection	Technological integration	ID	Visual inspection	Technological integration	ID	Visual inspection	Technological integration	ID	Visual inspection	Technological integration
1277	0	0	192	0	0	209	0	0	384	0	1	633	0	0	229	0	0	257	2	2
494	4	4	391	3	2	749	0	0	608	0	1	1511	0	0	449	4	4	3148	3	1
498	0	1	1090	0	0	245	2	2	628	0	1	1517	0	1	105	0	0	268	0	0
4217	4	3	380	3	3	854	0	2	1572	0	0	1571	0	0	527	0	0	215	2	2
469	0	1	246	0	0	3115	3	3	181	0	0	3193	3	2	361	0	0	164	0	0
1258	0	0	356	3	3	183	0	1	665	0	0	259	2	2	635	0	0	694	0	1
490	0	1	4231	4	3	1567	0	1	1077	0	0	154	0	0	1342	0	1	208	0	1
636	0	1	321	0	0	1525	0	0	204	0	0	1479	0	0	252	0	1	436	4	4
230	2	2	425	4	4	1437	0	1	360	0	0	437	4	3	602	0	1	1221	0	0
456	4	3	487	4	4	348	0	0	409	0	1	609	0	0	711	0	0	1559	0	0
667	0	0	1574	0	0	489	0	1	654	0	0	1512	0	0	919	0	0	1005	0	0
230	0	1	1499	0	0	895	0	1	1414	0	0	245	0	0	410	0	0	1463	0	0
514	0	0	4114	4	4	3188	3	2	359	3	3	3108	3	3	879	0	1	4233	4	4
911	0	0	695	0	0	76	0	0	4106	4	4	394	0	1	1230	0	1	488	0	1
721	0	0	138	0	0	552	0	1	572	0	1	1281	0	0	1173	0	0	2120	2	2
513	0	0	244	0	1	445	4	4	611	0	1	256	2	2	1386	0	0	439	0	0
4	0	1	792	0	1	692	0	1	596	0	1	1189	0	0	687	0	0	79	0	0
4207	4	4	777	0	0	1370	0	1	316	0	0	1462	0	0	453	0	0	387	0	0
3135	3	2	351	0	0	356	0	0	1278	0	1	97	0	0	28	0	0	300	0	0
723	0	0	468	0	0	599	0	1	666	0	0	4216	4	2	2119	2	2	767	0	0
53	0	0	4150	4	4	1422	0	1	5	0	0	1497	0	0	4178	4	4	4244	4	1
682	0	1	109	0	0	548	0	1	1078	0	1	1518	0	1	894	0	1	285	2	2
237	2	2	78	0	0	891	0	1	3101	3	3	1302	0	0	302	0	0	443	4	4
2114	2	2	1457	0	1	1201	0	0	4219	4	3	373	3	3	282	0	0	271	0	0
255	0	0	807	0	0	3106	3	3	323	3	3	491	0	0	153	0	0	3162	3	2
464	0	0	152	0	1	543	0	1	1314	0	0	0	0	0	4218	4	3	1301	0	0
1507	0	0	318	0	1	23	0	0	657	0	0	571	0	1	338	3	3	607	0	1
394	3	3	376	0	1	1061	0	0	861	0	0	793	0	0	386	0	0	311	3	3
193	0	0	2101	2	2	4205	4	4	4230	4	3	850	0	1	377	0	0			

Figure A1. Comparative statistical analysis table showing Visual Inspection VS Technological Integration with their respective sample IDs.

Appendix B. Data Management Model

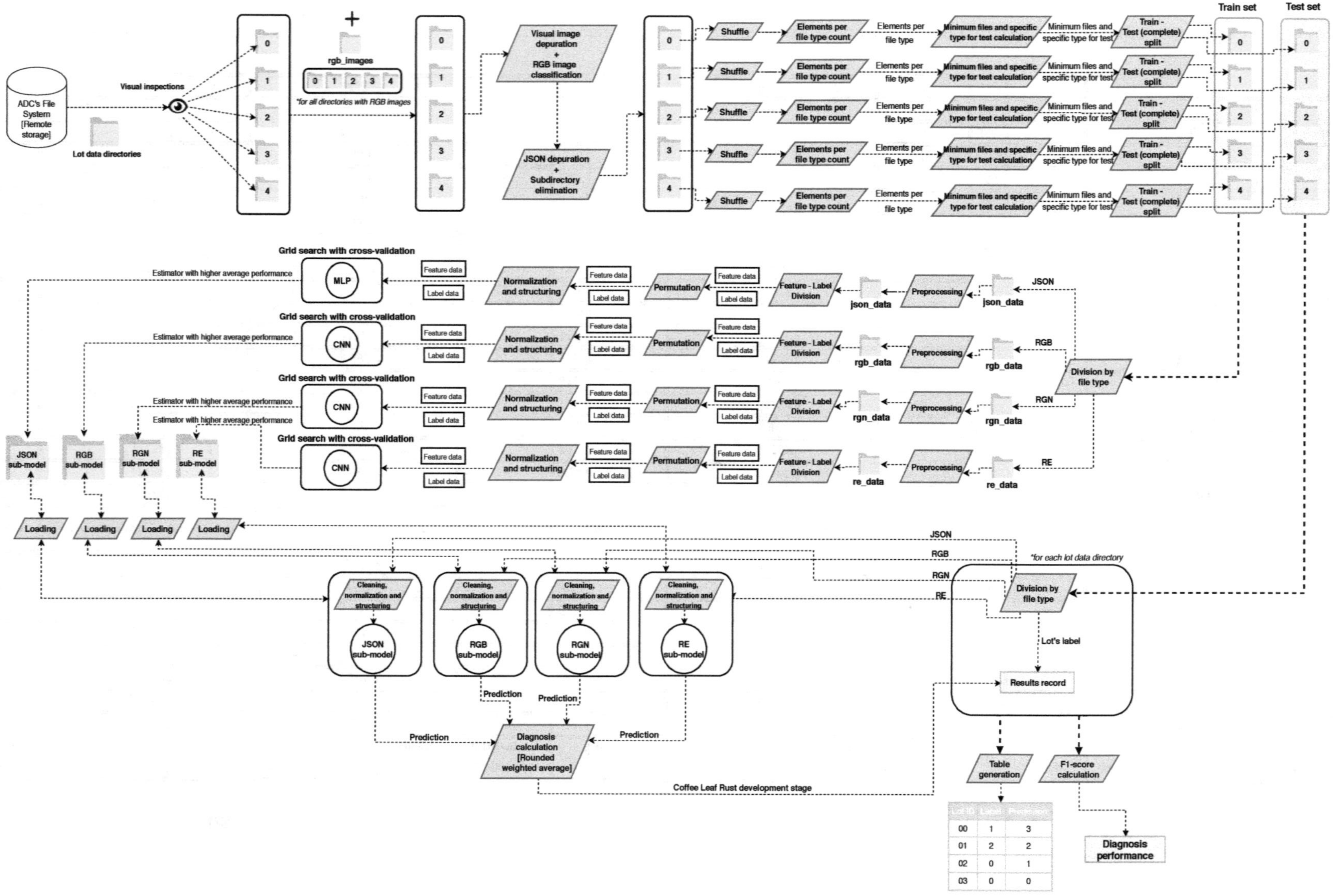

Lot ID	Label	Prediction
00	1	3
01	2	2
02	0	1
03	0	0

Figure A2. Detailed Data Management Model.

C. IoT Platform Dashboard

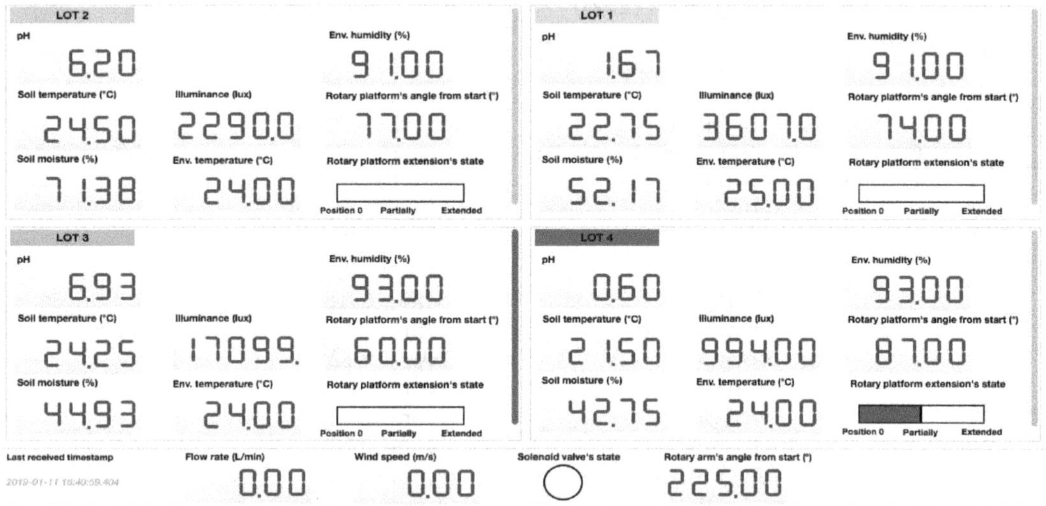

Figure 3. Implemented IoT platform real-time dashboard.

D. Submodels' Architectures

Table 1. JSON submodel's architecture.

N	Layer	Output Shape	# of Parameters
1.	Input Layer	(None, 6)	-
2.	Fully Connected	(None, 16)	112
3.	Batch Normalization	(None, 16)	64
4.	Activation	(None, 16)	0
5.	Fully Connected	(None, 64)	1088
6.	Batch Normalization	(None, 64)	256
7.	Activation	(None, 64)	0
8.	Dropout	(None, 64)	0
9.	Fully Connected	(None, 32)	2080
10.	Batch Normalization	(None, 32)	128
11.	Activation	(None, 32)	0
12.	Dropout (rate = rate/2)	(None, 32)	0
13.	Fully Connected	(None, 4)	132
14.	Activation	(None, 4)	0

Table 2. RGB submodel's architecture.

N	Layer	Output Shape	# of Parameters
1.	Input Layer	(None, 96, 128, 3)	-
2.	Convolutional2D (Kernel = (5, 5))	(None, 92, 124, 18)	1368
3.	Batch Normalization	(None, 92, 124, 18)	72
4.	Activation	(None, 92, 124, 18)	0
5.	Max Pooling (pool = (2, 2))	(None, 46, 62, 18)	0
6.	Convolutional2D (kernel = (5, 5))	(None, 42, 58, 36)	16,236
7.	Batch Normalization	(None, 42, 58, 36)	144
8.	Activation	(None, 42, 58, 36)	0
9.	Max Pooling (pool = (2, 2))	(None, 21, 29, 36)	0
10.	Convolutional2D (kernel = (3, 3))	(None, 19, 27, 54)	17,550
11.	Batch Normalization	(None, 19, 27, 54)	216
12.	Activation	(None, 19, 27, 54)	0
13.	Max Pooling (pool = (2, 2))	(None, 9, 13, 54)	0
14.	Dropout	(None, 9, 13, 54)	0
15.	Flatten	(None, 6318)	0
16.	Fully Connected	(None, 512)	3,235,328
17.	Batch Normalization	(None, 512)	2048
18.	Activation	(None, 512)	0
19.	Dropout	(None, 512)	0
20.	Fully Connected	(None, 128)	65,664
21.	Batch Normalization	(None, 128)	512
22.	Activation	(None, 128)	0
23.	Dropout (rate = rate/2)	(None, 128)	0
24.	Fully Connected	(None, 5)	645
25.	Activation	(None, 5)	0

Table 3. RGN and RE submodels' architectures.

N	Layers	Output Shape	# of Parameters
1.	Input Layer	(None, 128, 96, 3)	-
2.	Convolutional2D (kernel = (5, 5))	(None, 124, 92, 18)	1368
3.	Batch Normalization	(None, 124, 92, 18)	72
4.	Activation	(None, 124, 92, 18)	0
5.	Max Pooling (pool = (2, 2))	(None, 62, 46, 18)	0
6.	Convolutional2D (kernel = (5, 5))	(None, 58, 42, 36)	16,236
7.	Batch Normalization	(None, 58, 42, 36)	144
8.	Activation	(None, 58, 42, 36)	0
9.	Max Pooling (pool = (2, 2))	(None, 29, 21, 36)	0
10.	Convolutional2D (kernel = (3, 3))	(None, 27, 19, 54)	17,550
11.	Batch Normalization	(None, 27, 19, 54)	216
12.	Activation	(None, 27, 19, 54)	0
13.	Max Pooling (pool = (2, 2))	(None, 13, 9, 54)	0
14.	Dropout	(None, 13, 9, 54)	0
15.	Flatten	(None, 6318)	0
16.	Fully Connected	(None, 512)	3,235,328
17.	Batch Normalization	(None, 512)	2048
18.	Activation	(None, 512)	0
19.	Dropout	(None, 512)	0
20.	Fully Connected	(None, 128)	65,664
21.	Batch Normalization	(None, 128)	512
22.	Activation	(None, 128)	0
23.	Dropout (rate = rate/2)	(None, 128)	0
24.	Fully Connected	(None, 4)	516
25.	Activation	(None, 4)	0

References

1. Esposito, M. *Driving the Sustainability of Production Systems with Fourth Industrial Revolution Innovation*; White Paper; World Economic Forum: Colony, Switzerland, 2018.
2. OECD Directorate for Science, Technology and Industry. *ISIC REV. 3 TECHNOLOGY INTENSITY DEFINITION: Classification of Manufacturing Industries into Categories Based on R&D Intensities*; OECD: Paris, France, 2011.
3. Chen, Y.; Zhang, C.; Wang, S.; Li, J.; Li, F.; Yang, X.; Wang, Y.; Yin, L. Extracting Crop Spatial Distribution from Gaofen 2 Imagery Using a Convolutional Neural Network. *Appl. Sci.* **2019**, *9*, 2917. [CrossRef]
4. Liu, J.; Zhang, X.; Li, Z.; Zhang, X.; Jemric, T.; Wang, X. Quality Monitoring and Analysis of Xinjiang 'Korla' Fragrant Pear in Cold Chain Logistics and Home Storage with Multi-Sensor Technology. *Appl. Sci.* **2019**, *9*, 3895. [CrossRef]
5. Lee, J.W.; Kim, S.C.; Oh, J.; Chung, W.J.; Han, H.W.; Kim, J.T.; Park, Y.J. Engine Speed Control System for Improving the Fuel Efficiency of Agricultural Tractors for Plowing Operations. *Appl. Sci.* **2019**, *9*, 3898. [CrossRef]
6. Zhou, C.; Ye, H.; Xu, Z.; Hu, J.; Shi, X.; Hua, S.; Yue, J.; Yang, G. Estimating Maize-Leaf Coverage in Field Conditions by Applying a Machine Learning Algorithm to UAV Remote Sensing Images. *Appl. Sci.* **2019**, *9*, 2389. [CrossRef]
7. National Coffee Association USA. The Influence of Coffee Around the World. 2015. Available online: https://nationalcoffeeblog.org/2015/06/15/the-influence-of-coffee-around-the-world/ (accessed on 8 May 2018).
8. SCAA. *SCAA Protocols Cupping Specialty Coffee*; Specialty Coffee Association of America: Santa Ana, CA, USA, 2015.
9. CropLife Latin America. Roya del cafeto. 2018. Available online: https://www.croplifela.org/es/plagas/listado-de-plagas/roya-del-cafeto (accessed on 19 January 2019).
10. Rivillas, C.; Serna, C.; Cristancho, M.; Gaitan, A. *La Roya del Cafeto en Colombia: Impacto Manejo y Costos del Control*; Avances Tecnicos Cenicafe: Chinchiná, Colombia, 2011.
11. Carvalho, C.R.; Fernandes, R.C.; Carvalho, G.M.A.; Barreto, R.W.; Evans, H.C. Cryptosexuality and the Genetic Diversity Paradox in Coffee Rust, Hemileia vastatrix. *PLoS ONE* **2011**. [CrossRef]
12. The Observatory of Economic Complexity (OEC). Colombia (COL) Exports, Imports, and Trade Partners. 2018. Available online: https://atlas.media.mit.edu/en/profile/country/col/ (accessed on 23 January 2019).
13. Federación Nacional de Cafeteros. Estadisticas Historicas. 2018. Available online: https://www.federaciondecafeteros.org/particulares/es/quienes_somos/119_estadisticas_historicas/ (accessed on 10 January 2019).
14. Guzmán, O.; Gómez, E.; Rivillas, C.; Oliveros, C. Utilización del procesamiento de imágenes para determinar la severidad de La Mancha de Hierro, en hojas de café. *Cenicafé* **2003**, *54*, 258–265.
15. Martinelli, F.; Scalenghe, R.; Davino, S.; Panno, S.; Scuderi, G.; Ruisi, P.; Villa, P.; Stroppiana, D.; Boschetti, M.; Goulart, L.R.; et al. Advanced methods of plant disease detection. A review. *Agron. Sustain. Dev.* **2015**, *35*, 1–25. [CrossRef]
16. Lobitz, B.; Beck, L.; Huq, A.; Wood, B.; Fuchs, G.; Faruque, A.; Colwell, R. Climate and infectious disease: Use of remote sensing for detection of Vibrio cholerae by indirect measurement. *Proc. Natl. Acad. Sci. USA* **2000**, *97*, 1438–1443. [CrossRef]
17. Su, N.Y. Remote Monitoring System for Detecting Termites. U.S. Patent 6,052,066, 29 September 1998.
18. Mirik, M.; Norland, J.E.; Crabtree, R.L.; Biondini, M.E. Hyperspectral one-meter-resolution remote sensing in Yellowstone National Park, Wyoming: I. Forage nutritional values. *Rangel. Ecol. Manag.* **2005**, *58*, 452–458. [CrossRef]
19. Mulla, D.J. Twenty five years of remote sensing in precision agriculture: Key advances and remaining knowledge gaps. *Biosyst. Eng.* **2013**, *114*, 358–371, doi:10.1016/j.biosystemseng.2012.08.009. [CrossRef]
20. Calvario, G.; Sierra, B.; Alarcón, T.E.; Hernandez, C.; Dalmau, O. A multi-disciplinary approach to remote sensing through low-cost UAVs. *Sensors* **2017**. [CrossRef]
21. Goel, P.K.; Prasher, S.O.; Landry, J.A.; Patel, R.M.; Bonnell, R.; Viau, A.A.; Miller, J. Potential of airborne hyperspectral remote sensing to detect nitrogen deficiency and weed infestation in corn. *Comput. Electron. Agric.* **2003**, *38*, 99–124. [CrossRef]

22. Ortega-Huerta, M.A.; Komar, O.; Price, K.P.; Ventura, H.J. Mapping coffee plantations with Landsat imagery: An example from El Salvador. *Int. J. Remote Sens.* **2012**, *33*, 220–242. [CrossRef]
23. Bolaños, J.A.; Campo, L.; Corrales, J.C. Characterization in the Visible and Infrared Spectrum of Agricultural Crops from a Multirotor Air Vehicle. In Proceedings of the International Conference of ICT for Adapting Agriculture to Climate Change, Popayán, Colombia, 22–24 November 2017; pp. 29–43.
24. Piedallu, C.; Cheret, V.; Denux, J.; Perez, V.; Azcona, J.; Seynave, I.; Gégout, J. *Etudier les Variations Spatiales de NDVI pour Caractériser les Contraintes Environnementales Limitant la Vitalité des Forêts de Montagne et de Méditerranée*; CAQSIS; INRA: Clermont-Ferrand, France, 2018; pp. 1–17.
25. Chemura, A.; Mutanga, O.; Dube, T. Remote sensing leaf water stress in coffee (Coffea arabica) using secondary effects of water absorption and random forests. *Phys. Chem. Earth Parts A/B/C* **2017**, *100*, 317–324. [CrossRef]
26. Hamuda, E.; Glavin, M.; Jones, E. A survey of image processing techniques for plant extraction and segmentation in the field. *Comput. Electron. Agric.* **2016**, *125*, 184–199. [CrossRef]
27. Camargo, A.; Smith, J.S. An image-processing based algorithm to automatically identify plant disease visual symptoms. *Biosyst. Eng.* **2009**. [CrossRef]
28. De Melo Virginio Filho, E.; Astorga, C. *Prevención y Control de la Roya del Café: Manual de Buenas Prácticas para Técnicos y Facilitadores*, 1st ed.; CATIE: Turrialba, Costa Rica, 2015; p. 67.
29. Avelino, J.; Zelaya, H.; Merlo, A.; Pineda, A.; Ordoñez, M.; Savary, S. The intensity of a coffee rust epidemic is dependent on production situations. *Ecol. Model.* **2006**, *197*, 431–447. [CrossRef]
30. Haddad, F.; Maffia, L.A.; Mizubuti, E.S.; Teixeira, H. Biological control of coffee rust by antagonistic bacteria under field conditions in Brazil. *Biol. Control* **2009**, *49*, 114–119. [CrossRef]
31. Jackson, D.; Skillman, J.; Vandermeer, J. Indirect biological control of the coffee leaf rust, Hemileia vastatrix, by the entomogenous fungus Lecanicillium lecanii in a complex coffee agroecosystem. *Biol. Control* **2012**, *61*, 89–97. [CrossRef]
32. Azfar, S.; Nadeem, A.; Basit, A. Pest detection and control techniques using wireless sensor network: A review. *J. Entomol. Zool. Stud.* **2015**, *3*, 92–99.
33. Dargie, W.; Poellabauer, C. *Fundamentals of Wireless Sensor Networks: Theory and Practice*; Wireless Communications and Mobile Computing; Wiley: Hoboken, NJ, USA, 2010.
34. Chaudhary, D.; Nayse, S.; Waghmare, L. Application of wireless sensor networks for greenhouse parameter control in precision agriculture. *Int. J. Wirel. Mob. Networks (IJWMN)* **2011**, *3*, 140–149. [CrossRef]
35. Piamonte, M.; Huerta, M.; Clotet, R.; Padilla, J.; Vargas, T.; Rivas, D. WSN Prototype for African Oil Palm Bud Rot Monitoring. In Proceedings of the International Conference of ICT for Adapting Agriculture to Climate Change, Popayán, Colombia, 22–24 November 2017; pp. 170–181.
36. Bhardwaj, A.; Di, W.; Wei, J. *Deep Learning Essentials: Your Hands-On Guide to the Fundamentals of Deep Learning and Neural Network Modeling*; Packt Publishing: Birmingham, UK, 2018.
37. Goodfellow, I.; Bengio, Y.; Courville, A. *Deep Learning*; Adaptive Computation and Machine Learning; MIT Press: Cambridge, MA, USA, 2016.
38. Patterson, J.; Gibson, A. *Deep Learning: A Practitioner's Approach*; O'Reilly Media: Sebastopol, CA, USA, 2017.
39. Sulistyo, S.B.; Wu, D.; Woo, W.L.; Dlay, S.S.; Gao, B. Computational Deep Intelligence Vision Sensing for Nutrient Content Estimation in Agricultural Automation. *IEEE Trans. Autom. Sci. Eng.* **2018**. [CrossRef]
40. Sulistyo, S.B.; Woo, W.L.; Dlay, S.S. Regularized Neural Networks Fusion and Genetic Algorithm Based On-Field Nitrogen Status Estimation of Wheat Plants. *IEEE Trans. Ind. Inform.* **2017**. [CrossRef]
41. Sulistyo, S.B.; Woo, W.L.; Dlay, S.S.; Gao, B. Building a Globally Optimized Computational Intelligent Image Processing Algorithm for On-Site Inference of Nitrogen in Plants. *IEEE Intell. Syst.* **2018**. [CrossRef]
42. Fuentes, A.; Yoon, S.; Kim, S.C.; Park, D.S. A robust deep-learning-based detector for real-time tomato plant diseases and pests recognition. *Sensors* **2017**. [CrossRef]
43. Picon, A.; Alvarez-Gila, A.; Seitz, M.; Ortiz-Barredo, A.; Echazarra, J.; Johannes, A. Deep convolutional neural networks for mobile capture device-based crop disease classification in the wild. *Comput. Electron. Agric.* **2019**. [CrossRef]
44. Chemura, A.; Mutanga, O.; Dube, T. Separability of coffee leaf rust infection levels with machine learning methods at Sentinel-2 MSI spectral resolutions. *Precis. Agric.* **2017**, *18*, 859–881. [CrossRef]
45. Chemura, A.; Mutanga, O.; Sibanda, M.; Chidoko, P. Machine learning prediction of coffee rust severity on leaves using spectroradiometer data. *Trop. Plant Pathol.* **2018**, *43*, 117–127. [CrossRef]

46. Mollazade, K.; Omid, M.; Tab, F.A.; Mohtasebi, S.S. Principles and Applications of Light Backscattering Imaging in Quality Evaluation of Agro-food Products: A Review. *Food Bioprocess Technol.* **2012**, *5*, 1465–1485. [CrossRef]
47. Pulido, H.G.; De la Vara Salazar, R.; González, P.G.; Martínez, C.T.; Pérez, M.d.C.T. *Análisis y Diseño de Experimentos*; McGraw-Hill: New York, NY, USA, 2012.
48. Numpy.org. NumPy. 2018. Available online: http://www.numpy.org/ (accessed on 19 January 2019).
49. Pandas.pydata.org. Pandas: Powerful Python Data Analysis Toolkit. 2018. Available online: https://pandas.pydata.org/ (accessed on 19 January 2019).
50. Scikit-learn.org. Scikit-learn. 2018. Available online: https://scikit-learn.org/stable/ (accessed on 19 January 2019).
51. Keras.io. Keras: The Python Deep Learning library. 2018. Available online: https://keras.io/ (accessed on 19 January 2019).
52. Tensorflow.org. TensorFlow. 2018. Available online: https://www.tensorflow.org/ (accessed on 19 January 2019).
53. Burkov, A. *The Hundred-Page Machine Learning Book*; Andriy Burkov: Quebec, QC, Canada, 2019.
54. scikit-learn Developers. sklearn.model_selection.GridSearchCV. 2019. Available online: https://scikit-learn.org/stable/modules/generated/sklearn.model_selection.GridSearchCV.html (accessed on 19 January 2019).
55. scikit-learn Developers. 3.1. Cross-Validation: Evaluating Estimator Performance. 2019. Available online: https://scikit-learn.org/stable/modules/cross_validation.html (accessed on 19 January 2019).

PERMISSIONS

All chapters in this book were first published in MDPI; hereby published with permission under the Creative Commons Attribution License or equivalent. Every chapter published in this book has been scrutinized by our experts. Their significance has been extensively debated. The topics covered herein carry significant findings which will fuel the growth of the discipline. They may even be implemented as practical applications or may be referred to as a beginning point for another development.

The contributors of this book come from diverse backgrounds, making this book a truly international effort. This book will bring forth new frontiers with its revolutionizing research information and detailed analysis of the nascent developments around the world.

We would like to thank all the contributing authors for lending their expertise to make the book truly unique. They have played a crucial role in the development of this book. Without their invaluable contributions this book wouldn't have been possible. They have made vital efforts to compile up to date information on the varied aspects of this subject to make this book a valuable addition to the collection of many professionals and students.

This book was conceptualized with the vision of imparting up-to-date information and advanced data in this field. To ensure the same, a matchless editorial board was set up. Every individual on the board went through rigorous rounds of assessment to prove their worth. After which they invested a large part of their time researching and compiling the most relevant data for our readers.

The editorial board has been involved in producing this book since its inception. They have spent rigorous hours researching and exploring the diverse topics which have resulted in the successful publishing of this book. They have passed on their knowledge of decades through this book. To expedite this challenging task, the publisher supported the team at every step. A small team of assistant editors was also appointed to further simplify the editing procedure and attain best results for the readers.

Apart from the editorial board, the designing team has also invested a significant amount of their time in understanding the subject and creating the most relevant covers. They scrutinized every image to scout for the most suitable representation of the subject and create an appropriate cover for the book.

The publishing team has been an ardent support to the editorial, designing and production team. Their endless efforts to recruit the best for this project, has resulted in the accomplishment of this book. They are a veteran in the field of academics and their pool of knowledge is as vast as their experience in printing. Their expertise and guidance has proved useful at every step. Their uncompromising quality standards have made this book an exceptional effort. Their encouragement from time to time has been an inspiration for everyone.

The publisher and the editorial board hope that this book will prove to be a valuable piece of knowledge for researchers, students, practitioners and scholars across the globe.

LIST OF CONTRIBUTORS

Rajendra P. Sishodia and Ram L. Ray
College of Agriculture and Human Sciences, Prairie View A&M University, Prairie View, TX 77446, USA

Sudhir K. Singh
K. Banerjee Centre of Atmospheric & Ocean Studies, IIDS, Nehru Science Centre, University of Allahabad, Prayagraj 211002, India

Maria Casamitjana
Corporación Colombiana de Investigación Agropecuaria—AGROSAVIA, Bogotá 250047, Colombia
Geography Department, University of Girona, Plaça Ferrater Mora, 1, 17004 Girona, Spain

Maria C. Torres-Madroñero
Research Group on Smart Machine and Pattern Recognition, Instituto Tecnológico Metropolitano, Calle 54A No. 30 - 01, Barrio Boston, Medellin 050012, Colombia

Jaime Bernal-Riobo
Corporación Colombiana de Investigación Agropecuaria—AGROSAVIA, Bogotá 250047, Colombia

Diego Varga
Geography Department, University of Girona, Plaça Ferrater Mora, 1, 17004 Girona, Spain

Antonio Fernández-López, Antonio Ruiz-Canales and Manuel Ferrández-Villena-García
Engineering Department, Miguel Hernandez University of Elche, 03312 Orihuela, Spain

Daniel Marín-Sánchez
Computer Science and Systems Department, University of Murcia, 30100 Murcia, Spain
Google, Brandschenkestrasse 110, 8002 Zürich, Switzerland

Ginés García-Mateos
Computer Science and Systems Department, University of Murcia, 30100 Murcia, Spain

José Miguel Molina-Martínez
Food Engineering and Agricultural Equipment Department, Technical University of Cartagena, 30203 Cartagena, Spain

Sajad Sabzi and Razieh Pourdarbani
Department of Biosystems Engineering, College of Agriculture, University of Mohaghegh Ardabili, Ardabil 56199-11367, Iran

Davood Kalantari
Department of Mechanics of Biosystems Engineering, Faculty of Agricultural Engineering, Sari Agricultural Sciences and Natural Resources University, Sari 48181-68984, Iran

Thomas Panagopoulos
Research Center for Spatial and Organizational Dynamics, University of Algarve, Campus de Gambelas, 8005 Faro, Portugal

Sergio Vélez, Enrique Barajas, José Antonio Rubio and Rubén Vacas
Instituto Tecnológico Agrario de Castilla y León (ITACyL), Unidad de Cultivos Leñosos y Hortícolas, 47071 Valladolid, Spain

Carlos Poblete-Echeverría
Department of Viticulture and Oenology, Faculty of AgriSciences, Stellenbosch University, Private Bag X1, Matieland 7602, South Africa

José Alberto García-Berná, Brahim Benmouna, Ginés García-Mateos and José Luis Fernández-Alemán
Department of Computer Science and Systems, University of Murcia, 30100 Murcia, Spain

Sofia Ouhbi
Department of Computer Science and Software Engineering, CIT, United Arab Emirates University, Al Ain 15551, UAE

Andrew Clark and Joel McKechnie
Remote Sensing Centre, Queensland Department of Environment and Science, GPO Box 2454, Brisbane, QLD 4001, Australia

Marica Franzini and Vittorio Casella
Department of Civil Engineering and Architecture, University of Pavia, Via Ferrata, 3, 27100 Pavia, Italy

Giulia Ronchetti and Giovanna Sona
Department of Civil and Environmental Engineering, Polytechnic University of Milan, Piazza Leonardo da Vinci 32, 20133 Milan, Italy

Dolores Parras-Burgos, Daniel G. Fernández-Pacheco and Thomas Polhmann Barbosa
Departamento de Estructuras, Construcción y Expresión Gráfica, Universidad Politécnica de Cartagena, C/ Doctor Fleming s/n, 30202 Cartagena, Murcia, Spain

Manuel Soler-Méndez
Grupo de I+D+i Ingeniería Agromótica y del Mar, Universidad Politécnica de Cartagena, C/Ángel s/n, Ed. ELDI E1.06, 30203 Cartagena, Murcia, Spain

Fanxiu Chen, Bin Zhang, Liming Zhang and Fanzhen Meng
School of Science, Qingdao University of Technology, Qingdao 266033, Shandong, China

Endong Wang
Sustainable Construction, The State University of New York, Syracuse, NY 13210, USA

Pedro José Blaya-Ros, Víctor Blanco and Rafael Domingo
Dpto Ingeniería Agronómica, Universidad Politécnica de Cartagena (UPCT), Paseo Alfonso XIII, 48, E30203 Cartagena, Spain

Fulgencio Soto-Valles and Roque Torres-Sánchez
Dpto Automática, Ingeniería Eléctrica y Tecnología Electrónica, Universidad Politécnica de Cartagena (UPCT), Campus de la Muralla s/n, E30202 Cartagena, Spain

David Velásquez
I+D+i on Information Technologies and Communications Research Group, Universidad EAFIT, Carrera 49 No. 7 Sur - 50, Medellín 050022, Colombia
Department of Data Intelligence for Energy and Industrial Processes, Vicomtech Foundation, Basque Research and Technology Alliance (BRTA), 20014 Donostia-San Sebastián, Spain
Department of Computer Science and Artificial Intelligence, University of Basque Country, Manuel Lardizabal Ibilbidea, 1, 20018 Donostia/San Sebastián, Spain

Alejandro Sánchez, Sebastian Sarmiento and Mauricio Toro
I+D+i on Information Technologies and Communications Research Group, Universidad EAFIT, Carrera 49 No. 7 Sur - 50, Medellín 050022, Colombia

Mikel Maiza
Department of Data Intelligence for Energy and Industrial Processes, Vicomtech Foundation, Basque Research and Technology Alliance (BRTA), 20014 Donostia-San Sebastián, Spain

Basilio Sierra
Department of Computer Science and Artificial Intelligence, University of Basque Country, Manuel Lardizabal Ibilbidea, 1, 20018 Donostia/San Sebastián, Spain

Index

S

T

U

V

W

Printed in the USA
CPSIA information can be obtained
at www.ICGtesting.com
JSHW051401091023
49903JS00006B/234